EUROPEAN TEXTILE RESEARCH:
COMPETITIVENESS THROUGH INNOVATION

Proceedings of the European textile research symposium 'Competitiveness Through Innovation'

organized jointly by
Comitextil and the
Commission of the European Communities

held in Luxembourg, 18 and 19 September 1985

EUROPEAN TEXTILE RESEARCH: COMPETITIVENESS THROUGH INNOVATION

Edited by

C. BLUM
Comitextil

and

J. G. WURM
Commission of the European Communities

ELSEVIER APPLIED SCIENCE PUBLISHERS
LONDON and NEW YORK

ELSEVIER APPLIED SCIENCE PUBLISHERS LTD
Crown House, Linton Road, Barking, Essex IG11 8JU, England

Sole Distributor in the USA and Canada
ELSEVIER SCIENCE PUBLISHING CO., INC.
52 Vanderbilt Avenue, New York, NY 10017, USA

WITH 36 TABLES AND 177 ILLUSTRATIONS

© ECSC, EEC, EAEC, BRUSSELS AND LUXEMBOURG, 1986
Softcover reprint of the hardcover 1st edition 1986

British Library Cataloguing in Publication Data

European textile research: competitiveness through innovation.
1. Textile industry—European Economic Community countries
I. Blum, C.   II. Wurm, J. G.
338.4'7677'0094   HD9865.A12

Library of Congress Cataloging in Publication Data

European Textile Research Symposium (1985: Luxembourg).
  European textile research.

  'Organization of the conference by: Commission of the European Communities, Directorate-General for Science, Research, and Development, Brussels, in co-operation with Comitextil'—T.p. verso.
  'Published for the Commission of the European Communities, Directorate-General for Telecommunications, Information Industries, and Innovation, Luxembourg'—T.p. verso.
  Summaries in French, German, and Italian.
  Bibliography: p.
  Includes index.
  1. Textile industry—Congresses.   I. Blum, C.   II. Wurm, J. G.   III. Commission of the European Communities. Directorate-General for Science, Research, and Development.   IV. Comitextil.   V. Commission of the European Communities. Directorate-General for Telecommunications, Information Industries, and Innovation.
VI. Title.
TS1300.E97   1985   677   86-19754

ISBN-13: 978-94-010-8421-5   e-ISBN-13: 978-94-009-4323-0
DOI: 10.1007/978-94-009-4323-0

Organization of the conference by: Commission of the European Communities,
Directorate-General for Science, Research and Development, Brussels, in co-operation
with Comitextil

Publication arrangements by Commission of the European Communities,
Directorate-General for Telecommunications, Information Industries and Innovation,
Luxembourg

EUR 10651

LEGAL NOTICE
Neither the Commission of the European Communities nor any person acting on behalf of the Commission is responsible for the use which might be made of the following information.

PREFACE

J. HONEYMAN

The second textile and clothing research and development programme (1982-1985) supported by the EEC was undertaken with the prime objective of strengthening the competitiveness of the Community textile industries by stimulating innovation.

From the generation of the research ideas through to the application of new technologies, the optimum conditions for scientific research demand highly trained and qualified people working in well-equipped laboratories.

The Commission, in collaboration with the textile industries working through Comitextil, participated in the selection of suitable topics and supported the implementation of the programme and the publication and dissemination of the results.

In all, 24 institutes in seven member countries carried out the programme which was divided into 35 separate but co-ordinated contracts. The costs were shared equally between the Commission and the participating institutes. The results were presented by selected specialists at the closing symposium held in Luxembourg on the 18 and 19 September 1985.

The technical sessions covered the four topics : garment physiology and construction, quality of knitted fabrics and articles, application of new spinning technologies in the wool industry, and the upgrading of linen. In addition, prominent industrialists and Commission officials presented papers concerned with the present situation and the prospects for textile manufacture in the Community. Possible future research topics were outlined and the need was stressed for an increase in the amount of research and development to be carried out.

This publication contains English versions of the papers presented at the symposium together with French, German, and Italian summaries. The important points made during the discussion periods are also included. In separate sections, details are given of each institute involved and of the part it played in the programme.

*   *   *

**Organisationsausschuss/Organizing committee/Comité
d'organisation/Comitato organizzativo:**

M. C. Blum, Comitextil
Ms J. Candries, Comitextil
M. J. Wurm, DG XII, CEC
Ms A. Donohoe, DG XII, CEC
M.C. David, DG III, CEC
M. E. Papastathopoulos, DG III, CEC
Dr. J. Honeyman, expert CEC
Prof. I. Meuris, expert CEC
Dr M. Bona, Città degli Studi di Biella
Dr D. Finlay-Maxwell, John Gladstone & Co Ltd
Dr R. Jeffries, Shirley Institute
M. A.J. Larsen, Comitextil
M. J. Stryckman, Centexbel
M. M. Van Lancker, Centexbel

Local organisation : M. R. Linster, DG IX, CEC

Registration and publication : M. D. NICOLAY, DG XIII, CEC

Institute/Institutes/Instituts/Istituti :

ATPUL, Le Plessis-Belleville
Bekleidungsphysiologisches Institut Hohenstein eV, Bonnigheim
CETIH, Paris
CELAC, Chaineux
Centexbel, Gent
Centexbel, Verviers
Città degli Studi SpA, Biella
Dansk Textil Institut, Taastrup
Department of Textile Technology, Manchester
Deutsches Wollforschungsinstitut, Aachen
HATRA, Nottingham
ITF-Boulogne-Billancourt
ITF-Maille, Troyes
ITF-Nord, Villeneuve-d'Ascq
Institut für Textiltechnik, Denkendorf
Institute for Perception TNO, Soesterberg
Institut technique du lin, Paris
Instituut voor Bewaring en Verwerking van Landbouwprodukten, Wageningen
Lanerossi SpA, Schio
LIRA, Northern Ireland
Shirley Institute, Manchester
Stazione sperimentale per la cellulosa, carta e fibre tessili vegetali
        ed artificiali, Milano
Tecnotessile, Prato
Textil-Forschung, Bielefeld eV
Vezelinstituut TNO, Fibre Research Institute, Delft
WIRA, Leeds

CONTENTS

OPENING SESSION

Welcoming address
    F. BODEN, Minister for Education and Youth, Luxem-
    bourg

Opening address
    J.H. LEACH, President of Comitextil

The outlook for the European textile industry
    P.A. RUTSAERT, Commission of the European Communi-
    ties

WELCOMING ADDRESS

F. BODEN
Minister for Education and Youth, Luxembourg

Mr Chairman,
Ladies and Gentlemen,

On behalf of my colleagues in the Council of Ministers for Research in the European Community and of the Luxembourg Government, it is an honour and a pleasure for me to welcome you most warmly to the European Textile Research Symposium in Luxembourg.

Current scientific research seems primarily concerned with meeting the challenges of high technology areas such as data processing, micro-electronics, artificial intelligence, telecommunications, biotechnology, and new materials. It may not be immediately apparent to those unfamilar with the problems associated with textile production how such scientific research fits into Community activities in the textile sector. Yet you are assembled here today to examine together the results of the EEC's Second Textile Research Programme, a programme which has involved more than 35 contracts with 24 institutions in seven Community countries. The researchers involved have looked into the problems of garment physiology and construction, new spinning technologies, quality control methods and ways of maintaining the high quality of knitted fabrics and garments, and the upgrading of linen products. This symposium reflects a desire to strengthen European co-operation in research and faced with the challenge of technological change, to improve collaboration between higher educa-tion establishments, research institutes and industry with a view to optimizing the transfer of research findings to industry.

The presence at the symposium of representatives from the research institutes and the textile industry and the joint analysis of the results obtained should ensure that research results will quickly be applied at the production level.

What is at stake here is the preservation and strengthening of the competitiveness of the major basic industry which textiles represent. The high standard of living in Europe means that the textile industry is forced to meet very strict quality requirements while at the same time coping with the production constraints imposed by the high labour costs in our countries. Innovation, whether with regard to product quality or design or in production technology, is a major asset in the effort to meet these challenges.

By pooling the know-how derived from a rich European tradition, by exploiting the innovative ideas of our researchers and by utilising the potential for incorporating new technology into the various production processes, the textile industry stands a good chance of successfully rising to the challenges with which it is faced.

Moreover, as the history of the introduction of the power loom amply testifies, since the onset of the industrial age the textile industry has been accustomed to the upheavals which technological innovations can bring about.

I am convinced that today's and tommorrow's symposium will play a part in helping to solve the problems with which you are faced and in forging closer links between the research sector and the textile industry.

One of the Community's primary objectives is the creation of an internal European market, open to the outside, which guarantees free movement of persons, goods, capital and services. This single large market within the Community will serve to facilitate scientific and technological co-operation between the Member States and enhance the competitiveness and productivity of the European economy.

It is my sincere wish, on the occasion of this symposium, that the research sector and the textile industry, which are the areas of particular concern to you, should benefit greatly from this single market and that this major sector of the economy should succeed in achieving lasting growth and maintaining employment opportunities.

Ladies and Gentlemen, I should like to conclude, in my capacity as Minister for Tourism, by also wishing you a very pleasant stay here in Luxembourg. You may be sure that I would be deligthed to see you return here in less formal circumstances, that is, as tourists with more time at your disposal to visit the beauty spots of Luxembourg and become more fully acquainted with our country.

In the meantime, I wish you every success in your symposium.

OPENING ADDRESS

J.H. LEACH
President of Comitextil

I thank the European Commission for its action in favour of research applicable to the textile sector. Indeed, the textile industry was the first industrial sector to benefit from specific Community programmes carried out on a shared-cost basis.

Comitextil was the first Association to put forward structured research projects at Community level, for which the industry was willing to put up 50 % of the necessary financing. In this way the industry showed its faith in technological progress and in Europe, and the European Commission showed its conviction that there is a future for the textile industry in the E.E.C.

Over the past few years, too many theorists have termed the textile industry a "sunset industry" : an industry that will inevitably decline in the developed countries and shift to developing countries. My view is different. I believe there will be a new dawn and we are going to illustrate this over the next two days.

These programmes have made it possible for researchers from the institutes in the various Member States to work together on projects of common interest, but also to reach satisfactory results, results of which you will be able to appreciate the real value.

This is not the least of the merits of this operation : as a result of coming into contact with each other, these specialists came to a better understanding and appreciation of each other's work, which had beneficial effects, not only for the projects themselves, but also in a wider context, as several projects were carried out beyond the scope of any public framework. There is probably no better means of making Europe into a reality.

I take advantage of this occasion to thank all those who have actively participated in this work, in particular for the tenacity they have shown in overcoming certain obstacles, including the language barrier and of travelling around Europe. In the short term these could be regarded as hindrances to effectiveness, but in the long term then should facilitate the construction of a technological Europe which should make it possible to call on the best qualified in a specific field and thereby to create new working partnerships and to optimize the results.

Finally, I wish to thank all those who have come here to learn about the results obtained.

It is not always easy for the director of a small firm, where he is the real orchestra conductor, as it were, to hand over the direction for a few days; neither is this easy for the person responsible for the production of an average-size firm. Competition is cut-throat and this requires the participation of everyone in the company. However, everyone here today must certainly share my view that a new revolution is dawning in our sector. Modern textile research reaches from fibre physics and composite structures to energy conservation, from heat transfer phenomena to mill management, from aerodynamics to material sciences.

The second textile research programme will only show a few of the facets of the revolution in progress. Scientific research itself is only one aspect, however essential, of this revolution. Innovations in textile technology can only be competitive and successful if they are the result of joint efforts that stretch from basic research to optimized design and from advanced manufacturing techniques to technically competent sales people. This is why every Community intiative in this field must be studied and carried out in close cooperation with the industry and in a favourable environement.

The technological process is in fact advancing at an uneven pace, and although textile processing itself and certain operations preparatory to fabric assembly have benefitted from progressive automation, the operation which makes a garment from our products is still dominated by salary costs.

There are also encouraging medium-term prospects in these areas, but while awaiting their realization it will be necessary to maintain a system for regulating imports into the EEC, which should largely take account of the willigness of our world partners to agree to the opening-up of their own markets which are generally closed to the outside.

I now introduce briefly the following two speakers.

In Mr Ph. BOURDEAU and Mr J. WURM, his deputy, we have encountered two officials who have been particularly attentive to the concerns of our sector. We await Mr BOURDEAU's address with great interest, as it should enligthen us regarding the way in which the problem has been perceived and dealt with by the Community authorities.

Mr P. RUTSAERT, whose responsibilities within the Directorate-General for Industrial Affairs led him to deal with textile problems for several years, will discuss this research in the context of concerns of the Community's textile industry. His perfect knowledge of the problems and his ability to describe them in a few words, will form an excellent introduction to the more purely scientific part of this Symposium.

THE OUTLOOK FOR THE EUROPEAN TEXTILE INDUSTRY

P.A. RUTSAERT
Commission of the European Communities

If we wish to appreciate the importance for the textile and clothing industry of the research effort which has gone into producing the results discussed at this symposium, and the importance of the new BRITE programme which is currently being launched, it would be useful to concentrate for a moment on the outlook for this industry. I shall try to describe as concisely as possible what is happening at present and to see, on this basis, what could happen in the future, since future trends are, after all, effectively determined by factors which have already been indentified.
    The points to which I would like to drawn your attention can be grouped under demand, supply, and the political bacground.

DEMAND
    The demand for textile and clothing products per person increases with income up to a certain income level, at which demand tends to reach a ceiling level which varies according to consumption patterns. For example, this is about 16 to 17 kg in the Community and around 22 to 23 kg in the United States.
    It is not surprising therefore that internal demand in countries with a long history of industrialization should be virtually static whereas it is increasing rapidly in the developing countries and particularly these which have recently become industrialized.
    Thus while consumption in the world as a whole has practically doubled over the last 20 years, it has increased by almost 150 % in the developing countries, largely as a result of increase in population, and by only 50 % in the industrialized countries, where the main increase took place in the period 1960 to 1970. Since the 1974 economic crisis, it has practically come to a halt.
    In the developing countries, on the other hand, consumption continues to rise. The market in what are currently regarded as the developing countries has caught up with the market in the industrialized countries, at least in volume, and will no doubt be a long way ahead by the year 2000.

SUPPLY
    When production techniques change, competitiveness may shift from one country to another, according to the relative availability of the various production factors.
    After the war, textiles and clothing were, relatively speaking, still highly labour-intensive industries and the labour was largely unskilled, so that even a Nobel prize-winner in economics could maintain, with sound economic logic, that these types of activity should tend to move towards the less developed countries where unskilled labour was more readily available.

This did in fact happen for some time, and at an ever-increasing rate, causing the industry in the industrialized countries serious problems of lost markets, first abroad and then at home. Over the same period, however, technical progress transformed whole sectors of the textile industry in the narrow sense (spinning, weaving, dyeing, printing) into an industry that was highly capital-intensive, which naturally restored the competitiveness of the more industrialized countries.

The clothing industry is always very labour-intensive, but the level of skill of this labour varies considerably with the type of product. Obviously, products requiring more unskilled lablour tended to be the first to shift to countries where wages are lower. Products of the middle and top of the range requiring more highly skilled labour and greater capital outlay, have better withstood the challenge. At the top of the range, luxury products, where labour accounts for almost all of the costs, have also tended to move to countries where even skilled labour is still relatively cheap. At present, labour in the developing countries, and particularly the newly industrialized ones, is rapidly becoming more skilled. These countries are therefore in a position to compete with the industrialized countries in more and more sectors of the market. However, technological progress has not passed the clothing industry by. Some operations, such as designing and cutting out, are or could be highly capital-intensive, whereas other operations, such as making-up, are still highly labour-intensive.

Progress is in sight, and, in some cases, we have already begun to make headway. We could concentrate first of all on the simplest products, and the paradoxical result would be that our industry would regain its competitiveness in the very sectors of the market that it has lost, whilst continuing to lose ground in those in which it has best maintained its position hitherto. We cannot be certain of this, however, if the nature of technical progress is such that capital may be substituted for labour in both short and long runs. On this way the industry in the industrialized countries could regain its competitiveness over a wide range of products, since international competition would then depend more on specialization or fashion than on factory costs.

THE POLITICAL BACKGROUND

Finally, a word about the political context, and trade policies in particular.

Economic theory takes free trade as a point of departure. Economic reality, however, presents a very different picture. The point is generally made, especially in connection with textiles, that the industrialized countries to limit the access of products from low-wage countries require to their markets. It is often forgotten that the markets of the industrialized countries are, generally speaking, the most open, whereas other countries elect basically to import only what they cannot make themselves. This prevents the industrialized and the industrializing countries from specializing in particular qualities or types of product. The newly-industrialized countries are thus permitted to concentrate more capital resources and skilled labour on a few export sectors that would not necessarily have been economic in an open economy. To this extent, the industrialized country may be forced to maintain restrictions on entry to their markets while in an open economy this would not be necessary.

The way in which we compete with newly-industrialized countries and the type of trade policy which the industrialized countries adopt towards them in future will depend very much on whether or not they continue to favour a closed economy, not only for textiles and clothing but, more generally, for all products for which there is a large market.

Concluding these thoughts on the outlook for the textile and clothing industry, I am afraid I have not been very definite about how your industry will evolve in future. Forecasting is an exercise which, by its very nature, needs to be repeated regularly. But I hope, at least, that I have left you in no doubt as to my conviction, and that of my colleagues, that your industry does have a future and, to repeat the words of the Chairman of Comitextil, a rich future.

SESSION A

GARMENT PHYSIOLOGY AND CONSTRUCTION

Introduction by the Session Chairman
R. JEFFRIES, Shirley Institute

Evaluation of textile and garment comfort

Construction and make-up of garments

Summary and Conclusions by the Session Chairman

Contractors' sheets

GARMENT PHYSIOLOGY AND CONSTRUCTION

R. JEFFRIES
Shirley Institute

This session describes research work in two areas : garment physiology and comfort, and fabric specification and garment construction in relation to garment manufacturing processes and garment style and aesthetics. These two areas were included in the Second Textile Research Programme of the EEC because of their importance to the textile industry and to the manufacturers and retailers of garments. There is no doubt that improvement in the comfort of fabrics and clothing is currently an important and developing theme with manufacturers and retailers, and therefore a major group of projects on this subject is most opportune at this time. Similarly, there is an increasing interest in the development of clothing of good style, quality, and fit and in the automation of garment manufacture; in both of these areas the specification, classification, and construction of fabrics and of garments made from them is clearly of key importance.

Work on five projects is described. In each case the approach has been to produce results and conclusions of direct and immediate value to the EEC textile and clothing industries, to assist these industries in the design of better products and in the development of improved production processes.

The first rapporteur describes the research on the complex subject of the comfort of garments in relation to the nature and properties of the fabrics, garments, and garment assemblies involved. Two types of comfort are examined : "thermophysiological" comfort (related to the maintenance of the thermal and moisture balance required to keep the body at a comfortable functioning temperature), and "sensorial" comfort (related to the tactile properties of the next-to-skin garments and the way in which they interact with the skin).

Three projects deal with thermophysiological comfort. In one of these, detailed information has been obtained on the thermophysiological comfort of garments and on the development of evaluation methods and guidelines. Models of human skin behaviour have been developed, and the results of the thermal and moisture vapour transfer measurements inserted into predictive formulae to enable the calculation of a "comfort vote".

This "comfort vote" correlates well with wearer experience, and thus replaces much of the need for expensive wearer trials. Another of the projects has been concerned with improvements in the thermophysiological comfort of rainwear. Investigations of the various types of water proof, water vapour permeable fabrics have been carried out, particular attention being paid to the combination of permeability and ventilation. A technique for investigating the effects of ventilation has been developed. Guidance on the design of rain-protective clothing is given. A third project is concerned with physiological measurements on the dissipation of body heat and sweat in relation to garments made of wool, cotton, wool blends and cotton blends.

The fourth project investigates the subject of the "sensorial" comfort of next-to-skin apparel. In this work it was first necessary to

carry out careful and extensive wearer trials to identify which are the
sensations and physical factors that determine the sensorial comfort of
these next-to-skin garments. Based on the results of these trials a range
of simple objective and subjective test procedures have been developed,
for measuring such factors as local irritation, prickle, tickle, local
pressure, cold feel, scratchiness, wet cling, fibre shedding, and static
electrical effects.

The second rapporteur describes work on the assessment and classification of the physical properties of fabrics in relation to garment
design, in order to assist in the optimisation of the design thus leading
to improved manufacturing processes and the production of comfortable and
aesthetically satisfying garments. The ability of fabrics to withstand
the strains and multidirectional flexing involved in garment manufacture
is investigated. This type of investigation requires the employment of
advanced techniques, which may not be suitable for routine use by garment
manufacturers.

Garment manufacturers have been involved in the verification of the
results of investigations. In a second part of this project, work has
been carried out on fabric/machine interaction problems in industry; the
monitoring of fabrics in relation to the automatic adjustment of the
processing machinery has been considered. Two specific topics have been
studies experimentally : fabric feeding in sewing (to identify fabric and
machine characteristics affecting the movement of fabric through the
machine), and in-line underpressing (to investigate the compromises
needed between production speed, pressing conditions, and physical
dimensions).

EVALUATION OF TEXTILE AND GARMENT COMFORT

K.H. UMBACH
Bekleidungsphysiologisches Institut e.V. Hohenstein

SUMMARY

To be competitive, modern clothing besides having good mechanical and technological properties and being of easy care must possess good comfort characteristics. In order to include wear comfort as a constructional parameter in the development of textiles and garments it is essential that there exists a method to measure and evaluate comfort. However, to be of practical use to the fabric and garment-making industry this evaluation, in contrast to the current approach, cannot be performed with time-consuming and expensive wear trials with human subjects. It is necessary that economical test procedures restricted in number are available which can be applied in the laboratory and which yield results which correlate with the experience gained while wearing the fabrics and garments.

Therefore, the aim of the project was to develop such test and evaluation procedures for textile and garment comfort and with examples to show how they can be used in the design of improved textiles and clothing. This is obviously of great benefit to EEC manufacturers, because it will help to accelerate the rate of product development in textiles and clothing and to reduce the possibility of products being marketed which for one reason or another are uncomfortable to wear and therefore result in consumer resistance.

Because wear comfort is a very complex sensation which must be approched from different angles, the project was divided into three parts which are reported in this paper. The work was carried out at the Hohenstein Institute, the Shirley Institute and the Institute for Perception.

EVALUATION AND OPTIMIZATION OF THERMOPHYSIOLOGICAL COMFORT

To evaluate thermophysiological comfort the five5-level system of analysis that was developed is shown schematically in Figure 1. On Level 1 of this sytem the textiles are tested as fabric layers using a thermoregulatory model of human skin (Skin Model) which, by means of a heated sintered metal plate with numerous pores through which distilled water can be pumped, simulates the heat and moisture exchange from man's skin (see Figure 2).

With the Skin Model "normal" or stationary wear conditions with the wearer only sweating moderately and with only water vapour on his skin surface can be simulated as well as non-stationary or transient conditions. The latter are characterized by increased sweating in intermittent pulses with not only water vapour but also liquid sweat on the wearer's skin.

The physiological quality of a fabric under these wear conditions can be expressed in specific quantities or indices. For the stationary situations water vapour permeability index i_{mt} (the ratio of water vapour resistance to thermal insulation) and water vapour absorbance F_i are relevant. The wear properties of a fabric under increased sweating of the wearer can be judged by a buffering index K_d, expressing the fabric's

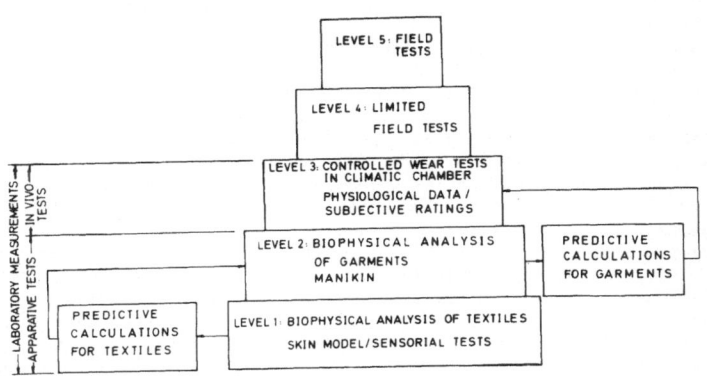

LEVEL 5: FIELD TESTS

LEVEL 4: LIMITED FIELD TESTS

LEVEL 3: CONTROLLED WEAR TESTS IN CLIMATIC CHAMBER

PHYSIOLOGICAL DATA / SUBJECTIVE RATINGS

LEVEL 2: BIOPHYSICAL ANALYSIS OF GARMENTS MANIKIN

PREDICTIVE CALCULATIONS FOR GARMENTS

PREDICTIVE CALCULATIONS FOR TEXTILES

LEVEL 1: BIOPHYSICAL ANALYSIS OF TEXTILES SKIN MODEL/SENSORIAL TESTS

LABORATORY MEASUREMENTS — IN VIVO TESTS
APPARATIVE TESTS — IN VIVO TESTS

Figure 1 : Five-level system for the analysis of the
thermophysiological properties of textiles and garments

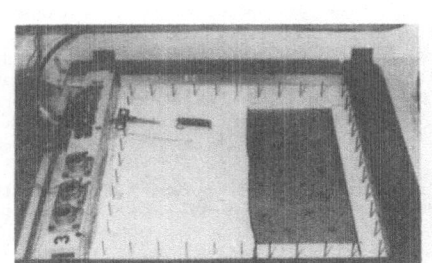

Figure 2 :

Left : Skin Model for the measurement
of the thermophysiological properties
of fabrics.

Right : Manikin "Charlie" for
the measurement of the thermophysiological properties of
ready-made garment ensembles.

capacity of controlling the humidity in the microclimate between skin and fabric after a sweat pulse, and by an index β_t, describing the fabric's temperature controlling capacity in the microclimate during non-stationary conditions. With heavy sweating a fabric can be judged with regard to thermophysiological comfort by the buffering indices K_f and the ability of water G, showing how well liquid sweet can be absorbed, and by the index F_1, describing the fabric's capacity to transport liquid sweet into the ambient air.

The specific quantities listed are inserted in the multicriterion diagram shown in Figure 3 by which the wear comfort of a fabric to be expected under stationary or non-stationary wear conditions can be judged. Because in practical use these conditions are occurring in continuous sequence, predictive formulae have been developed expressing a fabric's wear comfort under typical practical use conditions by a comfort vote CV ranging from "1" (very good) to "6" (completely unsatisfactory). These formulae have been developed separately for different garment types like underwear, T-shirts, trousers and sweaters. For underwear the following predictive formula holds :

$$CV = -8.418 \cdot i_{mt} - 0.5592 \cdot F_i - 2.3684 \cdot K_d - 6.734 \cdot \beta_T$$
$$- 6.7637 \, K_f + 0.0075 \cdot \Delta G + 13.4843$$

Numerous wear trials with subjects have shown that the comfort votes for a fabric predicted with these formulae are in very good agreement with the comfort sensations actually perceived by people during typical use conditions (see Figure 4).

Because the comfort evaluation of fabrics described is based on comparatively few laboratory measurements, it is now possible for the textile manufacturer to test and ensure good comfort characteristics of his products from the first stage of development. However, in a clothing ensemble many fabrics are combined and the air layers in the microclimate between the fabrics and adhering to the ensemble's outer surface are added. Consequently, the wear properties of the ensemble are determined not only by the fabrics' intrinsic physiological characteristics. Therefore, on level 2 of the system of analysis garments are tested with a life-sized Manikin representing a thermoregulatory model of man (see figure 2).

The Manikin is heated to man's skin and body temperature by electrical wires soldered to its inside and placed in a climatic chamber with controlled ambient climatic conditions. With the Manikin the total thermal insulation R_c of the clothing system can be measured directly as it is for the wearer in practical use. Because the Manikin is divided into 16 separately controlled heating circuits local thermal insulation of the garments at particular body sites can be determined.

It is essential that with the Manikin all the body positions (i.e. standing, walking or lying) can be simulated as they occur during the actual use of the clothing. Therefore, the Manikin can move its arms and legs by attached bars driven by an electric motor. Also the influence of convection and ventilation caused by the body movements of the wearer on the wear comfort of a clothing ensemble can be evaluated.

In order to determine the water vapour resistence R_e of clothing in the different body positions or movements corresponding to the thermal insulation values directly measured with the Manikin, a thermodynamic model is used in which the water vapour resistances of the ensemble's fabrics are inserted, as measured with the Skin Model.

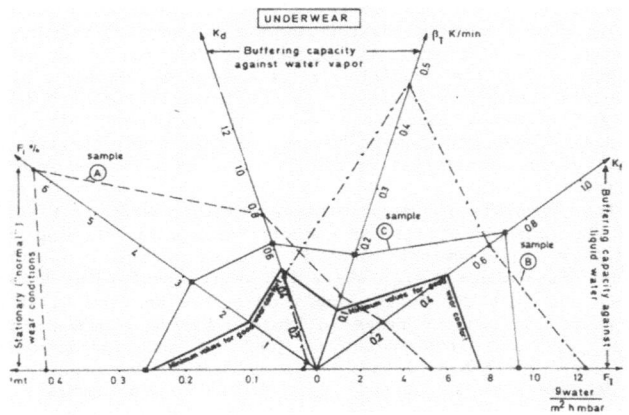

Figure 3 :

Multicriterion diagram for the
evaluation of the thermophysiological comfort of fabrics.

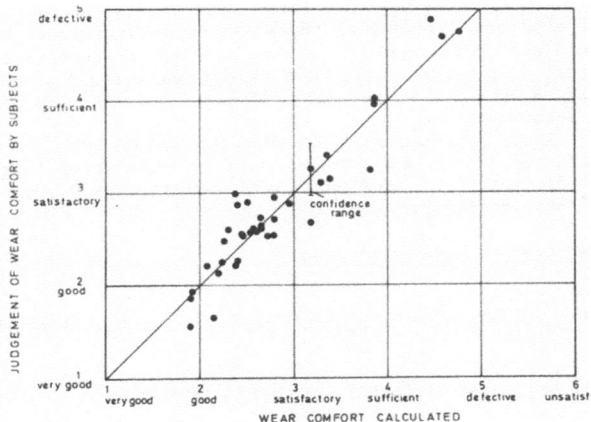

Figure 4 :

Correlation between predicted
and actually perceived wear
comfort of fabrics.

Extensive experiments have shown that the R_c and R_e values of clothing gained with the movable Manikin are in good agreement with the values actually found in controlled wear tests with subjects.

The actual benefit for the textile and garment industry of the Manikin measurements is given by the predictive model which was improved and enlarged during the project. It can be used for describing a clothing ensemble's thermophysiological comfort characteristics under all possible wear and climatic conditions which in their entirety could never be realized in wear trials with humans. In this model the fabric and garment quantities are directly inserted as measured with the Skin Model and the Manikin.

The principle of the predictive model is shown in Figure 5. With the model the range of utility or comfort range of a clothing ensemble can be determined. This comfort range is limited on the one hand by a minimum ambient temperature T_{amin}, where the wearer performing a certain kind of activity is just not feeling cold or suffering from hypothermia. On the other hand the comfort range is limited by the maximum ambient temperature T_{amax}, where the wearer is just not sweating so much that he is feeling uncomfortable or that he is suffering from hyperthermia.

Because both T_{amin} and T_{amax} depend on the ambient humidity, the range of utility of a clothing ensemble is shown by a psychrometric chart, as for the example of a business suit given in Figure 6.

From this chart the garment maker can deduce whether with his construction he is actually meeting the climatic requirements dictated by the garment's field of application and he is directly informed how constructional changes, i.e. changes in pattern or garment compositions, are afflicting the clothing ensemble's range of utility.

On the other hand with the predictive model the wear comfort of a clothing ensemble under particular climatic and activity conditions can be directly calculated. This is possible because wear trials with subjects have shown that there exists a significant correlation between man's subjective comfort feeling and certain quantities of his body like rectal temperature T_{ref}, mean skin temperature T_{sf} and heart rate HR as well as humidity in the microclimate above the skin P_{mf} (in mbar) caused by skin wettness.

These corelations, universally established in the wear trials performed within the project, led to predictive formulae for the wear comfort perceived, expressed by a comfort vote WC, again out of a 6 step ballot, ranging from 1 to 6 :

Wear comfort under heat load :

$$WC_f = 1.52 \ T_{ref} + 0.31 \ T_{sf} + 3.1 \ k_f + 0.02 \ HR_f - 67.85$$

with k_f = comfort factor calculated from :

$$k_f = \frac{\overline{P_{Mf}} - P_a}{\overline{P_{sf}} - P_a}$$

with $\overline{P_{sf}}$ = water vapour pressure at skin surface

p^a = ambient water vapour pressure

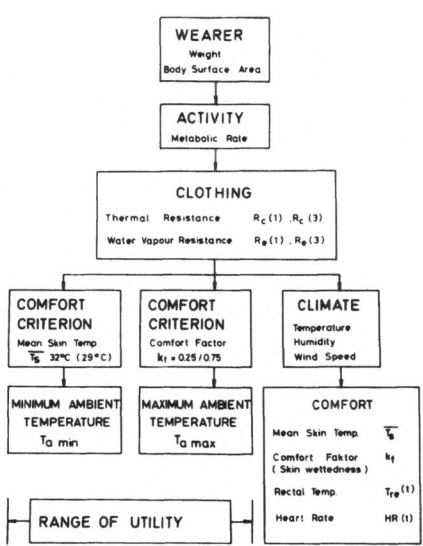

<u>Figure 5 : Principle of predictive calculations used in</u>
<u>the evaluation of the comfort characteristics of clothing ensembles</u>

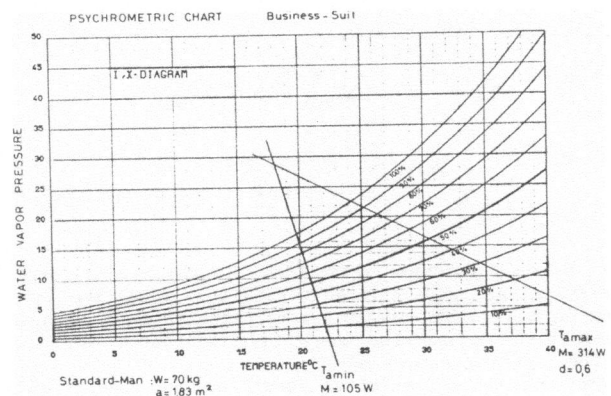

<u>Figure 6 : The range of utility of a business suit with the</u>
<u>wearer sitting (metabolic rate M = 105 W) or performing medium</u>
<u>heavy work (M = 312 W)</u>

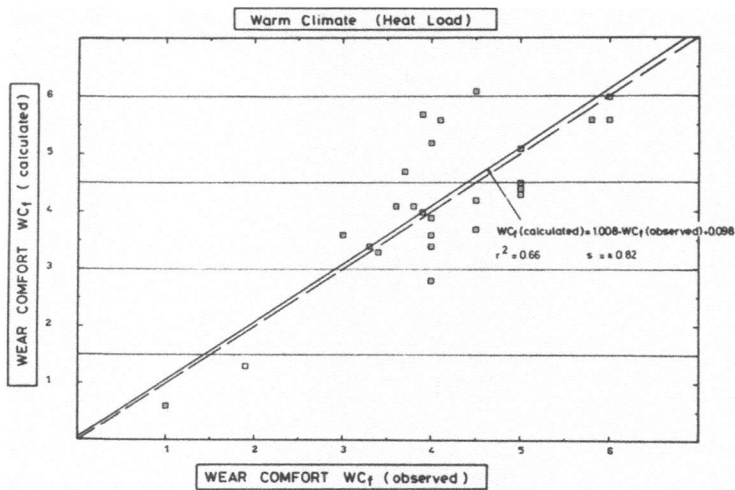

Figure 7 : Correlation between predicted and actually
perceived wear comfort of clothing ensembles

Wear comfort under heat loss :

$$WC_f = -5.61 \, \Delta T_{ref} + 0.6 \, \overline{\Delta T_{sf}} + 1.51$$

with ΔT_{ref} and $\overline{\Delta T_{sf}}$ = decrease in rectal and mean skin
temperature, respectively

All the body function quantities and the microclimate humidity in
these formulae can be calculated within the predictive model. As can be
seen from Figure 7 there exists a good correlation between the calculated
wear comfort votes and the subjective sensation of persons actually
wearing the clothing ensembles.

By reversing the predictive model, on the other hand, the thermal
and water vapour resistances can be calculated which either the fabrics
or the total clothing ensemble must possess in order to provide good wear
comfort under particular climate and activity conditions. Because these
values derived from the predictive model can be readily checked by Skin
Model and Manikin measurements, this system represents an effective and
reliable means by which the fabric and garment manufacturer can optimize
the thermophysiological wear comfort of his products for particular end-
uses.

Thus, the controlled wear trials with subjects in a climatic chamber contained as level 3 in the system of Figure 1 are in most cases unnecessary for the manufacturer. These wear trials should be only used to check the results of the predictive calculations for one particular wear situation. In case of agreement it can be concluded that the model's predictions with regard to other wear situations are also valid.

The controlled wear trials of level 3 are done in a climatic chamber with the test person performing a defined activity, i.e. walking on a treadmill (see figure 8). Objective body function and microclimatic quantities are collected on-line by computer from sensors attached to the subject's body (see figure). The subjective comfort sensation of the test persons is quantified by ballots and questionnaires.

In the development of clothing the wear tests of levels 4 and 5 of the system of Figure 1 should actually by restricted to a small number of items optimized systematically in their comfort characteristics through levels 1 to 3 of the system. In these wear tests which are run under the real ambient conditions of the clothing's field of application it can be seen whether parameters not included in the predictive calculations have some effect on the physiological comfort perceived by the wearer.

During the project the system of analysis developed for thermophysiological comfort was applied to design models of garments for specific end-uses which are optimized in their wear properties. The constructional guidelines deduced from these models are meant for direct application by the textile and garment industry, guaranteeing high quality products which particularly provide good wear comfort.

For instance, for knitted outerwear, guidelines can be set out concerning fabric parameters as well as fit and pattern of the garments which improve moisture transport properties via ventilation induced by the body movements of the wearer. Particularly, the importance of the textiles' air permeability for the resulting thermophysiological comfort of knitted outerwear could be assessed quantitatively.

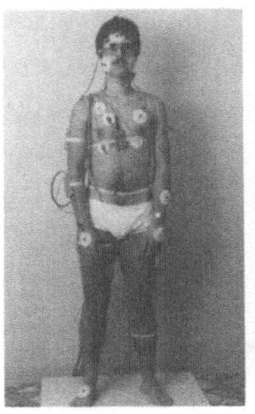 

Left : test person with sensor elements.   Right : test person walking on a treadmill.

Figure 8 : Controlled wear trials with subjects in a climatic chamber

In the field of sportswear it was found that with double-face constructions consisting of two different types of fibres material arranged in two combined layers better moisture transport properties and, thus, better wear comfort could be achieved compared to conventional one-layer constructions traditionally made from cotton. Especially in situations with the wearer sweating heavily by combining a conductive fibre, such as synthetics, worn next to the skin with a sorptive fiber (cotton, viscose or wool) used in the fabric's outer side an improved physiological functionality is resulting. This is effected by the ability of these double-face constructions to conduct liquid sweat away from the skin, storing it within the outer layer which is acting as a buffering zone. Thus, the microclimate next to the skin stays more dry, the fabrics cling less to the skin and a post-exercise chill following a period of heavy body activity is prevented more successfully.

For wet-weather protective garments, guidelines concerning textile parameters, particularly water vapour permeability, have been found which lead to clothing which is watertight but, nevertheless, possesses good thermophysiological wear properties. These guidelines have been fixed in a DIN Standard (DIN 61 539) and have been taken up by the textile industry to develop a new generation of wet-weather fabrics for occupa-tional-, leisure- and sports-wear. Already these products, showing much better wear properties than convention constructions, are selling well on the market, giving the European textile and garment industry an advantage over imported products from low-cost countries.

EVALUATION AND OPTIMIZATION OF SENSORIAL COMFORT

Sensorial comfort caused by the contact textile/sklin is essentially determining the physiological properties of fabrics. Poor sensorial characteristics will lead to poor wear comfort of a textile, even if it possesses good thermophysiological properties. However, the scientific knowledge of sensorial comfort has been rather incomplete. Therefore, the project reported has been performed with the following aims :

- to elucidate the physical factors and body sensations determining sensorial comfort or discomfort,
- to relate these factors and sensations to the physiological properties of fabrics and garments,
- to develop simple test methods for the quantitative evaluation of sensorial comfort,
- to formulate guidelines for the textile and garment industry to help in the design of sensorially comfortable fabrics and garments,
- to contribute background knowledge for a general theory of fabrics and garment comfort.

A first literature survey showed that little has been published of direct relevance to the subject of sensorial comfort and little is known about skin physiology in relation to sensorial sensations. Therefore, at first, a basic survey has been carried out including about one thousand people who have been questioned with regard to their attitudes and beliefs in relation to next-to-skin comfort. Parallel wear trials with 10 men and 10 women of various ages have been performed with 22 T-shirts consisting of the knitted and woven fabrics listed in Table 1. The test samples were carefully selected to cover a wide range of fibres and types of fabric construction.

NO	FIBRE CONTENT	FABRIC
		KNITTED
1	Polyester/cotton (65/35)	Honeycomb
2	PVC/Nylon (90/10)	Rib
3	Polypropylène/nylon (90/10)	String vest
4	Cotton/Nylon (90/10)	Evelet
5	Wool/polypropylene (50/50)	Rib
6	Polyester/viscose (50/50)	Interlock & Dropstitch
7	Wool	Interlock
8	P.V.C./Acrylic (85/15)	Interlock, brushed
9	Cotton in Polyester out	) interlock with) dropstitch, brushed
10	Polypropylene/nylon (90/10)	Rib
11	Superwash wool in Polyester out	Looped Interlock
12	Polyester	Rib
13	P.V.C./Acylic (85/15)	Evelet
14	Polyester	Interlock
15	Angora/Lamswool/Nylon (40/40/20)	Rib
16	Superwash wool	Rib, brushed, inside
17	Superwash wool	Rib
18	Polypropylene in Cotton out	Interlock, brushed
19	Polypropylene in Acrylic/Wool (80/20) out	Rib
21	Acrylic (Dunova)/Cotton/Nylon (60/30/10)	Rib
		WOVEN
20	Polyester	Plain
22	Viscose	Twill

Table 1
Test samples used for the wear trials

The detailed evaluation of the survey and the questionnaires answered by the subjects in the wear trials showed that the following factors determine sensorial comfort or discomfort :

- local fit
- local irration by garment labels
- prickle
- tickle
- fibre shedding
- wet cling
- initial cold feel
- scratchiness
- static electrical effects
- allergies

Particularly it was shown that fabric "handle" as determined by feeling the fabric with the hands is not the same as the feel of fabrics against the skin at various body sites. The torso is more sensitive in detecting prickle and tickle than the hand. Consequently, the results of

the Kawabata system for objective measurement of fabric handle are not
directly related to the sensorial comfort felt by the wearer of a
next-to-skin fabric in actual use.

To measure and determine quantitatively the different relevant
sensorial factors the test methods described below have been developed :

Local fit

A too tight fit of a garment can override all other discomfort
sensations. By using a specially developed pressure transducer the wear
trials showed the pressure applied to the body by localised areas of
garments constructed according to current patterns and typically worn
next to the skin (see Table 2). Also pressures that can be tolerated in a
local area without causing discomfort were determined (see Figure 9).
These critical values can be directly related to objective measurements
exerted by stretching the test samples on the local fit tester shown in
Figure 10. Thus, discomfort caused by local fit can be prevented.

		Mean mmHg	Range
FEMALE			
Bra at chest	Front	140	70 – 280
	Back	50	20 – 100
	Strap	40	10 – 100
Briefs at hips	Front	40	10 – 100
	Side	60	20 – 200
Tights at waist	Front	30	20 – 60
	Side	60	20 – 80
Under skirt at Waist	Front	50	20 – 140
	Side	80	40 – 220
Skirt at waist	Front	30	20 – 60
	Side	70	20 – 140
MALE			
Trousers at waist	Front	50	20 – 80
	Side	110	60 – 180
Socks calf level	Front	60	10 – 120
	Side	40	20 – 60

Table 2
Pressures exerted on the body by garments

Local irration by garment labels

Considerable discomfort can be caused by the edges of garment
labels, particularly if they are folded and heat sealed or heavily
resinated. A visual comfort evaluation can be based on electron
microscope photographs at 80 x magnification as shown in Figures 11 and
12. Whereas the woven label of figure 11 is comfortable, the heat-sealed
labels of this figures with their sharply jagged edges are causing acute
discomfort. In figure 12 again the corners of the folded woven label

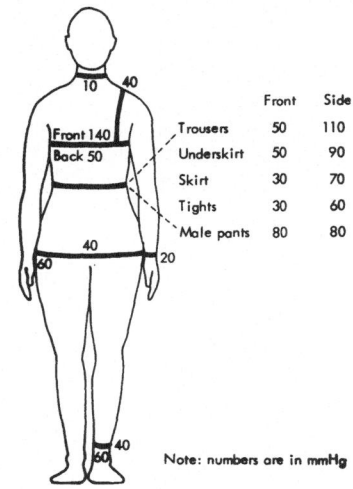

BODY MAP OF LOCAL FITTING AREAS OF GARMENTS

	Front	Side
Trousers	50	110
Underskirt	50	90
Skirt	30	70
Tights	30	60
Male pants	80	80

Note: numbers are in mmHg

Figure 9 : Tolerated pressures caused by garment fit at various body sites

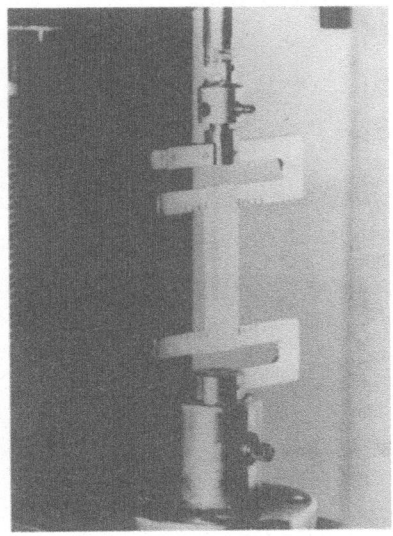

Figure 10 : Testing of garment fit with a modified Instron Tensile Tester

Figure 11

Upper left : woven label / Upper right : poorly cut heat sealed edge.
Lower left and right : heat sealed edge

Figure 12 :

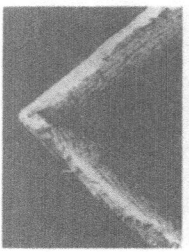 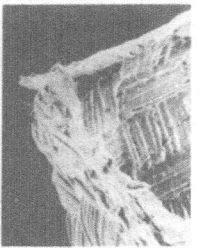

Upper left : woven label / Upper right : heat sealed edge
Lower left and right: heat sealed edge at x 80 and x 180 magnification,
respectively.

causes no discomfort, whereas the wearer is irritated by the jagged and split corners at the folds of the heat-sealed labels.

Prickle

Fabrics containing wool are prone to cause prickle sensations when more than 3.5 % of the fibres have a diameter over 30u and when 0.6 % of the fibres are over 40u in diameter. The diameter of the fibres can be determined by well-known methods.

Tickle

Garments are especially likely to tickle when the body is slightly sweating and the skin is slightly damp. Whether a fabric tends to cause tickle can be judged by photographs showing the fibre ends protruding from a fabric's surface.

Shedding of loose fibres

Fibres separated from the fabric during wearing and clinging to the skin cause sensorial discomfort. An instrument has been devised in which fabric samples are shaken in a defined manner and the loosened fibres collected. The amount of loose fibres is then assessed against photographic standards.

Wet cling

Wet cling is experienced when the body is perspiring profusely. This cling is especially "tacky" when the skin is covered with damp sweat residues. The apparatus developed to measure wet cling quantitatively is shown in Figure 13. The force required to pull a strip of the sample off a horizontally arranged Perspex cylinder is determined with the test fabric either dry or damp after exposure to high relative humidity or to moisture. Tacky cling is simulated by coating the cylinder with artificial sebum. The wear trials performed have yielded a good correlation between these cling values and the sensation of clinging perceived by the test subjects.

Figure 13
Wet cling tester

28

Initial cold feel

Initial cold feel is caused by a rapid transfer of heat from the body to the garment when it is first put on. Even though this effect lasts only for a few seconds, it can influence the wearer's reaction and judgement of the fabric.

The thermal impression meter shown schematically in Figure 14 is used to evaluate whether a fabric tends to cause an initial cold feel. A fabric is bought into contact with a heated copper block and the maximum drop in temperature of the block yields a measure of the cold feel in practical use. This fabric property is mainly determined by the surface structure and only to a minor extent by the fabric's fibre composition.

THERMAL IMPRESSION METER (TIM)

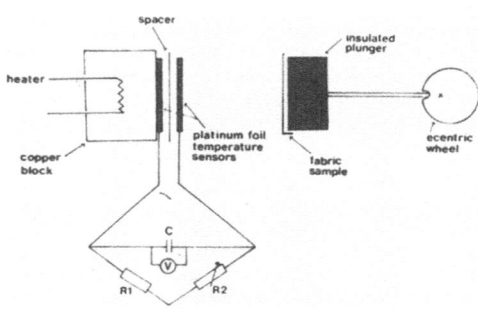

v voltmeter

Figure 14 : Principle of the measurement of initial cold feel caused by a fabric during its first contact with skin.

Scratchiness

The wear trials suggest that scratchiness is of less importance than the properties already discussed. It can be determined by measuring the friction of the sample against a material having the same friction coefficient as human skin. Tests on forearms of subjects resulted in a very fine grade of paper used for polishing metal surfaces beo,g ised as a skin simulant.

Static electrical effects

Static electricity can cause sensorial discomfort in 3 ways :

- spark discharge on removal of a garment,
- clinging of garments to each other or to the body while worn,
- skin irritation due to spark discharges.

The charging behaviour of garments can be determined on persons by removing the sample quickly from next to the skin and throwing it into a Faraday cylinder with an ambient humidity of 33 % r.h. and 50 % r.h. The parameters measured ar the charge on the garment and its polarity, and the voltage on the body.

In these tests the T-shirt samples containing PVC fibres produced the highest charges on body and garment. Also samples which were either entirely made from polyester or having a polyster face caused shocks to some of the test persons. However, they were not regarded as uncomfortable. The lowest static charge was found on the samples containing either a cellulosic component (cotton or viscose) or wool.

The clinging behaviour of a fabric caused be static can be determined by measuring its adhesion to a metal plate after the material has been rubbed against the skin of a person.

Allergies

A survey disclosed that the number of reported alleric reactions to clothing is small and most of these cases have occurred when people in the making-up industry are handling large quantities of fabrics rather than from individual items. Not surprisingly, in the wear tests performed within this project no allergic reactions of the test subjects have been observed.

On the other hand it has been found that, contrary to public belief, formaldehyde freed from the textile is no larger source for allergic reactions than fibre harshness, the fabric's surface roughness or residues of dyes, resins, softeners, etc..

Thus, in accordance with the aims of the project test methods have been developed by which it is possible to measure quantitatively the sensorial comfort of fabrics and garments. Furthermore, the basic knowledge obtained of the interaction between certain fabric parameters and sensorial sensations has led to guidelines which help the industry to design fabrics and garments worn next to the skin, having optimal sensorial wear comfort.

IMPROVEMENT OF PROTECTIVE CLOTHING FOR BETTER WEARABILITY AND COMFORT

Basically protective clothing has to give protection against hazards, a,d jave good thermophysiological wear properties.

If wear comfort is not sufficiently good, protective clothing will not be accepted and - as the experience shows - consequently not be worn, even if this means the neglect of standing rules. The aim of this study was to show principle means to improve the comfort of protective clothing without impeding its protective function. As an example wet weather protective clothing has been chosen, because the protection against wettedness is an often needed necessity.

The physiological problem with this type of clothing consists in the demand to be watertight, however, simultaneously to possess a water vapour permeability. Without sufficient permeability in warm climates or during physical activity of the wearer, sweat evaporation from the skin, and consequently the evaporative heat flux, is impeded, unbalancing the thermoregulatory heat exchange between the body and the ambience. The consequences are decreased performance of the wearer and unpleasant wetting of the underwear, often leading to severe post-exercise chills.

Basically with this type of clothing good physiological properties can be achieved by improving the water vapour permeability of the fabric and/or by improving garment ventilation via an exchange between "microclimate" and environmental air through garment openings due to body movements or to outer wind movement.

To draw up an inventory of rain wear fabrics currently used in work and sport wear (e.g. garments used for cycling and yatching) suitable test methods have been defined to evaluate water waterfproofness, water vapour and air permeabilities in a way that provides information relevant to practical wear conditions.

To confirm adequate waterproofness conventional hydrostatic head and shower tests alone are not enough, because they do not simulate well enough the mechanical strain that a fabric undergoes in practical use. Therefore a funnel fold test has been developed, where a circular sample is folded twice and filled with a aqueous solution of a colourant, showing leaks at the folds as coloured spots. Also a blow through test has been introduced, replacing the collecting dish of the conventional shower test by a water absorbent tissue of which the weight gain during prolonged rain is determined.

To measure water vapour permeability the sweating hot plate (Skin Model) could be used. But because for a fabric manufacturer this method is somewhat too laborious to apply for instance in quality control a simpler device has been sought. In using different control dish methods defined in various national standards it was shown that the test results were strongly influenced by the water vapour pressure or relative humidity at the fabric's inner surface (see Figure 15) which explains the discrepancies often found between the various test methods currently used.

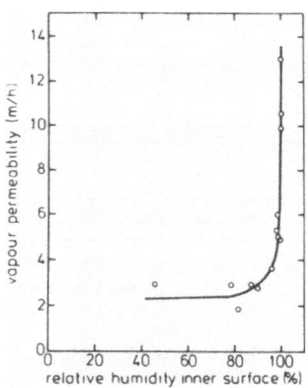

Figure 15
Dependency of a fabric's water vapour permeability
on relative humidity.

For the specific application on rainwear fabrics the most promising method for the determination of water vapour permeability turned out to be a tissue test, modifying DIN 53 122 by replacing the air over the sample by a soaked and squeezed tissue.

Air permeability was determined according to DIN 53 887. According to Figure 16 for some of the fabrics investigated air permeability is enhancing water vapour permeability, but on the other hand it is incompatible with waterproofness (see Figure 17). To be sufficiently shower proof the hydrostatic head has to be at least 40 cm. To be sufficiently watertight fabrics for work clothes have to possess a hydrostatic head of more than 130 cm (DIN 61 539), for particular sportwear (i.e. yachting clothing) the hydrostatic head has to be even higher. From Figure 17 it can be seen that fabrics fail this test if they have an air permeability of over 31 m^2s.

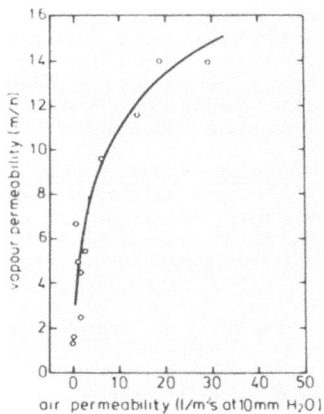

Figure 16
Correlation between air permeability and
water vapour permeability (at 65 % r.h.)

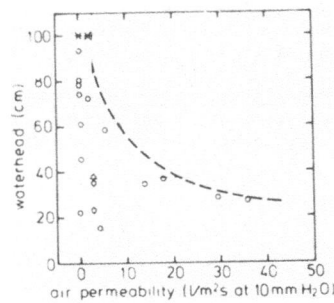

Figure 17
Correlation between air permeability and
hydrostatic head

The characteristics of some fabrics used for wet weather garments are now described :

Fabrics covered with a microporous film (laminates)
Fabrics covered with a microporous coating
These films or coatings possess micropores with a size much larger than water vapour molecules but much smaller than water drops. To be sufficiently watertight the film or coating must be water repellent. The water vapour permeability of these types of fabrics is good, in the case of some of the laminates even similar to "normal" garment fabrics.

Fabrics covered with a continous film
In contrast to the former fabric types the film has no pores but is hydrophilic, passing water vapour on a molecular basis, but not liquid water.

Like the microporous films and coatings these fabrics possess a good water vapour permeability combined with a waterproofness which meets the requirements for sportswear and work clothes.

Fabrics made of hydrophilic microfibres (fibre fineness ∿ 0.2 dtex)
The high density of these fabrics leads to a pore size at the crossings of the threads which is so small that liquid water penetration is prevented, resulting in a hydrostatic head large enough for use in most types of sportswear. Neverthless, water vapour permeability of these fabrics is very good.

Coatings with pores created by spark erosion by means of high voltage
With this technique the number of pores is far below the range needed for sufficient water vapour permeability. Also waterproofness is not good enough, not even for a shower proofness.

Multilayer fabrics
For moderate demands a fabric which by itself is not really waterproof can be made so to a certain extent when applied in a multilayer assembly, particularly one made of swelling fibres. However, such a swelling fabric loses its water vapour permeability when wetted.

Fabrics with water repellent finishes
Their waterproofness for most end-uses is not enough. Besides, the finish has to be renewed after each laundering or dry cleaning process.

Fabrics with a water vapour impermeable continuous coating
These fabrics though completely waterproof possess, no or negligible water vapour permeability, insufficient for physiological purposes.

*   *   *

To summarize, only the first four types of fabrics fulfil the demands of wet weather protective clothing to be simultaneously watertight but water vapour permeable. However, to decide, whether their water vapour permeability is really high enough for practical demands, wear tests with four subjects in a climatic chamber have been performed.

The tests have been carried out with two rain suits of identical pattern but differing in fabric (impermeable PVC-coated polyamide and semipermeable microporous coating on polyamide). As can be seen from Figure 18 the suits' pattern allowed practically no ventilation and additionally the subjects were wearing rubber gloves and boots and an impermeable hood and face mask. Thus, the influence of the fabrics' water vapour permeability alone on thermophysiological comfort could be evaluated free from other parameters.

The test climate was 14°C and 90 % r.h. In the climatic chamber the subjects cycled on a bicycle ergometer for 30 min. The work load was heavy, varying in degree and resulting in a metabolic rate up to 650 W.

Skin and rectal temperature, inner suit humidity, heart rate and the suits' outer surface temperature have been measured with sensor elements and the subjects' exhaled air has been analysed. Also the weight change of the test persons and clothing has been registered and the sensation of heat and humidity subjectively perceived by the test persons has been recorded by means of rating scales.

Figure 18
Subject with clothing outfit worn in the wear trials

The test results show that under the conditions chosen (cold/humid climate, heavy work load) the advantage in heat load on the person resulting in the semipermeable fabric against the impermeable sample is comparatively small. This is due to an increase in dry heat loss in the impermeable suit caused by an increase of the suit's inner and outer surface temperature, as a result of water vapour condensation by which condensation is released (see Figure 19).

34

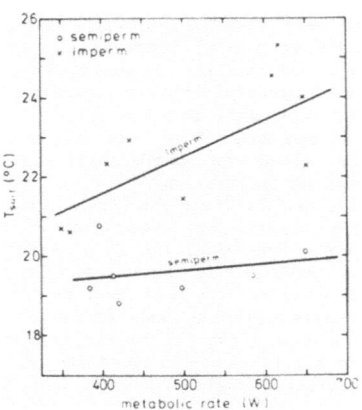

Figure 19
Outer suit surface temperature for the rainwear
tested as a function of metabolic rate

As can be deduced from Figures 20 this increased dry heat loss
(347 W as against 253 W with a 500 W metabolic rate) is to a large extent
compensating the lower evaporative heat loss in the impermeable suit
(14 W instead of 128 W). However, microclimate humidity, moisture accumulation and wetting of the underwear inside the impermeable suit were much
higher than in the semi-permeable suit, resulting in a somewhat better
comfort sensation in the latter.

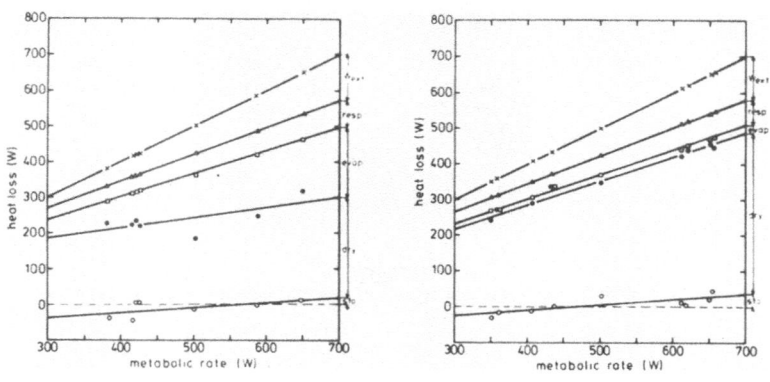

left : semi-permeable suit   right : impermeable suit

Figure 20
Cumulative representation of the components of the
heat balance in relation to metabolic rate

The results reported are only valid for the particular test situation chosen. In conditions of higher ambient temperature ($\sim$ 20°C) and lower work load the physiological advantage of a water-vapour permeable rainwear fabric is greater. However, the tests performed indicate that with regard to a physiological performance of rainwear as good as possible the fabric's water vapour permeability must be assisted by garment ventilation.

In order to determine which garment pattern will optimize ventilation, a quantitative test method was developed based on tracer gas technique. In this method a distributing and sampling harness of tubes for the tracer gas is fixed on the body beneath the outerwear (see Figure 21). The harness is divided between the upper and lower body. Thus, separate measurements for jackets and trousers are possible. Nitrogen is used as the tracer gas, the concentration measurements being effected with a mass spectrometer. By distributing the tracer gas under the garment until an equilibrium between inflow and ventilation is reached the latter can be determined as an absolute value in m^3/h.

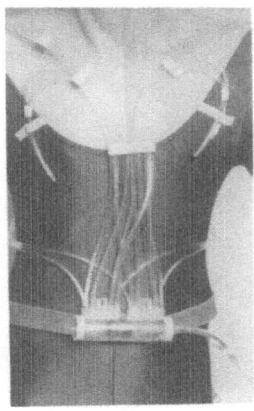 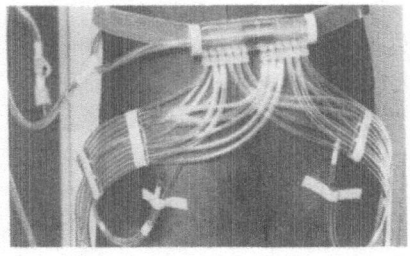

Left : harness for the upper body   Right : For the lower body.

Figure 21
Harness of tubes for the measurement of garment
ventilation by means of a tracer gas technique.

The tests performed with different garment models show that with vents constructed and arranged in a special pattern in a rain suit a ventilation rate caused by the pumping effects due to motions of the wearer up to $11m^3/h$ can be reached, noticeably improving the suit's moisture transport properties. With wind speeds over 2m/s the ventilation by external wind via these vents is effectively assisting the ventilation caused by body movements.

In conclusion, with regard to the fabric layers for wet weather protective garments the project has indicated that by combining different textile techniques fabrics can be developed with improved thermophysiological properties compared to the fabrics used to-day. For instance, such an innovation could be fabrics made of microfibres combined into suitable multilayer structures.

Additional application of the garment ventilation openings studied in the prototypes of the rainwear suits developed in the project can lead to improved protective clothing which really is combining good protection against environmental hazards with good wearability and comfort. For the European textile and garment industry this means an improvement of competitiveness in an important market section.

Besides this economic angle the project has enlarged the scientific basis for evaluating quantitatively the importance and influence of the. fabric water vapour permeability of fabric on the moisture transport properties of protective clothing as required for practical use.

CONSTRUCTION AND MAKING-UP OF GARMENTS

J. DESCHAMPS
CETIH

PRESENT SITUATION RELATING TO GARMENT DESIGN

Where clothing is concerned, purchase is most often guided solely by the appearance of the product. In the minds of the consumers, the features that are not visible in garments are assumed to be appropriate for their end-use, unless they stoically accept any deficiencies that may appear to them to be inherent in the desired product. Garment design and making-up are often the product of a succession of approximations, a large role being left to skills.

The operatives thus play an essential part in a making-up operation which relies greatly on their emperical knowledge of the raw materials, the performance they anticipate and the importance they attach to the various factors they take into account when making adjustments to their machines.

There is an ever increasing need for a more technical and scientific approach, as it is certain that a good number of companies will have to automate their production, so changing the role of the operatives whose technical expertise will have to be improved. It will be necessary to have the maximum possible power of defining the constructional characteristics of garments and the conditions for achieving them to produce the anticipated result directly.

The programme supported by the EEC comes just at the right time to remedy certain deficiencies and to provide the basis for rational action with reference to physiology and comfort, and to the design and make-up of garments. Physiology and comfort bring together a number of factors which may be summarised by analysing some of the functions of a garment.

- Physiological characteristics : "comfort"
  . fabric properties
  . freedom of movement in a garment

- Physiological characteristics : "appearance"
  . fabric)
) Aesthetic aspect : "fashion"
  . style)

- Physical aspect of the garment : "fit"
  . how the shape is created
  . selection of accessories
  . quality of make-up.

FUNCTION OF GARMENTS

Solely from the point of view of comfort, the significance of the various functions of a garment can vary according to circumstances and the subjects wearing the garments. Protection against climatic or mechanical influences does not in itself imply comfort : more subtle combinations of at least two other characteristics, physiological and psychological are also involved.

The physiological characteristic summed up by the expression "comfort" is closely related to fabric properties :

- involving transfers of air, moisture and heat between the individual and the environment,

- sensations tramsitted to the skin ranging from pleasant contact, through irritation right up to electric shock,

- or related to the freedom of movement of the garment the form of which may accentuate or attenuate certain properties imparted by the fabrics : air transmission, rain penetration, aggravation of perspiration or which may cause restriction of movement and even lead to organic disorders in the case of very closely fitting garments.

The psychological characteristic is synonymous with "aesthetic satisfaction". This involves the "appearance" of the garment and is where fashion is exploited. In the name of fashion, one person will accept garments considered uncomfortable by others who do not have the same psychological motivation. Here again, fabric and form recur, but from the aesthetic aspect. This aspect of fashion must be taken into consideration because this is what creates sales. Many cases are seen where there are flagrant contradictions between a comfortable garment and a fashionable garment. What did comfort matter to the smart set of last century when a lady would force her body into her corset to the point of collapsing in a faint ? Even today, fashion and comfort are often at variance. Some examples are provided by the rerouting from the original end-use of certain garments resurrected by fashion.

The form of the boiler-suit has been developed for the worker to be able to operate without his movements being impeded by restriction from his clothing. The fact that it is not divided at the waist reduces heat losses when the garment is used in winter on outdoor sites, but it does, however, have the drawback of not being totally suited to the physiology of everyday life. Fashion has overlooked this by adopting this type of garment for general use in sportswear and then for children's wear and women's wear by reducing the fullness necessary for body movement. For aesthetic reasons, comfort has been sacrificed for appearance.

The fashion for very closely fitting leather trousers follows approximately the same route. Originally this was designed for racing motorcyclists for protection from wind and in the case of falls. Leather is a material with sufficient impermeability to air to restrict heat transfer and enough friction resistance to soften a fall on the ground. The form of such garments should also fit sufficiently closely to the body to provide minimum resistance to wind and so not impede high speed.

Fashion has adopted this type of garment for town wear and its suitability for use has been distorted by the change of context. Although comfortable and well suited to the motorcyclist in the course of his activities, it has become a garment which does not allow air or perspiration transmission in heated draughtproofed offices. Nevertheless, smart men and women allow fashion phenomena to impose constraints which reason and their search for real comfort should forbid them to accept.

A third characteristic should be added to the two previous ones : the physical appearance and fit of the garment. This is dependent upon the actual construction including how the form is created, the selection of accessories, and the quality of make-up.

In satisfying consumer needs, the clothing industry must take into consideration simultaneously the demands of fashion and of the market, the current whims of its clientele, fit, comfort, and the adaptation of the form of the garment to the fabric selected for creating it. In addition, manufacturing costs and product quality must be correct if the firm is to remain competitive in the market.

PROGRESS OF WORK

Although fabric properties have a dominant influence on comfort, the form of the garment is equally important. However, the results obtained by any particular design, are frequently reached by common sense.

CETIH therefore concentrated its efforts on studying the fabric properties which it is necessary to take into account when designing a garment, especially if auxiliary aids are intended to be used at the design stage. At present the information supplied by fabric manufacturers is limited. It concerns essentially the composition and weight of the fabric. Literature sources, however, frequently give additional relevant information.

CETIH has also asked the British Clothing Centre to carry out that part of the programme involving characteristics to be considered during the actual manufacture, limiting the field of investigation to lockstitch seaming with simultaneous pressing. Thus, a clearer view has been obtained of the fabric characteristics to be considered in garment design. From this, deduction can be made of the cutting required to obtain the desired shapes and the equipment to be used in making-up and the machine settings can be predicted. These are the factors aimed at facilitating and simplifying making-up whilst achieving a more reliable result.

AIM AND SCOPE OF THE STUDY

This project has been limited to woven fabrics assumed to be free from faults. The aim of this study was to find means of enabling the industry to replace the successive approximation approach to garment development by a rational procedure leading quickly to the desired result. It must not be forgotten that the situation is such that a collection of several hundred garments has to be produced from a hundred or so fabrics in the space of a few weeks. This remit, the application of which is intended to be of economic advantage over the present approach and which will be shown to be an essential within the framework of general automation, requires awareness of the nature of the problems making this cautious approach necessary, their likely causes, and the factors to be assessed and controlled.

The study has been carried out as shown below.

PLAN OF THE STUDY :

- design :
  . definition of the problem
  . laboratory experiments
  . conclusions.

- making-up
  . problems related to making-up
  . conclusions

- future prospects.

STAGES OF THE STUDY

The principal stages are :

DESIGN

- Definition of the problem :
  . literature survey
  . enquiry in industry
  . identification of criteria by organising a joint experiment with industry
  . assessment and analysis of results by buyers

- Laboratory experiments :
  . study of distortions imposed in fabrics
  . characteristics of the fabrics concerned
  . seams for producing three-dimensional shape
  . identification and comparative study of suitable test methods
  . analysis of the variability of the results
  . case study of flared skirts

- Conclusion on the subject of design

MAKING-UP

- Problems related to making-up
  . machine/fabric interaction
  . guiding during sewing
  . study of control in fabric feed and influencing factors
  . possibility of integrating seam pressing.

- Conclusions on the subject of making-up

FUTURE PROSPECTS

GARMENT DESIGN

- Definition of the problem : Literature survey.
  The literature survey showed that numerous researchers had been interested in characterising the factors likely to be of influence within this study. Aspects which have been the subject of a number of publications and test methods are flat shear and bending stiffness. There is evidence of the difficulty presented by the characterisation of these parameters, the practical exploitation of which does not figure in the papers we have been able to obtain, with the exception of the work by Kawabata which involves "handle" rather than "sewability".

- Enquiry in industry.
  An enquiry has been carried out in industry aimed at discovering the criteria by which a particular fabric is associated with a particular form, ignoring any cost aspect. A particular fabric weight corresponds to a particular type of garment. This is the sole measurable value used. The other essential criteria are appearance, selected essentially on the basis of fashion, and "handle" which is assessed empirically by the maker-up by crumpling the fabric in the hand and sliding it between the fingers. Thus the evaluation of handle is

essentially subjective although a number of fabric properties such as softness, suppleness, stiffness, resilience, silkiness and drape are included.

    This enquiry revealed that technical criteria play only a very small role in fabric selection. Craft knowledge and experience acquired by operatives through successes and failures, has led to an informal un-expressed and even unconscious awareness which enables them to overcome selection problems. They can produce fabric/style combinations with an adequate success rate. Any search for less subjective methods has thus been neglected up to now and there is no common systematic procedure.

Identification of criteria by organising a joint experiment with industry
Organisation of the experiment

    The replies produced by the enquiry did not permit identification of the present means used in industry for taking account of fabric proper-ties in establishing garment patterns. An experiment was therefore organised to try and demonstrate the way in which fabric characteristics were allowed for by different makers-up who agreed to create "identical" models in the given fabrics. The garments specified were a coat, a costume (jacket and skirt), a dress and a skirt.

    The Committee for the coordination of the Fashion Industries (CIM) produced the style drawings of these four garment models in such a way that they incorporate reputedly difficult design and making-up assem-blies.

```
NAME : Amandine            REFERENCE : 310 101
PRODUCT GROUP :            INSTITUTED within contract
   Two-piece costume       between EEC and CETIH
SIZE OF MODEL : 40         DATE : 14/02/85
SIZE RANGE : /
```

Straight jacket, one-button, straight set-in sleeves. Sharp pointed lapels with raised top-stitching at 20 mm with .....

Figure 1

COMPANY : C.R.I.T.H.   NAME : AMANDINE   MODEL REFERENCE : 310101

	DEFINITION	REFERENCE	WIDTH DIMENSION	CONSUMPTION	INSTRUCTIONS / CONSTRAINTS	COMPOSITION LABELS
MATERIAL	Wool look fabric wool worsted / viscose P. & J. TIBERGHIEN - quality 400 g/m	101	90 cm	3.5 m	cut one way	wool 52 % viscose 48 %
TRIMMINGS						
LININGS						
SEWING THREADS						
FINISHING						

COMPANY : C.R.I.T.H.   NAME : AMANDINE   MODEL REFERENCE : 310101

BODY MEASUREMENTS CHART

	SIZE 40	TOLERANCES	OBSERVATIONS
Bust	90 cm	± 0.5	Use manikin CLEO, version 70, size 40
Waist		± 0.5	
Seat	95 cm	± 0.5	
Length 7th neck vertibra to waist	40 cm	± 0.5	

REFERENCE MEASUREMENTS FOR FINISHED GARMENT

Shoulder drop	14 cm	± 1.0	
Sleeve-hole depth	20 cm	± 1.0	
Jacket length centre back	65 cm		

Figure 2

Technical specifications have been prepared for each product, comprising a style drawing and summary description of the end-product for design purposes (Figure 1), a material data sheet and a body measurements chart (Figure 2) defining the size of the body to be clothed (a standard mannequin was given to each person so that the experiment was performed by everyone under identical conditions. Also provided were some preference measurements (length, depth of arm-hole etc..) for the finished garment as a target for the garment required, so that the products of each participant did not differ too greatly and could be compared.

Fabrics

The fabrics were selected from amongst those habitually used in making-up the type of garment in question to provide a range of four or five fabrics of sufficiently varied characteristics (weight, thickness, stiffness, drape...). These are shown in Figures 3 and 4.

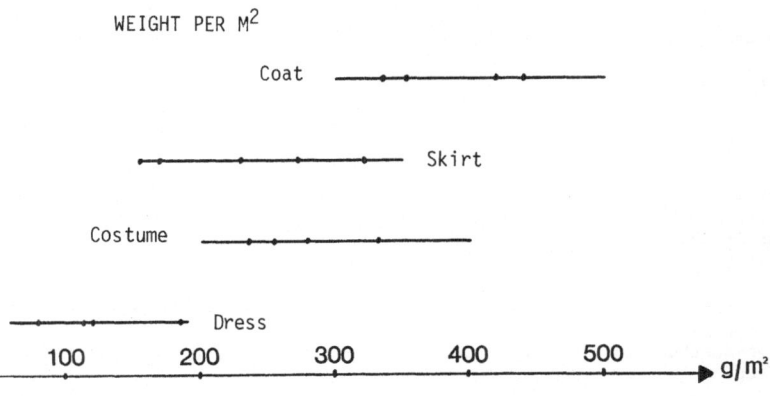

Figure 3

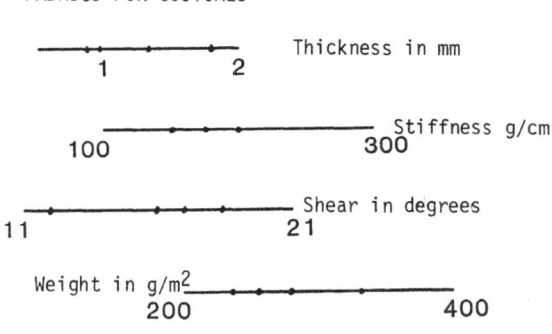

Figure 4

Makers-up

Each of the garments was produced in the range of selected fabrics by several different makers-up. The latter were chosen on the basis of their usual product range to form a panel as representative as possible of the making-up industry. A few stylist/designers were added to the panel.

Practical making-up

Each maker-up having the technical specification of the product, the standard mannequin and the various fabrics, produced the patterns and prototypes of the same model (sketch) in different fabrics (see schema). Latitude was allowed for modifying the patterns from one fabric to another so as to attain an equivalent result (appearance, three-dimensional shape, "hang" of the garment).

The garments produced were then analysed.

Participants : 4 to 5 makers-up per model

Garments produced for the study : 12 coats
                                              12 costumes (jacket-skirt)
                                              18 skirts
                                              20 dresses

Fabrics : 4 to 5 fabrics per model.

Results and conclusions

The results were analysed at two levels :

- comparison of garments made by the same maker-up with different fabrics.
- comparison of garments made with the same fabric by different makers-up.

In order to obtain a true economic perspective, these assessments were made not only by makers-up and specialists in clothing manufacture, but also by product managers and buyers from large stores and central purchasing authorities representing consumers.

Coats

No maker-up modified his patterns for the various fabrics used. ALl the accessors agreed in concluding that two fabrics were very suitable for the model selected, the third less so (particularly because of its appearance), and the fourth was considered by the buyers to be too stiff. There are no basic differences in appearance or pattern production between the different creations. The differences relate to points of detail : collar width, presence or absence of overedging and form of pocket flaps or cuffs. (Figure 5)

Costumes

The fabrics selected for the costumes were all considered to be satisfactory despite their varying physical properties. The costumes differed one from another essentially in the acuteness of cut of the collar, in the lapels of varying width and in the angle of slope of the pocket flaps. These differences do not significantly change the general appearance of the garments, the style of which remains essentially the same (Figure 6)

NO MODIFICATION OF
PATTERNS

FEW DIFFERENCES BETWEEN
MAKERS-UP

Shawl collar : width,
everedging

Cuffs : fold-back or
inserts.

Figure 5
Coats

NO MODIFICATION OF
PATTERNS

SLIGHT DIFFERENCES BETWEEN
MAKERS-UP

Collar : shape and width

Cut : shape

Pocket flaps : slope

Figure 6
Costumes

Skirts

The majority of makers-up did not modify the patterns because of the cloth. However, two of them used modified patterns for one fabric (the thickest and stiffest). One of them in this case even replaced the pleats intended to give fullness to the garment by tucks. Pleats or tucks : the overall difference in fullness is not significant. The spokesmen for the consumers received this modification favourably although the fabric was not considered to be suitable for producing the model selected. However, one of the fabrics selected which was unanimously considered to be too fine and soft by the makers-up for producing the model specified, was judged by the buyers to give the best result.

The differences found between the various skirts relate essentially to the form of the corselet, with varied accentuation of the pointed shape. The three-dimensional shape of the different garments is variable, but the skirts of a more "hobbled" character with a stronger barrel form appear to the buyers to correspond best to the style required for the model (Figure 7).

FEW MODIFICATIONS TO PATTERNS

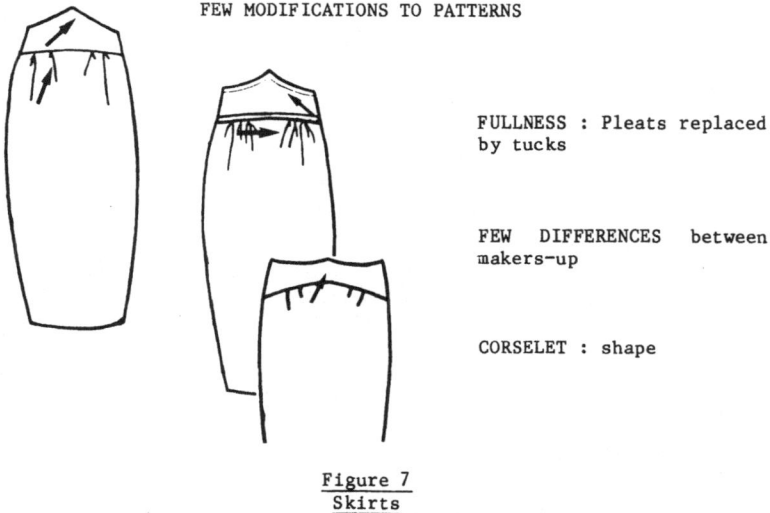

FULLNESS : Pleats replaced by tucks

FEW DIFFERENCES between makers-up

CORSELET : shape

Figure 7
Skirts

Dresses

Three of the four fabrics selected were judged unanimously by the makers-up to be satisfactory for producing the model, although their properties varied. Reservations were expressed in the fourth case. The majority of makers-up did not consider it necessary to modify the garment patterns because of the fabric. Just one of them changed the sleeve form to impart a particular style to the models. For the two fabrics with lower shear characteristics, the sleeve head of the garments has pleats whilst in the two other cases, the three-dimensional shape is created by introducing a biassed seam on the tope of the sleeve.

The dresses are the garments which vary most between makers—up. These differences arise essentially in the interpretation of the three-dimensional shaping in the shoulders, in the waist, in the hips and at the base of the garment (figure 8).

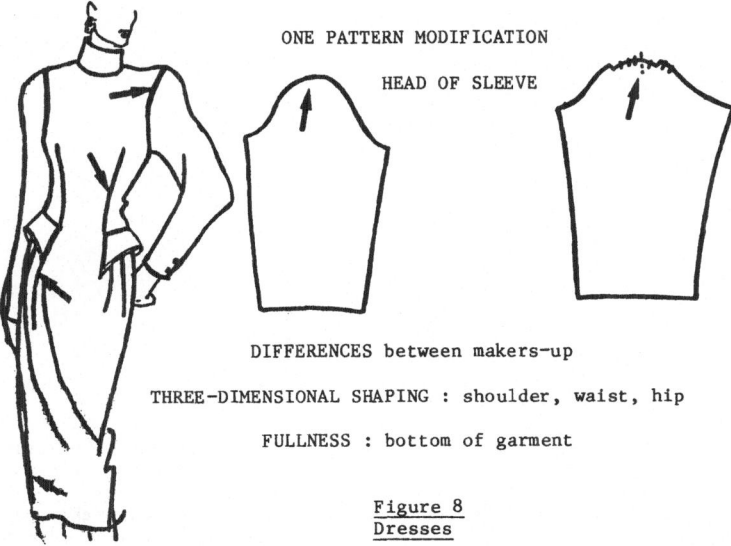

ONE PATTERN MODIFICATION

HEAD OF SLEEVE

DIFFERENCES between makers—up

THREE-DIMENSIONAL SHAPING : shoulder, waist, hip

FULLNESS : bottom of garment

Figure 8
Dresses

RESULT OF THIS EXPERIMENT

The production of these garments shows that the design and making—up of the same product from a style drawing and technical data sheets by different makers—up in most cases does not result in basically dissimilar garments, even though differences of detail do exist. Fullness is not, however, always interpreted in the same way, marking the influence of classical style or fashion from company to company. Changes of fabric rarely cause changes to be made in patterns. Differences in fabric properties are only taken account of if their variations are very great. The craft knowledge of the operatives enables them to solve any problems which arise. Generally, they adjust their patterns according to the cloth which will present the most problems, so that changes will not have to be made for other fabrics. This simplifies organisational and manufacturing problems, perhaps to the detriment of a more elaborate design.

FACTORS TAKEN INTO CONSIDERATION INTUITIVELY BY CLOTHING MANUFACTURERS IN DESIGN

Looking at the results of the experiment carried out, one may wonder what are the characteristics taken into consideration intuitively by makers—up based on their trade experience, and what should be examined in the future when, in designing and creating garments, the "craftsman" will be using information technology tools and automated equipment. Some of the factors appearing in the table below must be taken into account at the design stage.

Table I
Factors to be taken into account at the design stage

Properties	Pattern development	Making-up method & equipment	Garment aesthetics	Physical Wearer Comfort
Fraying	x	x	x	
Seam slippage	x	x	x	
Seam pucker	x	x	x	
Stretch	x	x		x
Bagging	x		x	
Shrinkage	x			x
Skew	x		x	
Dimensions of checks	x		x	
Eveness of checks	x		x	
Skin irritation	x			x

Fraying

This relates to the spontaneous easy release of warp or weft yarns from a piece of fabric along a cut edge resulting in the formation of a fringe. This fault must be taken into consideration in designing the model in order to allow sufficiently large seam values. The making-up equipment must provide for either overlocking or adhesive sealing of cut edges in order to preserve the aesthetics and solidity of the garment. Similarly, the cutton holes and piped pockets must be reinforced before cutting by the application of a hot-melt adhesive.

Seam slippage

Seam slippage occurs in a fabric when low tensile stress caused by simple movements of the garment wearer causes the threads held by the sewing thread to come away from those of the body of the fabric, thus forming an unsightly gap. This undesirable characteristic needs to be taken into consideration in the pattern development. If the slippage is excessive, then the fabric will be unsuitable for clothing purposes. If it is less pronounced it will be possible to find means of counteracting it by using these fabrics in models of sufficient fullness to keep wear stresses low, by increasing the freedom of movement allowed in the garment or by providing a more suitable lining which will withstand the stress instead of the outer fabric, by providing adequate seam values, and by using types of seaming which reduce the slipping tendency.

Seam pucker

A seam which wrinkles up after sewing is said to "pucker". This fault is of considerable detriment to the garment aesthetics. The fault arises from the structure of the fabric and from the way the seam is made. The tightness of the fabric structure, its "set", is the product of the number of its constituent warp and weft threads per centimetre and their thickness. It is possible to calculate the maximum limit of set based on the fabbric weave. When this is reached there is no more free space between threads, and insertion into the structure to the yarns

necessary for the seam will cause the fabric to become undulated in a way
that cannot be remedied by pressing. The execution of several rows of
stitches close together (double needles or overlocking) will accentuate
this waviness. This factor must be taken into account when the product is
being designed to avoid seams comprising multiple rows of stitches
parallel to the warp or weft direction of the fabric. In making-up, the
sewing threads adopted must be the finest possible, the number of
stitches per centimetre reduced as far as possible, and the stitch used
be of such a type that the interlacing of sewing threads occurs outside
the layers of seamed fabrics (chain stitch).

Stretch
The stretch of fabrics in a direction parallel to their constituent
yarns is generally low, but some fabrics containing elastomeric yarns are
designed to have greater stretch properties. The garments produced with
these fabrics are intended to fit close to the body and the stretch of
the fabric must be taken into account during development of the
patterns. Suitable sewing equipment is also necessary if unsightly
puckering is to be avoided at the seams due to extension occurring during
sewing.

Bagging
Bagging results from the incomplete elastic recovery of a fabric
after an imposed extension. This is seen especially in close fitting
garments at points where the fabric is subjected to repeated stresses
(knees, elbows etc..). This fault which spoils the aesthetics of the
garment may be allowed for during garment design, either by the use of a
more appropriate lining or by creating a fuller garment. In these ways
the stress is reduced as far as extension is concerned.

Shrinkage
The shrinkage occurring in a fabric during the successive pressing
treatments needed during its conversion into garments must be allowed for
at the design stage in order to increase its dimensions and retain sufficient freedom of movement, thus producing a finished garment of adequate
size to provide wearer comfort. The same applies if the shrinkage occurs
during normal care operations such as laundering or drycleaning.

Skew
A fabric is said to have "skew" when the weft threads are not
perpendicular to the warp threads. When this fault is of sizeable proportions, it spoils the balance of the garment and damages its aesthetic
properties. The influence of skew on appearance is much greater in
fabrics with visible weft, especially in check patterns, than in plain
fabrics. Skew, or more precisely the difference in skew between the new
fabric and the washed fabric, is also responsible for balance faults in
garments, in particular trouser legs which corkscrew after laundering.
This is an especially unsightly fault occurring in jeans and similar
cotton trousers. The development of this fault takes this course. The
loomstate fabric having considerable skew is straightened by stentering
and the garment is cut from this low-skew fabric. At its first laundering, the fabric relaxes and tries to resume its initial configuration
close to that of the initial skew. If the difference between the
straightened state and the laundered state is excessive, the fabric
rotates to find its natural balanced position.

Dimensions of checks

When designing pleated skirts in check fabrics of contrasting colours, it is important to take account of check size to calculte the leat dimension to be adopted, so that after pleating, the garment is of uniform shade with the tops of the pleats in all cases located in the same zone of the check pattern. It can be made very difficult or even impossible to produce a multiple lay for cutting garment panels. The same applies to matching the checks in the different pieces during garment assembly. The only option for using this kind of fabric is not to make a match, but this can spoil the aesthetics of the garments.

Skin irritation

Some woollen goods are harsh and unpleasant to the touch and may prove to be irritant. Their use must be avoided for garments coming into direct contact with the skin. This factor must be considered when designing the garment and a sufficiently thick lining placed between the skin and the fabric. In the case of children's wear, such as trousers, lining may not suffice. Children generally refuse categorically to wear anything "scratchy" and avoidance of such fabrics is recommended.

Laboratory experiments

Study of imposed deformations in the fabrics

The results of the enquiry and practical experience suggest that, with the exception of extreme cases, makers-up do not consider the ease with which fabrics can be moulded when they are designing a model. This finding needed closer investigation. Moulding may be divided into that imposed in fabrics by cutting pieces to that they adopt the shape of the body to be clothed and that formed spontaneously under the influence of the weight of the fabric, this being the source of flares. (Figure 9)

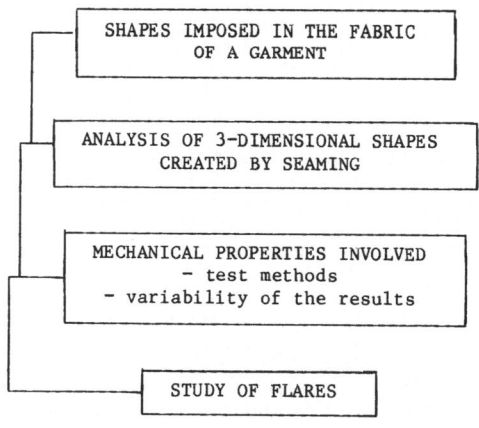

Figure 9
Laboratory experiments

51

Observation of the shaping imposed in a fabric at different loca-
tions in a garment allows the corresponding moulding of the fabric to be
noted, made more obvious by the use of a loosely woven fabric.

This phenomenon is particularly marked at the top of the sleeves and
at the bust but less marked at the waist. (Figures 10 and 11).

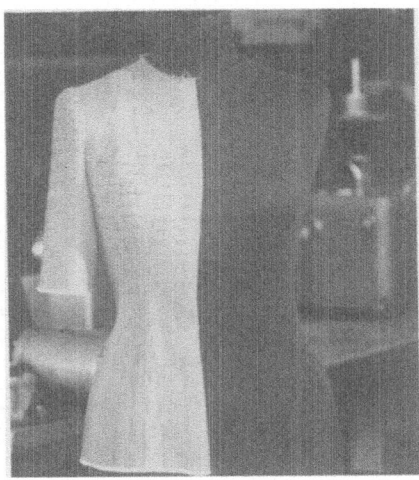

Figure 10

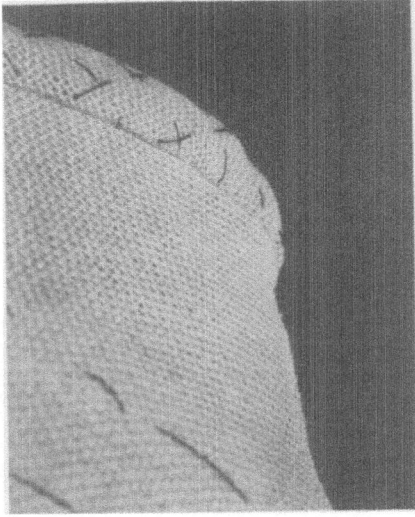

Figure 11

Seams for the creation of three-dimensional shape

As knitted fabrics have been intentionally excluded from this study, as also has the creation of shape by setting a fabric on a former (heat mouling), three-dimensional shape must be created by cutting out and then the seaming the garment pieces. All the seam configurations may be summarised by the diagram (figure 12). They differ in the values given to :

R : finite for infinite
L : equal or unequal
α , ß : variables

accepting that the right and left portions are part of the same piece of fabric or two pieces assembled into one fabric panel.

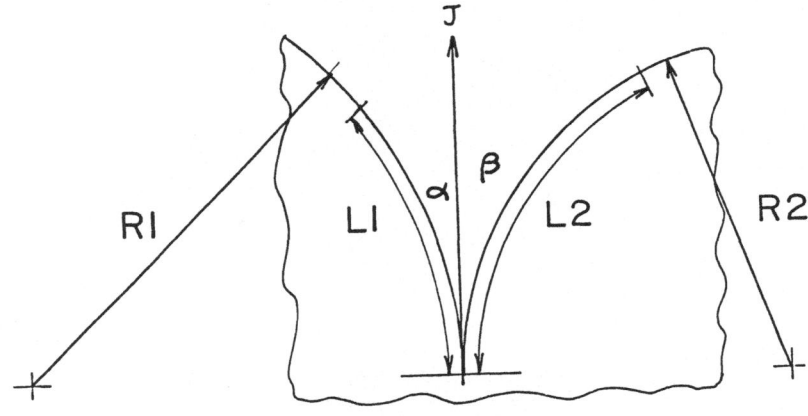

Figure 12

By forming an equation with the values of these geometric elements which allow a given theoretical shape to be obtained and with the bending and shear properties of the relevant fabric, it is possible to define the limits of the various possible seams.

The seams which enable three-dimensional shapes to be created in fabrics may be reduced to four principal types, allowing a more system-atic analysis to be carried out.

Assembly of two curves of identical radius and length at the intersection of whose tangents an angle and greater than zero is formed.

This case covers all darts and assemblies of two identical elements by a non-rectilinear seam.

Assembly of two curves of identical radius but different length

The small differences in length are distributed by take-up, intended to mould the rounding of the back on the shoulder or to allow the turn-back of a collar, a flap or a cuff of a garment. Any large differences are most frequently distributed into tucks in order to create a specific three-dimensional shape for the garment.

Assembly of two curves of different radius and identical length

This type of seam covers the fitting of collars to the body of garments, some waistbands for skirts or trousers, seams for assembling the sides of these garments and certain forms of cut pieces.

Assembly of curves of different radius and length

The form representative of this category is the seam attaching the sleeve to the arm-hole with distribution of take-up.

Only those seams with elements of differing length (distribution of take-up) and with radius equal or unequal require shear of the fabric threads along the seam.

TABLE II

ELEMENTS ASSEMBLED		LOCATION IN GARMENTS
LENGTH	RADIUS	
$L_1 = L_2$	$R_1 = R_2$	Darts
$L_1 \neq L_2$	$R_1 = R_2$	Shoulders - cuff Collar - yokes
$L_1 = L_2$	$R_1 \neq R_2$	Collar - side of skirt - Trousers - shoulder strap
$L_1 \neq L_2$	$R_1 \neq R_2$	Waist - Base of sleeves Sleeve attachment

MECHANICAL PROPERTIES INVOLVED

Two properties allow fabrics to assume approximately spherical shapes : bending stiffness and shear.

Bending stiffness is the ability of fabrics to bend under the influence of their own weight or an external force. This governs the "three-dimensional shape" of the garment, its aesthetic properties, and its comfort. "Weighting" allows it to be modified locally. Conversely, excessive stiffness is more difficult to correct. It complicates making-up aperations (difficult to mark pleats, impossible to open seams which need to be compensated by artifices such as overedging and makes it difficult for the fabric to pass through the folding guides.

The overall appearance may be changed by weighting certain parts of garments, especially by accentuating the difference in stiffness between warp and weft. This procedure has a very distinct influence on comfort. The bending property by itself only allows the fabric to adopt the forms of the bodies to be clothed at the expense of having a number of cut pieces, which despite everything would lead to acutely angled joints if the fabric was unable to exercise its shear properties. This is what makes apparel fabrics different from industrial fabrics, paper, or plastic sheets (Figures 13 & 14).

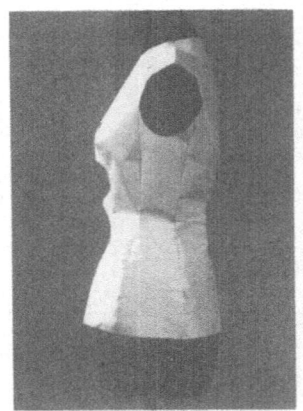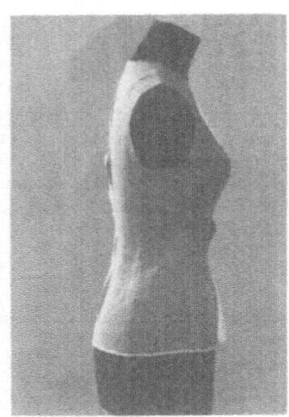

Figure 13

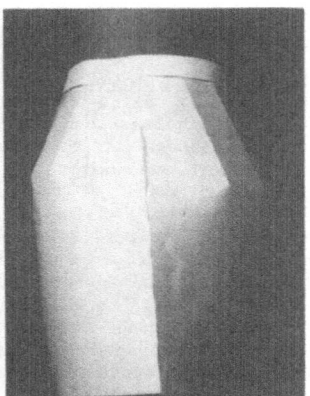

Figure 14

55

Shear

Shear stiffness may be defined as the way in which a fabric resists any action attempting to change the angle formed by the intersection of the warp and weft threads when made to pivot around their points of intersection. Various types of apparatus are described in the literature, the most recent being the Kawabata system which carries out tests automatically. Generally, this gives fabric rankings in the same order as less sophisticated methods.

The shearing property enables a three-dimensional shape to be obtained with a limited number of cut pieces. It also allows the fabric to be squeezed together and taken up without unwanted creases appearing, a principle used in attaching sleeves. This property also influences wear comfort as it partly governs the ease with which the garment is able to follow body movements. The importance of the shear characteristics in creating a three-dimensional form may be demonstrated by attempting to make a sphere firstly with a woven wire cloth and then with a sheet of paper. What is possible with the woven wire cloth cannot be achieved with the sheet of paper without cutting out, although the latter is much more flexible. The essential difference between the two is the total inability of the sheet of paper to shear, a property essential for simultaneous bending in two directions. (Figures 15, 16 and 17)

Figure 15

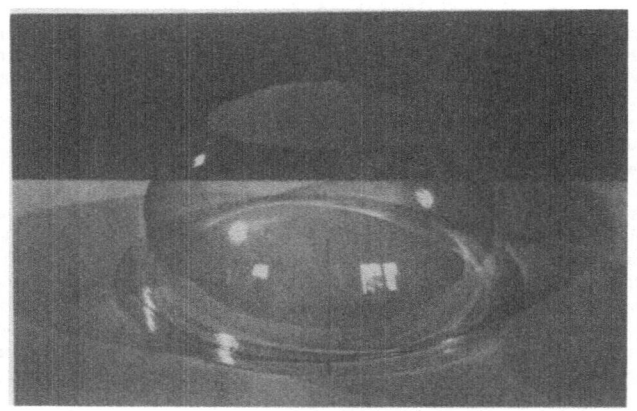

Figure 16

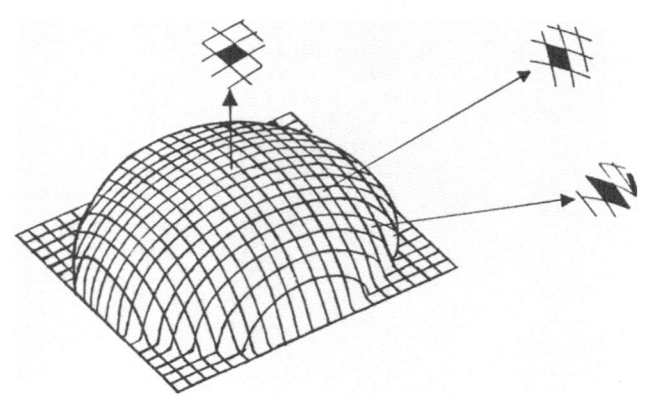

Figure 17
Warp and weft bending + Shear

This shear deformation corresponds to a change in orientation of the interlaced structure. The squared structure of the woven fabric becomes lozenge shaped and it is this interplay of diagonals which with a material of low extensibility, a woven fabric, enables the shape of a portion of a sphere to be assumed with a limited number of cut pieces.

Identification and comparative study of test methods

The methods described in the literature do not allow measurement of the simultaneous bending in several directions and the resultant shear, properties which are essential in the production of garments. The simultaneous evaluation of stiffness and shear by a device consisting of a hemispherical form and a series of rings of increasing diameter permits a rapid ranking of fabrics to be obtained. The fabric is placed on the hemisphere so that the centre of the fabric disc coincides with the summit of the sphere. It is found that the fabric rests perfectly on part of the sphere and then hangs around it in the form of waves, only touching it at certain points. The fabric rests perfectly on the sphere in so far as it is allowed by its shear properties and its ability to bend in all directions, and determination of the limit of contact can give a general assessment of the ability of the fabric to assume a three-dimensional form and thus be converted into a garment. To facilitate measurement, the limit of contact is sought by placing on the sphere-fabric assembly a succession of rings of increasing diameter. The limit is reached when the fabric assumes a wave form below the ring whilst the part above the ring is in perfect contact with the supporting sphere. (Figure 18)

Figure 18

58

The unwanted creases can only be re-absorbed by the removal of the material giving the flare formation beyond the contact zone, or by inserting "darts". This disadvantage of a test carried out on a sphere lies in the fact that with a slight variation in position of the interlaced structure (a change in latitude of some kind), there may be a corresponding important change, although small in absolute value, of the dimension of the diagonals. The diameter of the rings defining the contact zone of the fabric cap should vary by a millimetre at a time. The method can be used for a rapid assessment of the shaping propensity of a given fabric.

It is also possible to apply a tensile test to a bias-cut specimen to evaluate the length variation of the diagonals under a given load.

Variability of the results

The numerous tests performed on a whole range of fabrics with the objective of defining the range of each characteristic (which also enabled the industrial experiment plans to be evolved) were repeated in order to assess the variability of the results. As with many other textile properties, the variability is great. Factors to be added to the inherent variability in fabric structure include significant variation due to finishing treatments, as illustrated by the following example. The subjects are two identical fabrics, one of which has been subjected to SIROSET treatment :

TABLE III

	UNTREATED	TREATED
Bending stiffness	26.8	18.3 g f cm^2/cm
Shear	128.0	72.0 g f/cm degree

Is there any certainty that finishing treatment will be identical from one piece to another ? Do these differences in fact have any influence on garment construction ? Without disputing the value of theoretical models, these results tend to show that only tests performed on the finished fabric can provide useful information.

CASE STUDY - FORMATION OF FLARES IN SKIRTS

The whorl shape assumed by the fabric at the base of some skirts is difficult to predict at the pattern development stage. Consequently the creation of several prototypes is often necessary for the maker-up to develop models of this type, incurring extra expense. An investigation has been undertaken to see if the shapes can be predicted and if the fabric properties have any influence.

Method of creating flares

The study covered the two most common methods of creating flares. In the first, darts were inserted around a basic skirt. (Darts at waist level create the need for additional fullness at the base of the garment, thus imparting three-dimensional shape). In the second, each skirt panel was constructed at an angle, the waist and base lines of the garments in such cases corresponding to two arcs of concentric circles. By varying the angle at which the skirt is constructed, garments of varying fullness are created.

Fabric selection
The fabrics selected for the experiment were chosen in collaboration with weavers from a range suitable for making this type of skirt.

Method of measurement
As preliminary tests had shown the great similarity of form of the profiles of the one skirt at different levels, the measurements were made only at the bases of the skirts which were photographed from above. The curve areas were determined by weighing, following reproduction on paper of known constant weight per unit area.

Results and conclusions

Reproducibility of the number and form of flares
The study demonstrated the fluctuating way in which flares form, depending on fabric movements, and the impossibility of determining any preferential configuration. For one skirt, the number of flares may vary fluently from 4 to 8 and there are parallel changes in their form. The area of the projected curves fluctuates equally to a large degree for the same garment. Measurement of this area does not allow study of the development of skirt form as a function of patterning or of the fabric used.(Figures 19, 20 and 21)

HISTOGRAMMES DES NOMBRES DES GODETS

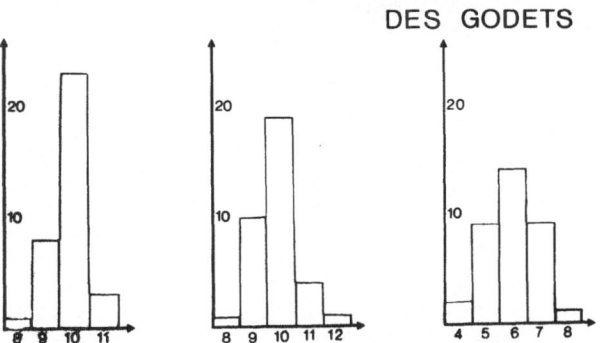

Figure 19
Histrograms of the numbers of flares

60

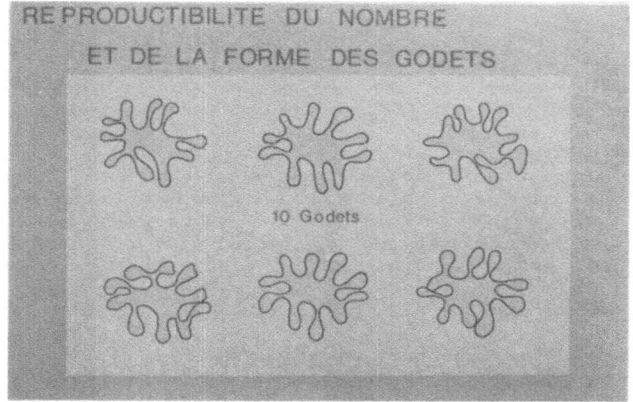

Figure 20
Reproducibility of the number and form of flares -10 flares

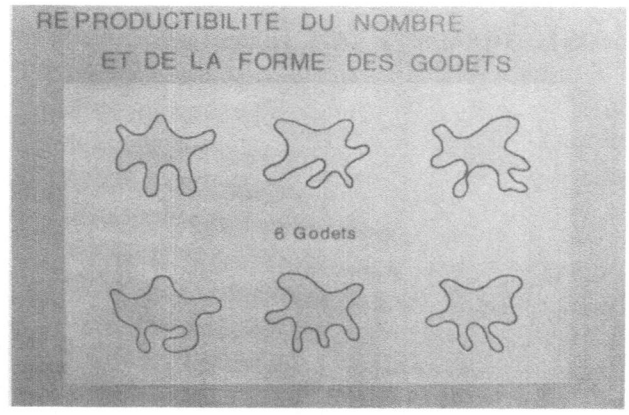

Figure 21
Reproducibility of the number and form of flares - 6 flares

Evolution of the number of flares according to fabric properties

Investigation of the number of flares formed in skirts produced with different fabrics (bending stiffness between 20 and 380 mg.cm) and identical patterns, does not allow detection of any significant influence of the nature of the fabric on flare formation, given the random character of flare location. However, when using fabrics of very high bending stiffness (of the order of 2500 mg.cm) which in current practice are considered unsuitable for the garment being considered, the number of flares drops significantly (5 instead of 10), with the appearance and comfort of the garment changing entirely.

Evolution of the number of flare according to patterns

Patterns have a major influence on flare formation. The number of flares formed depends on the fullness of the skirt at hem level. (Figure 22).

Number of flares

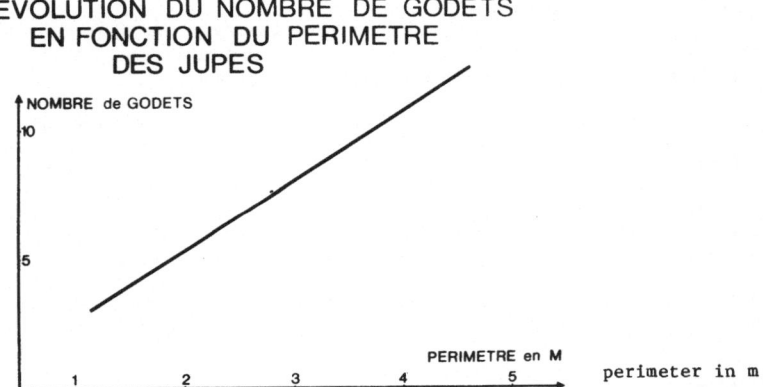

Figure 22
Evolution of the number of flares as a function
of skirt perimeter measurement

In a publication following our first report giving these results, Japanese researchers reached identical conclusions. For a given size and length, the experimentally obtained equation :

N = Number of flares = 2.6 x hem periphery (in m) (equation 1)

gives a good fit for this phenomenon. It may be reformulated to include the following :

T : waist circumference   )
H : skirt length   ) expressed in metres
$\emptyset$: the angle at the centre in which the skirt is recorded,
    expressed in degrees,

which enables it to be summarised as follows :

N : 2.6 T + 0.0454 H $\emptyset$ (equation 2)

Industrial application of the results

Using equation (2) the maker-up can determine the angle for cutting his patterns to attain the desired number of flares, thus avoiding actual trial garment production, it being understood that because of fabric movements the number of flares will not remain constant but will vary around the predicted value.

Conclusion relating to design

This study provides answers to hypotheses relating to the stiffening of fabrics which would destroy their essential characteristics : suppleness, bending in several directions, and shear.

If prediction of cutting out requires advances expensive methods, these will be outside the scope of makers-up. If several hours of testing by trained staff are needed to give information enabling a garment to be planned from a given piece of fabric, the procedure is bound to fail because the characteristics of the next piece will be different.

In the case of the textile industry, efforts must be directed towards improved regularity in the basic properties of fabrics including skew, cloth width, shade variation, and dimensional stability.

In the clothing industry, an improved knowledge of cloth geometry is needed so that satisfactory shape may be created without repeated re-adjustment. Some modifications reveal a standard of perfectionisms rather than of necessity.

CHARACTERISTICS HAVING AN INFLUENCE ON MAKING-UP

AIM OF THE PROJECT

The purpose of the work undertaken was to study certain fabric-machine interactions in the making-up of garments. At present it is the operative who makes any machine adjustments he considers useful for compensating fabric properties and machine performance. It sometimes takes two weeks for a trained operative to handle efficiently a new fabric. This explains why few factors have been quantified.

Rationalisation of current methods and automation of production imply that the factors to be controlled have been identified and that it is possible to measure them. Fabrics are the source of 12 % of problems created by the introduction of automatic equipment in the clothing industry. Rational solutions are seldom found.

Problems arise at all stages of production.

Scope of the study
The study related especially to the guiding of fabrics during sewing
and to the possibilities of combining pressing (seam opening) with
seaming. (Figure 23)

COMMON FABRIC/MACHINE INTERACTION PROBLEMS

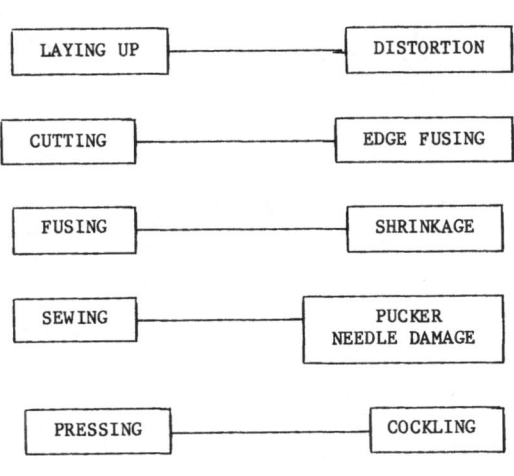

Figure 23

FABRIC FEED IN SEWING
This part of the study was aimed at identifying and quantifying the
fabric characteristics which affect fabric movement through the folding
guides and other areas over which they are passed. Although the literature is rich in descriptive accounts, few results of measurements have
been published. In passing through the guides, the fabrics are subjected
to compression forces, which above a certain limit, cause them to buckle.

Bending
The folding guides cause the fabric to bend, sometimes in zones
where the edges are steeply curved. Their efficiency is influenced by the
fabric's resistance to bending and on the hysteresis of this property.

Friction
Account must also be taken of frictional forces in passing through a
folder, but fabric stiffness often masks this characteristic. A simple
test that allows evaluation of the tendency of a fabric to be deformed
during sewing consists of driving a disc-shaped piece of fabric on a
surface of known roughness in a rotary movement imparted by a circular
impeller placed at the centre or on one radius of the disc. The conditions under which the cloth buckles are recorded : the diameter of the
impeller and its position in relation to the centre.(Figure 24 and 25).

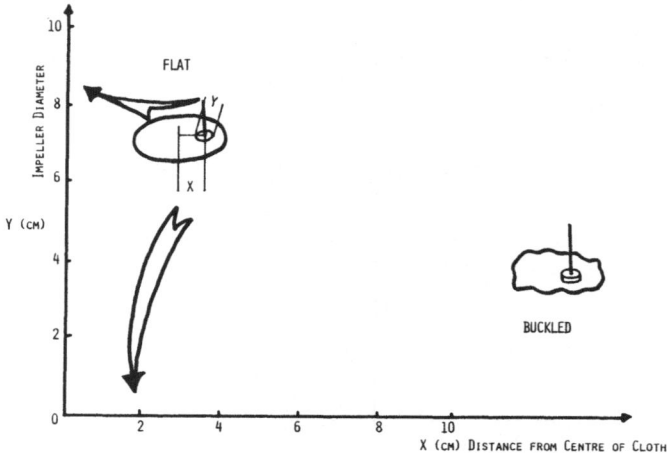

Figure 24

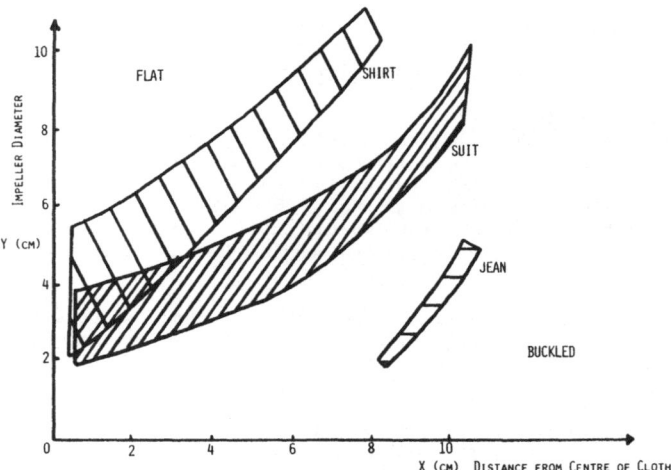

Figure 25

65

Control of feed
Feed irregularities cause variations in stitch length. The dozen or so factors influencing feed have been the subject of investigation and measurement.

Special test apparatus
For automation purposes, it is important to know how a fabric is fed without any action being taken on the part of the operative. A special sewing test that has been developed for this purpose involving only the fabric/feed interaction provides information on stitch density variation, fabric extension in several directions, displacement of thicknesses, and variations in stitch balance. The test, which enables the effects of each adjustment to be compared, consists of sewing the edges of two superposed circular pieces of fabric held together by one stitch at their centre. Frictional forces are eliminated by a cushion of air between the lower disc and the machine bed. A strain gauge was connected to a microprocessor to make a detailed study of the movements of the feed dog and of the presser foot. This enabled the degree of penetration of the feed dog into the fabric to be determined bearing in mind that the sharpness of the teeth of the dog may vary. (Figure 26)

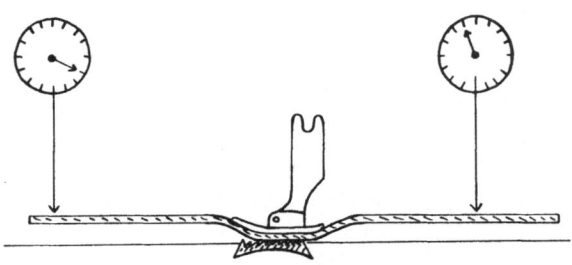

Figure 26

Results of the feed experiments
Stitch density is influenced by the speed of sewing. It is more irregular at high speed and a loss of control occurs during acceleration with the fabric skidding. To regain control it is necessary to increase the pressure of the presser foot to be a level above the critical threshold. (Figure 27 & 28)

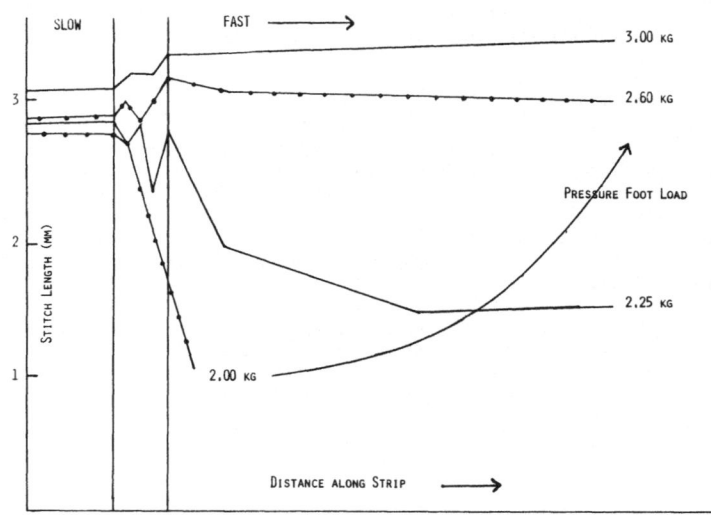

Figure 27

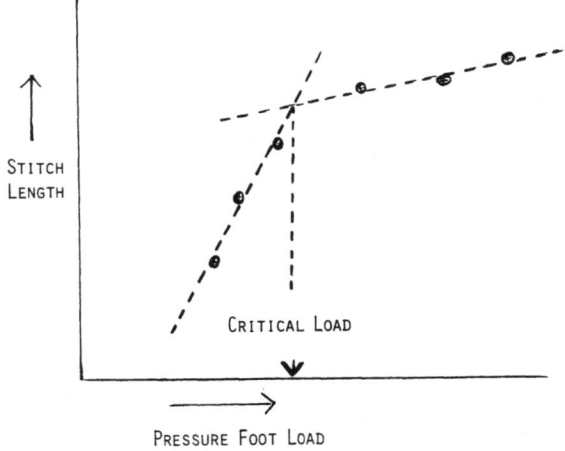

Figure 28

Influence of the fabrics

Their compressibility, extensibility and coefficient of friction are involved. Complex reactions such as fabric extension between the guide edge of the feed dog and the extension of the presser foot would require more elaborate studies. Bouncing of the presser foot contributes to loss of control of sewing. The compressibility of the fabric dampens the oscillatory motion of the bar of the presser foot. Measurement of the energy absorption on several fabrics subjected to an impact test have revealed anomalies in the cases of fine fabrics.

Conclusion relating to fabric feed

Control is dependent upon the following factors :

- Fabric : orientation, compressibility, thickness and extensibility.

- Feed dog : type, setting, movement and wear.

- Presser foot : load and inertia.

- Resistance exerted on the piece by the guides and the folders.

On-line pressing

The integration of seam pressing during sewing is economically justified by the reduction of the handling operations but it does demand that some compromises are made between production speed, pressing conditions and permitted physical dimensions, in order to allow a wide variety of fabrics to be handled.

In addition to its composition, fabric construction needs to be considered : if this is not symmetrica different, treatments may be required to attain good seam opening in various directions. Bending stiffness is another factor for consideration. Seam opening is more difficult with lightweight soft fabrics as the seam has a tendency to fold along one side of the seam line. The study has enabled actual data to be collected relating to on-line pressing of commercial fabrics under a variety of conditions.

Control of machine automation

At the present time, the influences of the various parameters are rarely described in terms other than "low values of ..." or "high values of ..." without quantification. The tolerable variations in parameters need to be quantified with regard to their influence on product appearance, without striving for a futile standard of precision for each measurable property. The selection of materials to be used on making-up machinery would certainly benefit more from tests of "associated parameters" than from mathematical calculations made from a weighty combination of a series of tests. The BCC fabric test for twist-buckling is one example of a simple effective test.

Several approaches are possible to automatic machine settings : the results of a range of selected tests may be used or machine control sensors may be incorporated. Improved methods of machine adjustment must be quicker and more precise than the present emperical methods.

Modification of machines, processes and fabrics

There is no set pattern of machines and procedures in clothing manufacture. They have been subjected to many changes in the course of the past twenty years, sometimes resulting from problems arising by the introduction of new fabrics, as in the case of the development of over-feed on laying-up carriages, with the aim of making them suitable for stretch fabrics.

Automation which demands a greater degree of precision and less variability in fabric properties and machine performance would justify an investigation being carried out to determine whether it is necessary to control environmental humidity in making-up workrooms. Where fabric development is concerned, the approach of the textile engineer may be shown to be of greater importance when the parameters of the garment production process are more precisely defined in a technical specification. Fibres and fabrics will thus be designed to meet specific conditions. One example is provided by fabric structure : experiments with a triaxially woven fabric have demonstrated the possibility of producing stable open-weave fabrics having a lower degree of anisotropy than conventional fabrics.

The fabric test programme conducted within this project has provided examples of measurable differences for friction, shear resistance and reaction to sewing when the same basic fabric is subjected to different finishing treatments. There is an ever increasing number of mechanical and chemical finishing methods and their influences on the process of clothing manufacture need to be monitored. Closer cooperation must be established between the maker-up and the fabric finisher so that improvement of the fabric properties does not affect their aesthetic and comfort qualities, or vice versa.

Conclusions

The investigation has provided answers to some questions relating to fabric/machine interaction. A sewing test developed within this project demonstrates the effects of factors influencing feed.

In order to attain uniform stitch density in a lockstitch machine with drop feed, the load applied by the presser foot must exceed a "threshold" value. This value varies in proportion to the resistance exerted on the fabric fed towards the needle.

Sewing performance, governed by a complex combination of machine characteristics and fabric properties, cannot be predicted if only certain factors are considered in isolation.

FUTURE PROSPECTS

Modernisation of clothing companies, their automation and the introduction at all levels of new techniques involving information technology, would appear at first sight to require an awareness and quantification of all the variables having any influence on garment design and garment manufacture. Detection and measurement of those properties having a fundamental influence on the conversion of fabrics into garments could bring about the avoidance of the repeated trials and readjustments which are frequently unavoidable during the design of garments.

The fabric performance criterion which, at the end of this investigation, appears to have the greatest influence in the conversion of fabric into garment (ability to pass from the initial "two-dimensional" state to the final "three-dimensional" state) is the combination of simultaneous multi-directional bending and shear. In view of the narrow range of variability of this criterion, when considering fabrics intended for the same end-use, demonstration of any differences is dependent upon precise measurements. Methods to be applied should make use of sophisticated techniques including photogrammetry and lasertechnology.

it is feared that such techniques could be too costly and difficult to put into practice for them to be accessible to a large number of makers-up capable of being carried out by the companies themselves.

This investigation has also produced the following findings.

Makers-up possess a command of techniques and intuitive knowledge of materials, sufficient for them not to have to make any changes to their patterns when fabrics vary.

In the creation of garments in which the creation of three-dimensional shape is of primary importance (as in flared skirts) the fabric properties have no perceptible influence on the number of flares when the fabrics used are suitable for this end-use.

With regard to making-up, it appears that a better knowledge of seam construction would reduce the role played by empiricism in machine settings and would permit criteria to be defined allowing the "self-adjustment" of machines according to the materials they have to work with. It is important to know the degree of complexity and accurancy required of test method especially as the variability of material properties is so great. Before devoting more resources to reducing this variability it is necessary to consider the resulting benefits and their value. It is essential that investigations should be preceded by a study of market requirements.

This does not in any way question the evolution of the textile and clothing sectors towards computer control and automation, but it may provide stimulus for thought to be given to the limits to be placed on them.

SUMMARY AND CONCLUSIONS BY THE CHAIRMAN OF THE SESSION

R. JEFFRIES
Shirley Institute

The session on garment physiology and construction was in two parts :
- evaluation of textile and garment comfort,
- optimisation of design and manufacturer of Garments.
A first conclusion of the session is that these two topics were well chosen, since both of the topics are relevant to the present and future needs of the textile and clothing industries of the Community if they are to remain innovative and competitive in the future.
The comfort of textiles and garments in all its aspects is an area of functionality that is of major interest to manufacturers and users of fabrics and clothing, and one that will grow in importance in the future, for several reasons. First, as the use of protective requirements become more exacting, so will the requirements for comfort become more demanding; second, there is an increasing consumer requirement for sportswear that is comfortable, particularly in situations where profuse sweating is present; third, there is an increasing requirement that even ordinary, everyday wear should be more comfortable.
In his report on the work on comfort, Dr. Umbach described research at the Hohenstein Institute, at Shirley Institute, and at TNO Institute for Perception on thermophysiological comfort, sensorial comfort, and the comfort of semi-permeable ("breathable") rainwear, respectively. The project has provided a sound basis of knowledge that will assist industry to design clothing that is more comfortable in a range of situations, reducing the need for length and expensive wearer trials.
It was concluded, however, that although excellent progress has been made, much remains to be done in this area. First, more background scientific knowledge is required on all aspects of clothing comfort. Some of this work must necessarily be aimed at achieving a still greater understanding of the physiological, psychological, and quasi-medical aspects of the problem; for instance, although much is known about pain thresholds on the skin, much less is known about the sensitivity of the skin to the very light contact of fabric during the wearing of a garment. More needs to be learned about the whole psychological concept of comfort. For instance, is comfort merely the absence of discomfort? Could there not be a more positive aspect to the comfort of clothing; for example, could textiles and clothing be designed that could induce positive feelings of well being? This possibility may be too far-fetched, but perhaps not! Second, new materials need to be developed that have still better comfort properties, perhaps materials that will help to produce the positive feelings of well-being rather than merely the lack of discomfort that was mentioned earlier. Third, new garment designs need to be developed, to utilise to the full the properties and potential of these new comfortable materials. New ways of ventilating clothing may also need to be devised to optimise the total comfort properties of the garment.
Dr. Umbach's contribution resulted in a good discussion that emphasized the value of the work on comfort to the textile and clothing

industries. The research done on the project assists industry to predict reliably the comfort bevahiour of clothing. Dr Manni described work supported by Lanerossi on the comfort, assessed by physiological measurements, of fabrics made from wool compared with wool blends, and of cotton compared with flax. Much interest was shown in the technical and scientific nature of protective outerwear that permits water vapour to pass but not liquid water. In reply to a further question, Dr Umbach said that other work at the Hohenstein Institute had investigated the effects of fabric finishes on comfort, but this work had not been reported at the Symposium because it was not part of the project (in relation to finishes, much depends on whether the finish is hydrophilic or hydrophobic). Much interest was shown in the discussion about the nature of the techniques used to measure comfort factors; these are the key to the whole subject of comfort.

In the second part of the session, Mr Deschamps described the work of CETIH and of subcontracted work at the British Clothing Centre on the optimisation of the design and manufacture of garments. Two major conclusions were reached. First, it was concluded that this subject is very important for the future of the textile and clothing industries. The relationship between the properties of textiles fabrics and the design and the manufacture of clothing needs to be much better understood if the future competitiveness and well-being of the industry is to be secured. This improved knowledge of the interface between textiles and clothing is essential to replace or at least to supplement the existing techniques for the design and manufacture of garments. These are based mainly on personal skill and experience, gained over many years of work, and although they have proved satisfactory up to the present time they will certainly not be sufficient on their own for the future. The new information on the textile/garment interface will certainly enable the quality and productivity attainable with existing methods to be increased but, more importantly for the future, will form an essential step in the direction of the automation of clothing manufacture and the increasing use of robotic systems in this area. The work described by Mr Deschamps is concerned with investigating the physical properties of textiles fabrics in relation to the optimisation of the design of garments that are functionally effective and comfortable and yet are of good style and aesthetics, and to making the clothing manufacturing process more amenable to the increasing use of automation and robotics (i.e. to understanding more fully the interactions between fabrics and garment manufacturing machinery).

The second conclusion arising from Mr Deschamps report that must be emphasised is that this work on the fabric/garment interface is in its very early stages. Investigations carried out on the project made an excellent contribution but a very large amount of work will be required on this very complex and difficult subject to provide the firm basis of knowledge that will be necessary in the future. The full collaboration of scientists, technologists and industry will be essential if success is to be achieved in this task. The excellent collaboration that Mr Deschamps obtained from industry is indeed encouraging; this collaboration must be continued and developed.

The two topics in the session on Garment Physiology and Construction are of key importance to the textile and clothing industries of the Community if they are to remain innovative and competitive.

S E S S I O N   A

GARMENT PHYSIOLOGY AND CONSTRUCTION

* * *

Contract 016-TEX-F - J. DESCHAMPS

Contract 020-TEX-I - E. MANNI

Contract 023-TEX-NL - W.A. LOTENS

Contract 034-TEX-UK - R. JEFFRIES

73

CONTRACTOR :BEKLEIDUNGS-
            PHYISOLOGISCHES INSTITUT e.V.
ADDRESS D - 7124 Bönnigheim

CONTRACT N° : 006-TEX-D

STARTING DATE : 1/10/1982

FINISHING DATE : 30/9/1985

PROJECT LEADER : Dr. K. UMBACH

TELEPHONE : 7143/ 27.10

TELEX : 724.913

TITLE OF PROJECT

GARMENT PHYSIOLOGY AND CONSTRUCTION

AIM OF PROJECT
Evaluation of Textile and Garment Comfort

SUMMARY OF RESULTS
The general objective of the project was to provide broad-based background information on the thermophysiological aspects of textiles and garments and to develop uncomplicated evaluation methods and guidelines which enable textile and garment manufacturers to improve the quality of their products with regard to wear comfort.

For the comfort evaluation of textiles, with a thermoregulatory model of human skin (Skin Model), methods have been developed to measure heat and moisture transport properties of fabric layers under "normal" wear conditions as well as under non-stationary conditions combined with heavy sweating of the wearer. The material-specific quantities determined with these methods are inserted into predictive formulae by which the thermophysiological comfort of the textiles, to be expected in different fields of use, can be calculated and expressed in a comfort vote. It has been ascertained that this calculated comfort vote is in very good agreement with the comfort sensations perceived by persons.

For ready-made garments similar predictive calculation models have been developed based on garment-specific thermal and moisture resistances determined by means of a life-sized, movable manikin.

With the predictive models a system is now available for the universal description and evaluation of the thermophysiological characteristics of clothing under specific climatic and wearer's activity conditions.

The significance and benefits of the project's results for the industry lie in the fact that time-consuming and expensive wear trials with humans became expendable. Now the textile and garment manufacturer can rely on comparatively uncomplicated laboratory tests to ensure good comfort characteristics of his products during the very first stages of a product development.

Further help to the industry is given by the fundamental guidelines deduced within the project which can be directly applied to optimize the wear properties of specific textiles and garments.

CONTRACTOR : CETIH   CONTRACT N° :016-TEX-F

ADDRESS : 14 rue des Reculettes   STARTING DATE : 9/6/1983
          FR - 75013 PARIS
                                           FINISHING DATE : 31/12/1985

PROJECT LEADER : J. DESCHAMPS   TELEPHONE : 1 / 45.35.24.01

                                           TELEX : 270.019

TITLE OF PROJECT

PHYSIOLOGY AND CONSTRUCTION OF GARMENTS

AIM OF PROJECT
Design of garments.

SUMMARY OF RESULTS

1. – Consideration of the physical properties of fabrics in garment design so as to obtain :

- optimum garment design with maximum avoidance of adjustment,
- easier making-up,
- production of garments of comfortable form and aesthetically satisfactory.

1.1. Classification of fabrics on the basis of physical properties.

Systems of classifying fabrics still relate to categories of end-use. They are essentially subjective and relate directly to visual or tactile information although they are sometimes biassed by simple measurable features such as cloth weight.

The automation of the making-up processes makes it necessary to have a more objective classification of fabrics based on measurable criteria which may be easily used by the majority in industry.

A survey has been carried out of the fabric properties at present considered more or less intuitively in the design and manufacture of garments and which are worthy of systematic investigation.

The properties most significantly influencing the ease of making-up and the production of high quality garments with comfort and aesthetics meeting the expectations of the designer have been investigated.

The maximum extent of their variations, covering all categories of garments, has been determined together with their variability within and between pieces. The purpose of this is to determine the variations which are significant.

1.2. Identification of the constraints imposed on fabrics and deformations suffered during making-up

Photographic study of changes in the structure of fabrics at the principal seaming operations (side seaming, shoulder seaming, yokes, sleeve setting seams and darts) has permitted identification of fabric features having predominant influence on the "making-up properties" of fabrics.

1.3. Attempt to find methods allowing industrial measurement of ease of cutting and multidirectional curvature of fabrics.

The capacities of fabrics to be cut and to be bent sumultaneously in several directions appear to be the dominant features affecting their ease of making–up and an overall measurement of these parameters has appeared to be desirable. Experimental methods have been developed to do this.

1.4. "In situ" evaluation and verification of fabric properties involved in their classification.

It appeared opportune to look at the influence of these fabric properties at the actual stages of design and manufacture in a clothing factory. Factories have made up skirts, and dresses, coats,jackets from a given specification and given fabrics, following their usual procedures, in order to check of the hypotheses evolved.

Assessments have been made of the results obtained (volume produced, garment comfort and aesthetics) by experts in the clothing industry and consumer representatives.

1.5. Conclusions

This study constitutes a preliminary approach to the problem of the influence of the physical properties of fabrics on garment design.

Positive design of the form sought has not yet been achieved either from the aesthetic or comfort aspect. Its attainment would be accelerated by better control of the uniformity of fabric properties supplied to the clothing factories.

2. – Fabric/Machine interactions in garment manufacture.

2.1. Introduction

Objectives of the work were
(a)   to review fabric/machine interactions in industry against the background of automation.
(b)   to carry out experimental studies on selected problems areas and,
(c)   to consider fabric monitoring as an aid to automatic adjustment of processing machinery.

2.2. Interactions

Each step in the production chain has its own interaction problems. The relative importance of different problems depend on the cost of the fabric and of labour in garment production. Two specific topics (2.3. and 2.4.) were selected for experimental study because they were seen as key areas for future development on which little recent work had been published.

2.3. Fabric feeding in sewing.

There is at present insufficient information to predict the limitation of machines (and fabrics) and their optimum settings. The aim was to identify fabric and machine characteristics affecting the movement of fabric through a sewing machine, including guides, folders and associated surfaces.
It was concluded that control of fabric in sewing depends on :
– fabric orientation, compressibility, thickness and extensibility,
– feed dog type, adjustment, movement and wear,
– presser-foot load and inertia,
– drag on workpiece by folders and other surfaces.

The threshold presser-foot load needed to provide stitch length control depends on the level of the drag on the fabric as it approaches the needle.

2.4. In-Line underpressing

There is economic justification for combining underpressing and sewing operations. Compromises between production speed, pressing conditions and physical dimensions are needed if a wide range of commercial fabric is to be processed by a single unit. Quantitative data were assembled for a range of fabrics.

2.5. Instrumentation

Two tests have been developed as aids to fabric selection and machine settings :
- A buckling test for assessing fabric limpness,
- and a circular sewing test to study feeding without the influence of the operator.

Feed dog movement was examined using a microprocessor analyser.

2.6. Conclusions and recommendations.

There is a lack of quantified published data on fabric/machine interaction problems. Fabric finishing can have a major effect : fabric anistropy is a factor that should be recognised more widely. Sewing behaviour depends on a complex combination of machine conditions and fabric properties and cannot be predicted confidently from a few isolated factors.

Industry is recommmended to ensure good communication (quantified wherever possible) between clothier, machine maker, garment technical designer and fabric manufacturer/finisher.

CONTRACTOR : LANEROSSI

ADDRESS : SCHIO / Italy

PROJECT LEADER : Dr. E. MANNI

CONTRACT N° : 020-TEX-I

STARTING DATE : 1/10/1982

FINISHING DATE : 31/3/1985

TELEPHONE : 69.33.60

TELEX

TITLE OF PROJECT

EFFECTS OF VARIOUS CLOTH GARMENTS DURING REST AND WORK

SUMMARY OF RESULTS

The influence of various cloth garments during rest and work was analyzed on young subjects of both sexes by recording skin temperatures (armpit, thigh, back, thorax) and cutaneous impedances (armpit, thigh) in different conditions of environment and season. The dresses were : pure wool, mixed wool, flax and cotton.

The muscular work consisted in going up and down two stairs (each one 20 cm high) for a period of 3 min, at a frequency of 24 cycles/min for a total of 14,000 kg.m for males and 11,000 kg.m for females. The room temperature of the indoor experiments was 22°C in winter, while for the experiments outdoor was 13° - 17°C. The summer temperature was 23.5°C.

The results were analyzed statistically by means of an Apple 3 computer. The statistical significance was calculated by analyzing the variance of type 2 x 5 (cloth type x time) with repeated measures. For example, the temperature was compared in the males wearing woollen garments and mixed woollen at times 0, 2, 10, 20, 30 min after the work. Such a type of statistical comparison was repeated for each temperature and impedance measure for the two types of winter and summer garments either in males or in females for a total of 36 comparisons.

Winter garments

In resting conditions and wearing woollen garments the skin temperature was higher than when wearing mixed woollen. Immediately after the muscular work of 3 min and afterwards there was a decrease in temperature of 0.3 - 0.8°C. However, in the majority of cases the noted differences were not statistically significant : wearing pure wool induced a minor decrease of skin impedance both indoor and outdoor. After muscular work the decrease of impedance was of 10 - 20 units in both sexes.

Although the subjects reported higher comfort with woollen dresses, the results were not significant.

Summer garments

The results obtained wearing flax and cotton dresses were more complex and less clear.

In some cases flax dresses protected more than the cotton ones, while in some other tests this was not the case. Also in this group of experiments the muscular work induced at the beginning a decrease in temperature (0.1 - 0.4°C) followed by an increase.

The differences between flax and cotton observed at the beginning of the test sometimes increased during and after work, while in others decreased up to the inversion. However, also in these investigations, the results were not statistically significant.

Concerning the influence of cotton and flax on sweating, a minor reduction of impedance was noted with dressing in cotton rather than in flax. The influence of work was not consistent and the results were not statistically significant.

79

CONTRACTOR : T.N.O.

ADDRESS : INSTITUTE FOR
PERCEPTION - P.O. Box 23
NL - 3769 ZG SOESTERBERG

PROJECT LEADER : Dr.W.A. LOTENS

CONTRACT N° : 023-TEX-NL

STARTING DATE : 1/10/1982

FINISHING DATE : 31/3/1985

TELEPHONE : 3463/ 14.44

TELEX

TITLE OF PROJECT

IMPROVEMENT OF PROTECTIVE CLOTHING FOR BETTER
WEARABILITY AND COMFORT

SUMMARY OF RESULTS

New tests have been developed :
- vapour permeability "tissue test"
                      "during rain"
- waterproofness "funnel test"
                 " blow through"
- ventilation "tacergas technique"

   In a co-operative effort of the Institute for Perception TNO, Ten
Cate Overall Fabrics and the Fibre Research Institute TNO, an investigation was carried out into the discomfort suffered by wearers of rainwear due to moisture and heat storage.
   An inventory available materials showed that the usual laboratory
techniques are insufficient to test fabrics for their waterproofness and
vapour permeability, when in protective clothing. Additional tests have
been developed. Only a limited number of fabrics show both good waterproofness and vapour permeability. There are four basic technologies :
microporous coating or films, hydrophilic films, microfibre wovens and
multilayer assemblies.
   Each technique has its pros and cons with respect to functionality
and economy. Special attention has been paid to the possibility of having
a fabric adapt its permeability to the needs of the moment : active
regulation. Major effort, however, has been devoted to the investigation
of the combined effect of permeability of the fabric and ventilation of
the garment to keep the wearer dry. By means of subject experiments in a
climatic chamber, applying a new tracer gas method for the determination
of ventilation, it has been shown that during moderately hard work both
optimal permeability and ventilation are needed to avoid moisture storage
in the clothing.
   The measurements support a mathematical model that helps explain the
importance of the many variables involved, such as air temperature and
clothing insulation. The project winds up with design rules for protective clothing, improving the wearability and comfort.

CONTRACTOR : SHIRLEY INSTITUTE

ADDRESS : Wilmslow road
Didsbury
GB - MANCHESTER M20 8RX

PROJECT LEADER : Dr. R. JEFFRIES

CONTRACT N° :034-TEX-UK (H)

STARTING DATE : 1/10/1982

FINISHING DATE : 30/9/1985

TELEPHONE : 61/ 445.81.41

TELEX : 66.84.17

TITLE OF PROJECT

GARMENT PHYSIOLOGY AND CONSTRUCTION : EVALUATION AND
OPTIMISATION OF SENSORIAL COMFORT

AIM OF PROJECT

1. To decide which are the sensations and physical factors that determine the sensorial comfort of next-to-skin garments, and to relate these sensations and factors to the physical properties of the fabrics and garments.

2. To develop simple objective or subjective methods for evaluating the various aspects of the sensorial comfort of fabrics and garments.

SUMMARY OF RESULTS

1. A detailed literature survey has been carried out. This has produced much information of indirect value to the project but confirmed that very little has been published of direct relevance to the subject of sensorial comfort. This emphasized the need for careful wearer trials to reveal which are the physiological and physical factors of importance in determining the sensorial comfort of a next-to-skin garment.

2. In the wearer trials, 20 members of the Institute staff have worn T-shirts made from 22 fabrics selected to cover the wide range of fibres and fabric types that are commercially available, or likely to be used, for underwear garments. The main conclusion is that the following are the main factors determining the sensorial comfort or discomfort of a fabric or garment :
   (i)   pressure on the body caused by tight fit,
   (ii)   irritation caused by sewn-in labels,
   (iii)   prickle, particularly around the neck and shoulders (caused by fibres of coarse diameter),
   (iv)   tickle, associated with fabric hairiness,
   (v)   the presence of loose fibre on the body and in the face,
   (vi)   wet cling,
   (vii)   initial cold feel,
   (viii)   static electrical effects when the garment is removed,
   (ix)   scratchiness.
   A major conclusion is that the "handle" of a fabric (as assessed by feeling the fabric with the hands) does not correlated with the sensorial comfort of the fabric when it is worn in garment form. Thus, such factors as roughness or smoothness, softness or stiffness, limpness or liveliness, which are important in determining the "handle" of a fabric, are not important in determining its sensorial comfort.

81

3. Methods of assessment or measurement of the factors (i) - (ix) listed in paragraph 2. above have been developed. Some of these tests are objective, some subjective or partially subjective. The fabrics studied in the wearer trials have been investigated with these new methods and techniques, in addition to a wide range of other tests and evaluations of a more established and standard nature, and correlations between the results of the wearer trials and the test results have been sought.

Information has also been obtained and collated on skin allergies produced on some wearers by certain types of fibre and fabric.

A major aim of the work is that the test methods, conclusions recommendations, and guidelines that will result from this work will so far as possible be simple and straightforward, and immediately usable by industry.

NEW SPINNING TECHNOLOGIES IN THE WOOL INDUSTRY

M. BONA
Città Degli Studi

For many years cotton spinning technology has been making steady advances thanks to the development of novel techniques which have partially replaced the classical spinning frame; the most important of these new techniques fall in the category of open-end systems, intially of the rotor type but more recently — at least in prospect — of the friction or air jet type.

It is understandable that these important development should primarily have involved the cotton sector and certain synthetics, both for objective technical reasons, these shorter, stronger fibres being easier to handle than wool, and because the research and development efforts of the textile machinery manufacturers have been mainly directed towards a potential replacement market at least ten time bigger than the wool market.

Accordingly, to promote the adoption of new spinning technologies for wool, with a view to reducing manufacturing costs or restraining their rise, it was necessary to organise a reasearch programme based on the real needs and with the bakcing of the industry; this programme was to concentrate on the behaviour of fibres subjected to the new non-conventional processes and tpo establish the mechanical and textile conditions for their eventual belonging to five EEC countries, had already made some progress on the initiative of certain members of the group when it received decisive and, above all, coordinated support from the launching and funding of the Second Community textile Research Programme.

Naturally, the first phase consisted in testing the various makes of rotor open-end spinning machines available on the market, without, however, neglecting other non-conventional solutions such as friction and hollow-spindle spinning.

In this phase the Community research programme produced some very interesting results, relating to both the carded and the combed product, whose practical importance is underlined by the fact that, thanks to the collaboration among several institutes, it was possible to test numerous types of machines.

However, it was gradually realised that the technical aspects of the problem should also include the preliminaries to spinning proper, and in particular the initial preparation of the fibres and carding.

As regards fibre preparation, the programme made it possible to propose various interesting modifications to the conventional processes used for virgin fibres, such as, for example, mechanical slub catching which may be accompanied by a simplified form of classical carbonising; naturally, special attention was paid to the use of recycled raw materials, which is of considerable economic importance to the European carded wool sector and poses the most difficult technical problems in relation to the subsequent operations.

With regard to the future use of new spinning technologies, whether starting from virgin or, in particular, reclaimed materials, one of the most important results of the programme had undoubtedly been the design, development to the prototype stage and practical testing of new types of cards which combine the advantages of the traditional wool and cotton systems, with a view to producing a homogenous, perfectly clean and open sliver for feeding into an open-end spinning machine. It is desirable for the manufacturers of both spinning machines and conventional cards to establish the necessary contacts with the research institutes to develop industrial machinery based on the new design concepts which could, moreover, also lead to interesting development in the field of conventional spinning.

A final aspect of the research programme concerned the examination of the physico-chemical behaviour of the wool during the manufacturing process, and indispensable requirement if optimum use is to be made of the new technologies and such disadvantages as the accumulation of dust and desposits in the rotors, which would prevent their economic exploitations, are to be avoided.

Among the characteristics of wool which make it more difficult to spin with the new technologies as compared with cotton, for example, one should mention its greater fragility and the need to employ additives, which must be carefully chosen and metered out, if it is intended to spin on open-end spinning machine.

Accordingly, special importance must be attached to the development in the course of the research programme of laboratory methods which it will be possible to use, in practice, for monitoring the manufacturing process and for selecting the wools best suited for use with the new technologies.

The three technical reports which follow will present the results of the research programme. They will deal with the three main topics mentioned above : preparation, spinning, and monitoring.

By way of a general conclusion, it should be stressed that, thanks to the excellent spirit of cooperation displayed by the institutes participating in the Community programme, a genuine basis for decisive developments in spinning technology in the wool sector has now been established, even if significant problems still remain to be solved, such as the translating of the new ideas into machines and industrial processes which will call for cooperation between the institutes and the textile machinery companies.

An important topic for future research has emerged from this work, namely the use of new yarns with a structure anyhow more or less different from the traditional one : this implies modifications to the processes downstream from spinning and the adaptation of the new products to the requirements of the market.

METHODS OF WOOL PREPARATION FOR BETTER UTILISATION OF RAW MATERIALS AND THE APPLICATION OF UNCONVENTIONAL SPINNING TECHNIQUES

P. ARTZT
Institut für Textiltechnik

SUMMARY

The results presented represent a summary of the coordinated research projects undertaken by CELAC (Belgium), WIRA (UK) and ITT (FRG). The results point the way to new opportunities for rationalising current methods of preparation for wool wastes. The reclaimed fibres can be used for end-products of higher qualith than is usual at present. The spinning process can be shortened by the application of unconventional spinning techniques. We call on industry to cooperate with the institutes in making practical use of the results.

*   *   *

By comparison with cotton, wool is a very costly raw material. For this reason it is necessary to utilise the maximum possible percentage of the wool used in the production of high quality end products, i.e. the proportion of waste must be kept to the minimum. The wastes occurring in wool processing may be divided into two categories of widely differing nature.

CARD WASTE

Card waste occurs as a by-product in the production of the wool sliver following wool scouring. During carding, a large proportion of impurities still remaining after scouring are removed. It is unavoidable for a certain percentage of fibres to be removed along with the dirt, these being predominantly long fibres.

WOOL NOILS

Wool noils consist predominantly of short fibres which are interspersed with a proportion of vegetable matter.

Both types of wastes are valuable raw materials for the spinning industry. The problem is the separation of the fibres from the contaminants. In the present state of technology, these wastes are generally carbonised, i.e. vegetable impurities are chemically oxidised. Disadvantages of this process are as follows :

1) High capital costs for the equipment
2) High energy consumption
3) Constant use of chemicals
4) Complex process control
5) Wool damage and fibre felting.

However, damage to the fibres can be reduced by appropriate super-
vision of the conditions of acidification and drying. Removal of water by
hydroextraction or by drying at moderate temperature is an important
factor in the conservation of the properties of the fibres. On the other
hand, neutralisation in alkalin medium may be the cause of fibre felting
which incurs fibre breakage on the card and is a major factor in nep
formation.

The aims of the projects were thus to investigate methods meeting
the following requirements :

1) Possibility of a simple separation of wool from contaminants without
   wet treatment or chemical processing.

2) Mechanical separation of short fibres which are not capable of being
   spun.

3) Preservation of the natural properties of the wool.

4) Modification of the carbonising process with the objective of reducing
   fibre felting.

Research on points 1 to 3 was undertaken by the Institut für Textiltechnik, Denkendorf and on point 4 by Celac, Chaineux.

Mechanical cleaning of wool noils
Attempts to clean wool wastes by mechanical means were first
performed on wool noils, which are not so heavily contaminated as card
waste, even though the wool may have a heavy burr content.

Investigations at ITF-Denkendorf on the mechanical cleaning of noils
were performed on a Cape wool of mean fineness 21 um. As the noil is
consistently finer than the top, the resultant mean fineness of the noil
was approx. 19um.

The first priority was to ascertain the properties of the noil. A
significant criterion for the subsequent use of a fibrous material is the
fibre length. This was therefore determined on the noil using the
measurement method for cotton (Digital fibrograph) as the stapel lengths
are similar (Figure 1).

For purposes of comparison, the table also gives the fibre length
distributions of a New Zealand wool noil of mean fineness 24 um and a
typical cotton waste because some distortion of the test results is
produced by the high crimp of the wool, the fibre is longer than determined by the Fibrograph method. Because the three fibre materials have
quite similar fibre length distributions, there is a chance that it might
be possible to spin wool noil by techniques now used for cotton in which
the percentage of short fibre does not have a particularly disturbing
effect. This is the case for rotor spinning technique especially and also
for friction spinning. Nevertheless, it is essential that the fibre must
be adequately cleaned without fibre damage.

Methods of cleaning studied
A prototype machine was designed based on cotton cleaning principles, as shown in the diagram of Fig. 2. The wool noil was fed via a
feed chute then feed rollers transferred the fibres to the opening
roller. After opening, a separator knife separated fibres from contaminants.

88

Fibre length measurement

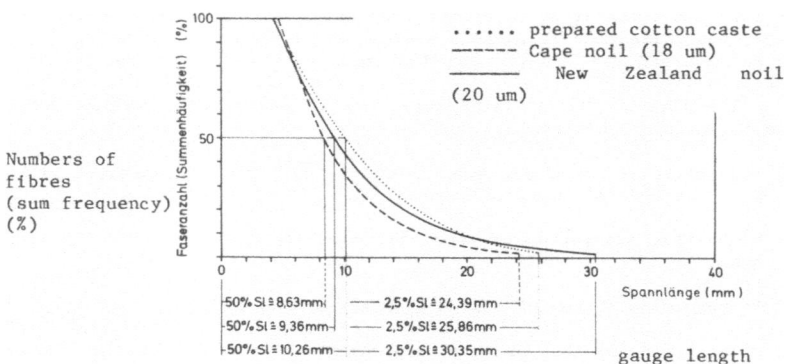

Numbers of
fibres
(sum frequency)
(%)

prepared cotton caste
------ Cape noil (18 um)
New Zealand noil
(20 um)

**Figure 1 : Fibre length distribution of wool noils
compared with cotton fibres**

Function diagram / cleaning machine

1. feed chute
2. feed rollers
3. pressure duct
4. knife
5. vacuum duct
6. pinned roller
7. condenser
8. doffer rollers
9. fan

A. opening zone
B. cleaning zone
C. dusting zone

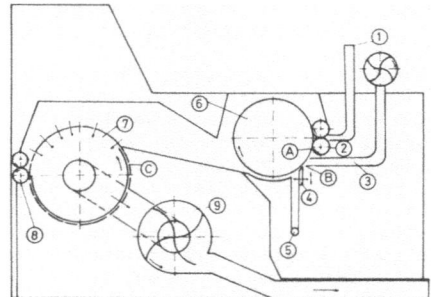

Figure 2 : Cleaning principle with roller feed

The cleaning action of the knife could be assisted by air pressure before the knife and vacuum after the knife, both of which could be switched on as required. The cleaned wool was then carried to a condenser for more thorough removal of dust and finally was taken off in web form by delivery rollers for further processing. The prototype machines as equipped with variable speed drives for feed rollers and cylinder in order to permit optimisation of the machine parameters. The cleaning efficiency which is important for evaluation of the process was checked by means of the Shirley Analyser. It is defined as :

$$R = \frac{\text{Wt. of dirt before} - \text{Wt. of dirt after}}{\text{Wt. of dirt before}} \cdot 100$$

R thus expresses the percentage of dirt that has been removed from the original dirt.

The parameters under investigation were :
1. clearance separator knife – cylinder
2. clearance separator knife – feed roller
3. angle of separator knife
4. air pressure
5. speed of cylinder rotation
6. production rate, kg/h.

The extensive trials with this prototype machine produced the following results :

1. For effective mechanical cleaning, the wool noil must have thorough pre-opening.
2. Mechanical pre-cleaning should free the wool noil from the coarsest contaminants. If coarse contaminants pass through to the sawtooth clothing of the cleaning machine, then the clothing becomes filled with dirt.
3. The mechanical stress on the wool fibre in pre-cleaning should be low.
4. In contrast to cotton, the dirt is more intimately attached to the fibre. Separation of the dirt and fibre therefore only occurs when the fibres are firmly held.

Because of the large clearance between the nip of the feed roller and the cylinder, however, there was no adequate combing out of the beard of fibres on the prototype machine. The level of cleaning attained with this system was therefore inadequate, at about 3-10 %. However, the trials pointed the way towards what steps must be taken to achieve better results with mechanical cleaning. In collaboration with a machine-making company a design has therefore been developed, based on cotton cleaning, which consists of pneumatic pre-cleaning followed by mechanical cleaning. This machine was adapted for cleaning wool noils.

The pneumatic cleaning stage (Figure 3) operates without any particular mechanical action. The opened wool noil is sucked upwards by the airstream. The heavier and coarse particles drop downwards counter to the airflow. This pre-cleaning greatly eases the task of the actual cleaning machine. The principle of the cleaning machine which has proved successful for cotton preparation is shown in fig. 4. Before the feed chute there is an opening unit. The wool is fed to the sawtooth cylinder via a feed tray which makes it possible to have a low nip clearance to match the fibre length.

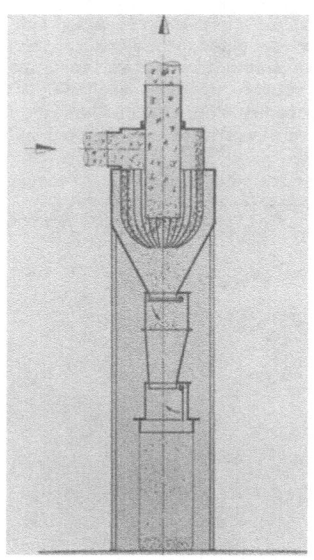

Figure 3 : Pneumatic cleaning stage

1. feed rollers
2. drum (preopening)
3. removal of coarse impurities
4. feed chute
5. taker-in cylinder
6. dish
7. drum
8. knife
9. grid
10. carder roller
11. trash roller
12. trash removal
13. duct for good fibres
14. condenser
15. suction extraction of material

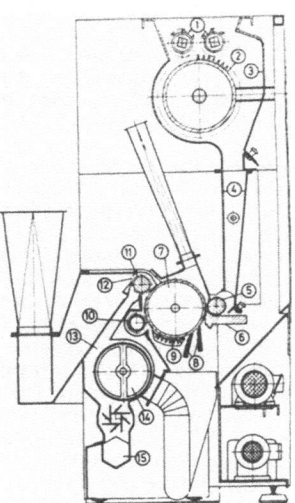

Figure 4 : cleaning machine for wool noils

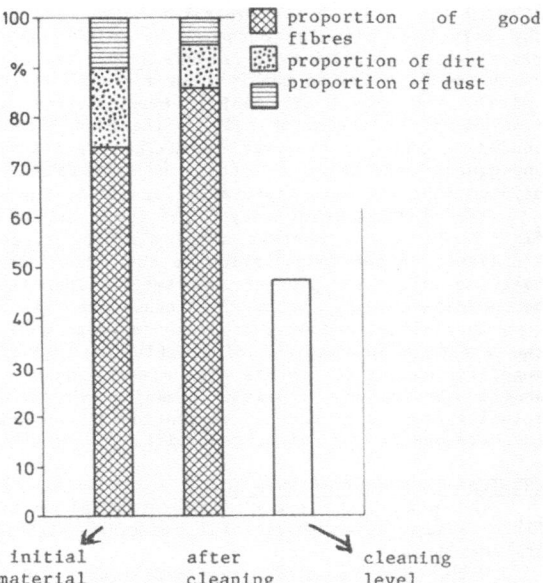

proportion of
good fibres,
dirt, dust

Figure 5 : cleaning efficiency on wool
noil with mechanical preparation

Figure 6 : visual comparison of wool noil
with and without cleaning

The sawtooth cylinder separates the dirt from the held fibre. The dirt is separated at the separator knives and at the grid. The heart of the principal cleaning stage is a rapidly rotating roller (6000 rev. min-1) which provides dirt removal across the full working width, as used on the rotor spinning machine. As a result of the high centrifugal force and the high level of opening, even smaller contaminant particles are removed at the saparator knives. The good fibres are sucked away by the condenser and dust removed there.The significance of dust removal for subsequent processing is often underestimated. Great attention must be paid to dust removal, especially for use of the fibres in rotor spinning.

The cleaning efficiency of an installation of this type is shown in fig. 5. The fibre content after cleaning rises to 86 %; the cleaning efficiency of the installation is thus about 50 % which means that about half the impurities present are removed. This is a very good result for a mechanical cleaning system. The total waste was 24 %.

In a direct comparison of the separate wool samples the influence of the cleaning process is distinctly seen (Fig. 6). The reduction in vegetable matter content is also clearly shown. It is clear that the aim should not be the complete cleaning of the wool before the actual spinning process as further separation of contaminant particles occurs on the card and on the spinning machine.

CHEMICAL TREATMENT OF WOOL NOILS

The aim of the experiments performed by CELAC is mainly to reduce fibre felting, even though this reduction might be accompanied by a loss of cleanliness.

With this in mind, the performance of the following has been compared :

1) the conventional carbonising process,
2) a carbonising process without beating before neutralising,
3) a carbonising process without beating and without crushing before neutralising.

Figure 7 outlines the various treatments in diagram form.

Testing of the properties of the materials at the various stages of processing and the properties of the end-products (fig. 8) has allowed the following conclusions to be drawn :

- short fibres are removed during the operations of crushing, beating and neutralising. This removal compensates for fibre breakage and is balanced by an increase in mean fibre length accompanied by a reduction in variation,

- the loss of short fibres is greater in the case of the lot subjected to crushing and beating. As we had envisaged, the consequence of beating is to separate the fibres and to promote the removal of the shortest of them during the neutralising process (start of neutralising, 16.2 mm and end of neutralising, 26,2 mm),

- the lot not subjected to crushing and beating together with the lot that was crushed but not beaten are more open (less felted) than is the case with the lot subjected to conventional neutralising. Measurement of standard bulk by the Drean and Schutz method (IWTO Tecnical Committee, January 1983) confirms the opinion of specialists on the more bulky appearance of lots that have not been beaten.

Figure 7 : diagram of carbonising tests

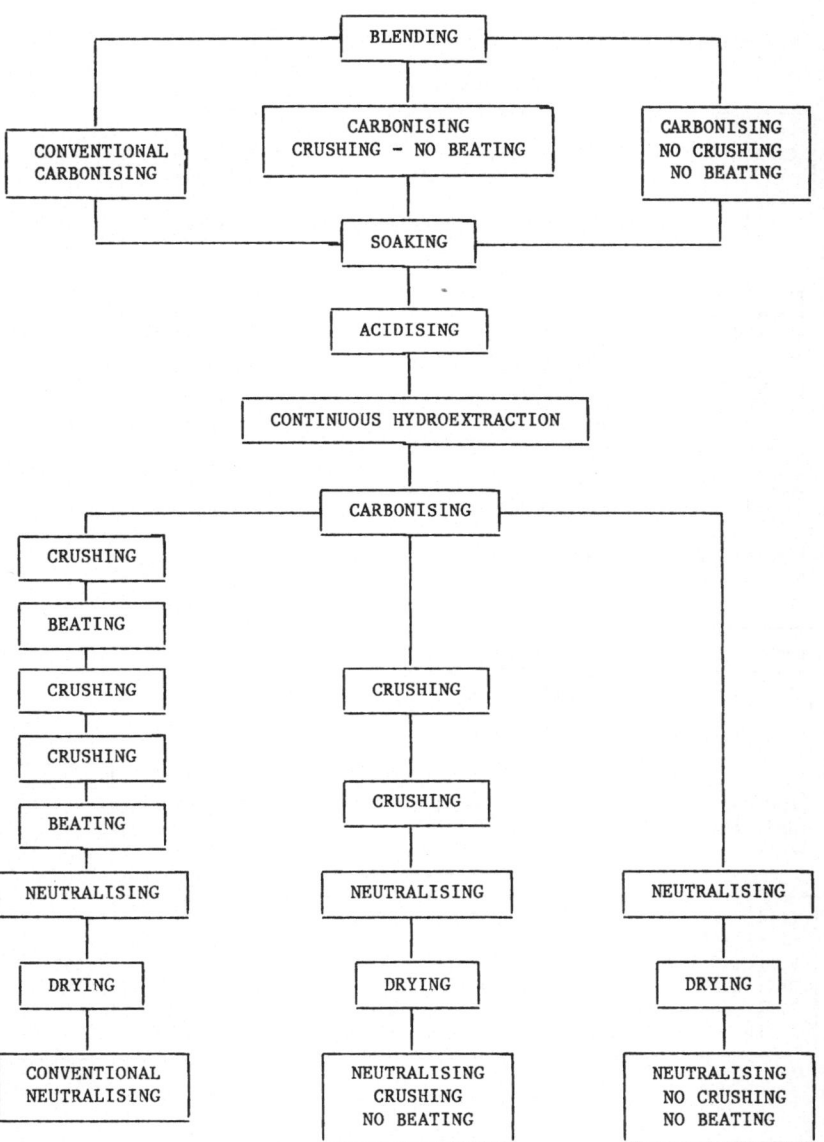

Figure 8 : Properties of materials at different stages
in the production cycle (physical parameters)

2 A : Length of fibres and standard bulk

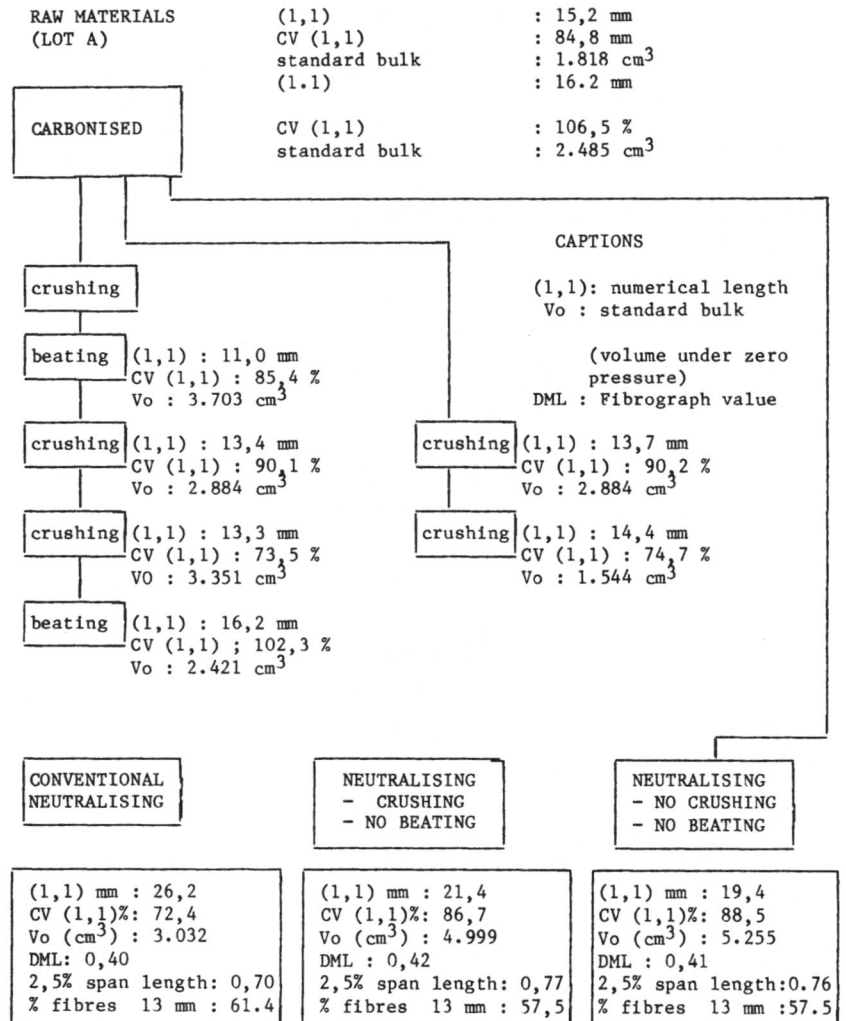

RAW MATERIALS   (1,1)   : 15,2 mm
(LOT A)   CV (1,1)   : 84,8 mm
                   standard bulk   : 1.818 cm^3
                   (1.1)   : 16.2 mm

CARBONISED   CV (1,1)   : 106,5 %
                   standard bulk   : 2.485 cm^3

                                      CAPTIONS

crushing   (1,1): numerical length
                                      Vo : standard bulk

beating (1,1) : 11,0 mm   (volume under zero
         CV (1,1) : 85,4 %   pressure)
         Vo : 3.703 cm^3 DML : Fibrograph value

crushing (1,1) : 13,4 mm   crushing (1,1) : 13,7 mm
         CV (1,1) : 90,1 %   CV (1,1) : 90,2 %
         Vo : 2.884 cm^3 Vo : 2.884 cm^3

crushing (1,1) : 13,3 mm   crushing (1,1) : 14,4 mm
         CV (1,1) : 73,5 %   CV (1,1) : 74,7 %
         VO : 3.351 cm^3 Vo : 1.544 cm^3

beating (1,1) : 16,2 mm
         CV (1,1) ; 102,3 %
         Vo : 2.421 cm^3

CONVENTIONAL   NEUTRALISING   NEUTRALISING
NEUTRALISING   - CRUSHING   - NO CRUSHING
                         - NO BEATING   - NO BEATING

(1,1) mm : 26,2   (1,1) mm : 21,4   (1,1) mm : 19,4
CV (1,1)%: 72,4   CV (1,1)%: 86,7   CV (1,1)%: 88,5
Vo (cm^3) : 3.032   Vo (cm^3) : 4.999   Vo (cm^3) : 5.255
DML: 0,40   DML : 0,42   DML : 0,41
2,5% span length: 0,70 2,5% span length: 0,77 2,5% span length:0.76
% fibres 13 mm : 61.4 % fibres 13 mm : 57,5 % fibres 13 mm :57.5

From the aspect of cleanliness, the lot not subjected to crushing and beating is distinctly more heavily contaminated than the lot neutralised in the conventional way. The neutralised lot that was crushed but not beaten contains a large amount of small straw fragments (number of straws/g : 532 against 159 in the case of the conventionally neutralised lot) but the weight percentage remains fairly low (0.86 against 0.41).

The experiments were followed through as for as spinning. Three carding systems were tried : a woollen card set, a cotton card, and a mixed card. The wools were blended with polyester staple fibres (1.7 tex, 38 mm) to produce a blend of 60 % noils with 40 % polyester.

Regardless of the method of carding adopted, the carbonised noil neutralised without beating produces a less neppy web (Fig. 9). Moreover, the difference in cleanliness between the conventional noil and the noil not subjected to beating is slight. THe results obtained at the spinning stage also confirm the advantage found for noils that have not been beaten. Comparative tests performed with different mechanical cleaning techniques do not permit the same standard of cleanliness to be reached, the fault removal efficiency with mechanical opening is of the order of 50 %.

Carding removes a further 50 % of the residual vegetable matter. For lots with a heavy fault content it appears that carbonising is a necessity. However, for noils with light burr content, mechanical opening may suffice.

EFFECT ON RECOMBING

At the ITT Denkendorf, the mechanically cleaned wool noil was blended with 40 % polyester (1.3 dtex) and processed on a cotton card without any problems. Finally attempts were made to recomb the card slivers both on a cotton comber and on a specially adjusted wool comb. It was demonstrated that recombing is possible without any major problems, especially on a cotton comber. Depending on fibre length distribution, short wool fibres were combed out to a greater or less extent, the blend ratio reversing from the initial 60 % wool/40 % polyester to 40 % wool/ 60 % polyester. Recombing positively improves the sliver quality and opens up completely new opportunities in the processing of wool noils into blend yarns.

Rotor-spun yarns 50 tex (Nm 20), and 36 tex (Nm 34) and ring-spun yarns 36 tex (Nm 34) were produced from the slivers that had been recombed and from those that had been only regilled. The yarn values obtained are given in fig. 9. The benefits of recomning on a cotton comber are clearly visible. A further result is demonstrated : mechanical precleaning of the wool noil, whereby about half of the dirt content is removed, when combined with recombing, is fully adequate for the production of high quality yarns. Any residual dirt is removed on the card, during the combing process and on the rotor spinning machine. Performance on the rotor spinning machine was very good.

A comparison of trials 2 and 3 (fig.9) demonstrates the significance of mechanical pre-cleaning in conjunction with recombing on the cotton comber. Without mechanical cleaning and recombing, yarn strength drops from 14.3. cN/tex to 7.8 cN/tex and yarn faults rise to an unacceptably high level. It is thus obvious that correctly designed mechanical pre-cleaning leads to the production of yarns of high quality with good performance on the spinning machine.

Attempts were also made to produce ring-spun yarns from these slivers. However, drafting problems arose on the roving frame. Consequently the ring-spun yarn (trial 5, fig. 9) was very irregular

(Uster CV 26 %) and neppy. This trial also demonstrated clearly that reprepared wool noils should be spun on the rotor spinning system. CELAC compared rotor spinning machines fitted with a selector and with an opening roller. These indicate that in wool processing the more gentle selector opening has the advantage because fibre abrasion is reduced and thus also fibre deposition in the rotor. This comparison is, however, rather academic as the rotor spinning machines involved are no longer being made commercially.

YARN DESCRIPTION	WITH RECOMBING		WITHOUT RECOMBING		
	on wool comb	on cotton comb			
	50 tex rotor yarn	50 tex rotor yarn	50 tex rotor yarn	36 tex rotor yarn	36 tex ring yarn
Mechanical cleaning yield good fibre (%)	76	76			
mixing with 40 % PES 1.3. dtex, 38 mm	x	x	x	x	x
overall yield (%)	57.8	54.9	92.7	92.7	92.7
count-related max. strength (cN/tex)	11.10	14.34	7.80	11.82	17.01
max. extension (%)	14.06	12.85	11.20	10.56	10.90
Uster CV (%)	15.75	15.76	17.20	16.43	25.96
Thin places/1000 m	10	12	41	27	437
Thick places/1000 m	84	54	138	37	617
neps / 1000 m	1200	172	1317	685	1424
test number	1	2	3	4	5

Figure 9 : Yarn properties as a function of sliver preparation.

As end-products of 100 % natural fibres are highly valued by the consumer, rotor yarns were spun from a blend of 50 % cleaned wool noils and 50 % cotton. Performance on the cotton card at a production rate of 60 kg/g (1.5 m working width) was very good. The slivers were drawn 1 x and then spun on the rotor spinning machine to a fineness of 30 tex. The yarn values are summarised in Fig. 10.

Performance up to this fineness (with about 150 end breaks/1000 rotor hours) was very good. Yarns of count 50 tex and 30 tex were made into shirtings and compared with a commercial product made from ring-spun yarns. The end-product was regarded as "of commercial marketable quality" both by the weaver and the shirt manufacturer.

Spinning system yarn count	OE rotor yarn 50 tex	OE rotor yarn 30 tex
count-related max. strength (cN/tex) max. extension (%) Uster CV (%) thin places / 1000 m thick places / 1000 m neps / 1000 m	6.1 4.26 16.29 85 170 64	5.8 3.69 17.95 174 220 147

Figure 10 : Yarn properties of rotor-spun yarn with
50 % cleaned wool noil and 50 % Sudan cotton

RESULTS OF THE MECHANICAL CLEANING AND PROCESSING OF CARD WASTE.

Card wastes are those occuring in wool combing plants. Depending on the wool origin and the scouring process, card wastes of 15 % to 35 % occur in carding. The mechanical condition of the card has an appreciable influence on the amount of waste produced. Important roles are played by the state of the card, clothing, the production speed and the machine setting. It is thus difficult to produce precise data of amount of wool waste. This must be taken into account in discussing the yield of mechanical cleaning by comparison with carbonising.

The trials were carried out with card wastes of 21.8 um and 28 um wools. Fibre length measurement by the WIRA method gave mean staple lengths of 41 and 42 mm with the longest fibres 140 and 200 mm respectively (Fig. 11). Card waste A from the 21.8 um wool was used without carbonising for optimisation of the mechnical cleaning process. A well known machines (fig. 12) for the mechanical cleaning of fibres was used here. An enclosed chamber in which a drum rotates is fed intermittently with fibres. The material remains in the cleaning chamber under vacuum for a specified period of time.

The drum operates in conjunction with worker bars which, may be set at variable clearance from the drum. The cleaning efficiency depends largely on :

1) the rate of throughout (Kg/h)

2) the time spent in the drum (sec)

3) the clearance of workers bar to drum, and

4) the tupe of grid adopted.

The machine setting must be matched to the input material in each case. There is the chance also of removing short fibres. This certainly then reduces the fibre yield but significantly longer fibres are obtained. It is especially important to try to reach a compromise between yield and fibre length. In our investigations, the following settings proved to be best :

-------- card waste A (21,8 um, mean staple length - WIRA method - 41 mm)

————— card waste B (28 um, mean staple length - WIRA method - 42 mm)

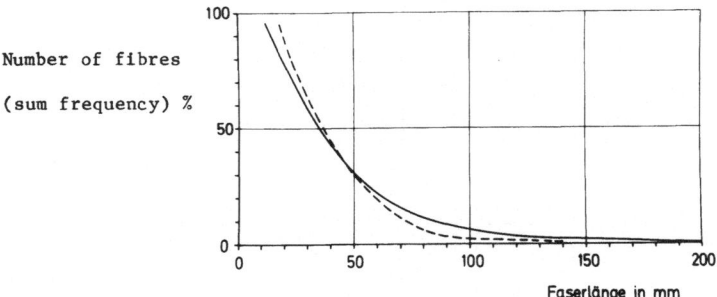

Number of fibres

(sum frequency) %

Fibre length mm

Figure 11 : length distribution of card waste fibres

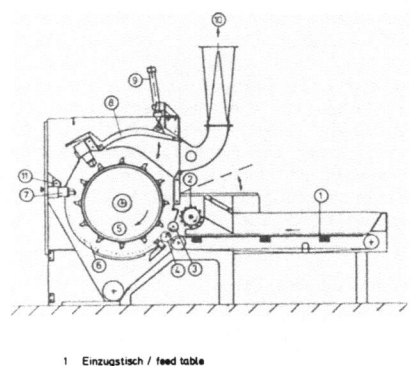

```
1   Einzugstisch / feed table
2   Druckwalze / pressure roller
3   Einzugswalze gefedert / feed roller, spring-loaded
4   Einzugswalze / feed roller
5   Tambour / main cylinder
6   Roste / grids
7   Arbeiterbalken / worker bar
8   Haube mit einstellbarer periodischer Bewegung /
    cover, with adjustable, intermittent movement
9   Druckluftzylinder / pneumatic cylinder
10  Materialabsaugung / material extraction
11  Staubabsaugung / dust extraction
```

Figure 12 : Cleaning machine for card waste

1) production rate : 200 kg/h
2) time spent in drum : 6 - 8 sec
3) clearance between worker and drum : as close as possible.

Under these conditions, the 21.8 um card waste was cleaned so thoroughly that it contained only 12.0 % dirt and dust (Fig. 13). Fig. 13 clearly demonstrates the progress made in cleaning. By the definition given previously, the cleaning efficiency is about 70 %. As more than 50 % waste occurs in this process, a machine of this type must be provided with a very good extraction system.

Expenditure on an extraction system must therefore be reckoned with in calculating the capital costs of a cleaning installation. The cleaned fibres were carded on a conventional wool card. The slivers were then gilled on intersections. In a parallel trial, a carbonised wool of the same fineness was processed. To compare the processing performance of the two fibres, ring-spun yarns were produced in counts 36 tex (Nm 28) and 500 tex (Nm2). Fig. 14 gives the spinning plan. When carbonised wool was processed in unadulterated form, problems immediately arose on the card. The carded web contained appreciably more neps than was the case with the wool prepared by the mechanical method. Application of an additive to the carbonised wool produced only a slight improvement of processing performance on the card. By comparison with a mechanically prepared wool, the wool was very "brittle". Waste on the card when processing carbonised wool was 10 % greater than with the mechanically prepared wool.

A particularly distinct feature was the difference in quality between the two fibre materials in recombing. The mechanically cleaned lot was recombed and produced a noil percentage of 26 %. Under the same conditions, as a result of the higher short-fibre content of the carbonised wool, the percentage noil was 37 %. This confirmed the speculation that the initially higher yield in carbonising can be partially attributed to a greater short-fibre content. A proportion of these short fibres is removed in mechanical cleaning and the remainder in recombing. The quality of the ring-spun yarns produced from the mechanically prepared wools was distinctly higher than the quality of the yarns produced from the carbonised wool.

The end-uses to which the two fibre raw materials can be applied therefore differ, the mechanically prepared card waste being more suitable as a raw material for the worsted spinner than the carbonised card waste.

Because of the fibre length distribution of the cleaned card waste it was not suitable for spinning on the rotor spinning system. Spinning trials were therefore performed on the wrap-spinning system. Fig. 15 shows the results of these trials which demonstrated that the wrap-spinning technique is eminently suitable for processing mechanically cleaned card caste.

With the wrap-spinning technique it is possible to spin with practically no end breaks. In particular, the wrap-spinning technique is thus an ideal process for spinning fibres of widely differing fineness, as the hairiness of wrap-spun yarns is significantly less than that of ring-spun yarns.

The subject of investigations by WIRA was the direct spinning of wool card slivers on the ITG 300 rotor spinning machine. A two-cylinder wool card with web divider was used in these trials. For processing on the rotor spinning machine in question, research has shown that the longest fibre should not exceed 82 mm.

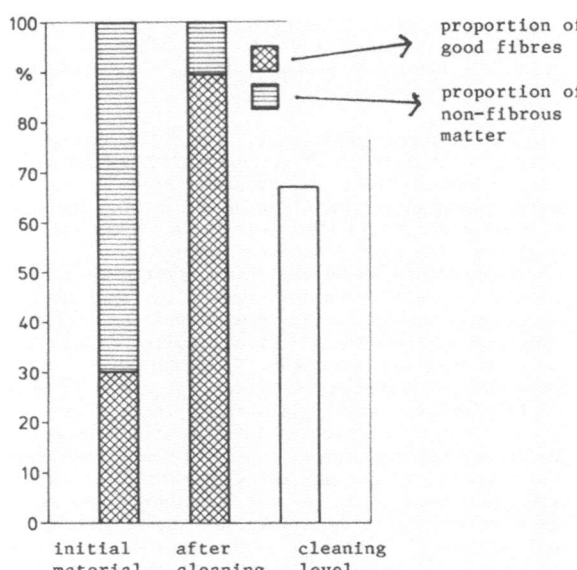

proportion of
good fibres,
dirt, dust.

Figure 13 : Cleaning efficiency on card waste
using mechanical cleaning

RAW MATERIAL	CARD WASTE			
	ring-spun yarn		ring-spun yarn	
	36 TEX	Nm 28	500 Tex	Nm 2
Mechanical cleaning yield (%)	32		32	
Carbonising		x		x
carding yield (%)	95	90	95	90
intersectings I-III yield (%)	98	95	98	95
Comb. noil (%)	20	37		
intersectings IV -VI	98	96		
Roving - Uster CV (%)	5,0	6,5		
YARNS VALUES				
. strength cN/tex	5,6	6,5		
. extension %	18,7	4,8		
. Uster CV %	16,8	18,0	8,0	11,0
. thin places /1000m	98	206		
. thick places/1000m	71	135		
. neps/1000m	74	124		

Figure 14 : Yarn properties of yarns from 100 % card waste

(1) : mechanical precleaning (2) : chemical precleaning	CARD WASTE A (21,8 um)							CARD WASTE B (28 um)		
	(1)			(2)				(1)		(2)
Yarn description	71 tex wrap	500 tex wrap	36 tex wrap	71 tex wrap	36 tex wrap	500 tex wrap	36 tex wrap	71 tex wrap	36 tex wrap	71 tex wrap
Mechanical cleaning	x	x	x					x	x	
Chemical cleaning				x	x	x	x			
Mixing with 50 % Trevira 350, 3.3 dtex, 60 mm		x				x				
YARN VALUES										
count-related max. strength (cN/tex)	6.6		5.6	6.3	7.9		4.8	6.3	3.5	5.6
max.extension (%)	11.7		18.7	11.0	14.1		9.3	11.3	5.7	8.2
Ulster CV (%)	13.2	8.0	16.8	12.3	16.1	10.8	18.1	16.4	26.1	16.3
Thin places /1000m	2		98	1	64		206	74	909	84
Thick places /1000m	7		71	4	56		135	30	465	44
neps / 1000m	6		74	8	16		124	11	147	8
weaving trail	x	x	x	x	x	x	x			

Figure 15 : Comparison of yarn values of ring-spung and wrap-spun yarns from mechanically cleaned and carbonised card waste.

This means that the fibres need to be carefully selected for the different end-use applications. Typical end-uses recommended for yarns spun from such wools include upholstery fabrics, blankets and carpets. Relatively coarse yarns are predominantly used in these sectors. The research results have demonstrated that the length of wools to be processed by the rotor spinning technique is limited for two reasons :

1. Mechanical design factors restrict fibre length to a maximum of 82 mm on the drafting unit of the ITG 300 rotor spinning machine.

2. If the yarns are used as carpet pile yarns, then the fibres longer than 80 mm produce a relatively poor surface appearance.

As the pile yarns for the products chosen were largely coarse yarns having a relatively high number of fibres in the yarn cross-section, no particular attention needs to be paid to wool fineness. There is no need for a drawing process between card and rotor spinning machine for yarns of this count. The results show that the same wool can be processed on the rotor spinning system as on the ring spinning system. This is important from the aspect of versatility in the spinning plant. As this technique of wool processing does not use combed tops but instead uses scoured wools, the vegetable matter content of the raw wools is of vital importance. The wool card used had a web crusher fitted for this reason.

The use of the double card gave very good fibre mixing which is a requirement for trouble-free performance on the rotor spinning machine. WIRA's experience was that a humid atmosphere is essential both in carding and in rotor spinning in order to achieve good processing performance and satisfactory yarn quality. The wool should be scoured to a residual fatty matter content of approx. 0.5 % which is adequate for processing on the card and on the rotor spinning machine. For the spinner it is important to know that there are no advantages at all in the application of further additives which results in unwanted deposits on the opening rollers and in the rotors.

The experiments also demonstrated that it was possible to spin all yarns with the same rotor, the same rotor speed, opening roller and opening roller speed, and the same doffing tube. Yarn twist, however, needed to be varied to meet the needs of the separate end-uses. Despite the somewhat lower strength of the rotor-spun yarns by comparison with ring-spun yarns, there were no disadvantages of any kind in weaving with regard to machine efficiency.

The results demonstrate that solid-shade velour carpets can be produced provided that some degree of "tuft definition" may be tolerated. Perhaps there will be a revival in carpet yarn production by rotor spinning.

TEXTILE WASTES IN THE E.E.C.*

INTRODUCTION

The objective of the survey on textile wastes in the E.E.C. carried out by WIRA was to determine the position of that part of the textile industry concerned with re-cycling textile wastes. The original conception was concerned mainly with wool, with some reference to open-end spinning. This has been extended within the terms of reference to include all types of fibres traded internationally as waste.

----

* This section was added after the lecture by Dr. Artzt in Luxembourg.

SUMMARY

   General findings, such as the amount of fibres available from the
making-up industry, show that   10–15 % of woven cloth appears as "new"
waste. Figures are available for the weights of woven cloth produced on a
world-wide basis and broken down into individual countries. Similarly,
"soft" waste from the hosiery trade can be treated in the same way, but
with lower waste figures due to fewer cutting operations. Other uses,
such as carpets, do not really lend themselves to re-cycling in the
general sense, but could be a source of either re-cycling for a new
product or as a source of energy. Recent work in italy shows that 48.000
tonnes per annum are available (all non-recyclable textile waste in the
Prato area).

   In the synthetic fibre area, primary producers of fibres made waste
up to 10 % of production, initially. Over the years, improvement in
production methods have reduced this to about 6 %, of which 2.5 to 3.5 %
is recycled "in-house". This means that there has been a drop to one
quarter of original supplies in this area.

   Regenerated cellulosics and mixtures of natural fibres have been
coped with by the industry for a considerable time. However, the difficulties associated with the newer mixed blends, particularly polyester
types, proved to be a serious problem. Traditional methods of collecting
and dealing with such materials led to almost insurmountable obstacles
particularly because of dye and fibre composition. Variations only in
areas where determined and intelligent sorting methods could be carried
out at an economic level has the trade been maintained.
In the E.E.C. only the Prato area in Italy can deal with modern recycled
fibres fabrics.

   Indications are that Prato is moving towards a general up-market
trend, with at least 40 % of processed materials being synthetics.

   With the introduction of statutory labelling requirements giving
country of origin and fibre content the problem of sorting could be
eased.

   Colour sorting is another major problem, particularly with modern
dyestuff involved. Re-dyeing or over-dyeing is difficult with mixed dyestuffs, which suggests that the problem of instrumental colour sorting
should be tackled as a major research project.

NEW TECHNIQUES OF CARDING AND SPINNING

J.P. BRUGGEMAN
I.T.F.-NORD

INTRODUCTION

The work reported here is a summary of the studies carried out jointly by : Dr Artzt of the Institut für Textiltechnik, Mr Grignet and Dr Knott of Centexbel, Mr Brach of Celac, Dr Blankenburg of the Deutsches Wollforschungsinstitut, Dr Cooke of Umist, Mr Windle of Wira, Mr Torelli of Tecnotessile and Messrs Pochen and Edme of ITF-Nord.

Open-end spinning is a method of producing yarn at speeds of about 4 times those on a conventional ringframe.

The cotton industry has been benefitting from this for the past 15 years and the wool textile industry could not remain indifferent to this technological development, although the open-end technique demands that raw material presentation is compatible with this method of spinning with respect to length, as a function of rotor diameter; fineness, as a function of the number of fibres in the yarn cross-section; and fault content, surface properties and additives.

The concept of arriving at a form of presentation for wool and its blends which is acceptable for the open-end process was developed quickly and collaboration between research workers and industrialists has resulted in processes and production lines closely resembling the cotton system.

Industry has become aware that this method of spinning is capable of complementing, if not replacing, woollen spinning, although it requires selection of raw materials and especially their treatment before spinning was described by Dr. Artzt. A quick survey has located :

- in Germany : 4000 rotors for wool or its blends,
- in England : 4000 rotors for fibre blends of length between 60 and 100 mm,
- in Belgium : 300 rotors,
- in France : 1200 rotors of the woollen type
  800 rotors for stretch-break converted combed wool,
- in Italy : 8418 rotors in woollen spinning processing wool waste but mainly synthetics.

These figures indicate the importance for industry of the work carried out by the research centres in the E.E.C.

The materials we have processed most frequently have been : noils, blends of the Prato or Mazamet type, stretch-broken converted wools and new wools for hollow spindle spinning.

Apart from stretch-break converting, which we shall come back to later, these do not present any particular difficulties in material preparation. Carding is the one operation which, starting with raw materials prepared by the processess described by Dr. Artzt, will produce an initial sliver of maximum possible cleanliness, free from contaminants including vegetable pieces and evenness of count with good homogeneity of composition.

Figure 1

In conventional woollen type spinning (Figure 1), the card delivers a web divided into slubbings which are fed to the ringframe. The open-end technique has given rise to the concept of omitting the tape condenser at the card delivery and replacing it with a delivery into one or more slivers intended for direct spinning on the open-end machine. This process, attractive for carpet yarn, possesses one drawback which is almost prohibitive as far as finer yarns are concerned, namely the absence of any system of count control between slivers and within a sliver. Furthermore, when using recycled materials, the card web is in the form of poorly openend fibrous clumps which represent an obstacle to open-end spinning. Industry and research have therefore turned to the use of the cotton type card (Figure 2) and also the duo-card which should permit an even sliver to be obtained with count compatible with open-end spinning (Figure 3).

Figure 2

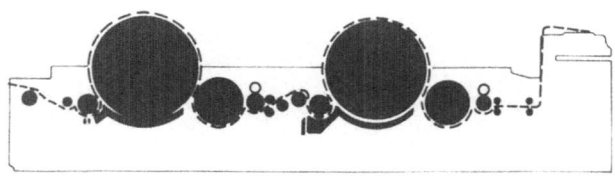

Figure 3

Although the results obtained were very encouraging they have revealed some deficiencies including excessive vegetable fault content, poor parallelisation of fibres, and a large amount of neps.

The various laboratories therefore asked machine makers and industry to study a new type of machinery for wool taking the into consideration sliver cleanliness, reduction in a nep content, opening of the material, fibre parallelisation, sliver evenness, and homogeneity of blends.

The basic idea was to ally the opening powers of the wool card's workers/strippers with the cotton card's powers of fibre separation and cleaning resulting from its carding flats.

Two prototypes of what we call "mixed cards" have appeared so far. One ine France, the SACM-T single-cylinder card, compprises 2 worker/stripper carding points and a section of 50 card flats, 12 of which are in operation (Figure 4).

Figure 4
Composition of a card set of the Prato type used in the preparation of carded recycled materials.

The other in Italy, is much larger and comprises one cylinder
equipped with roller carding points and a cylinder with card flats. The
sliver delivered from this unit is again acted upon by a third cylinder
also equipped with card flats.

Figure 5 shows a diagram of this first prototype compared with a
conventional Prato card set. The general principle of these two machines
has evolved from the same line of thought : whilst the workers/strippers
ensure initial opening and good blend homogeneity, the flats provide
thorough cleaning and good parallelisation.

Conventional Prato card set

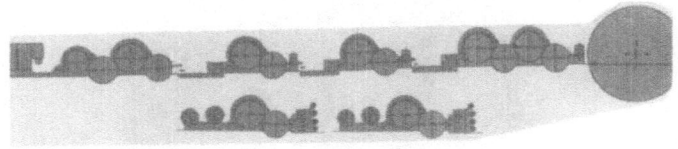

Prototype

Figure 5
Prototype of the new mixed card (rollers and flats)
designed and researched by Tecnotessile.

The results obtained show a distinct improvement in cleanliness and
blend homogeineity, and a reduction in neppiness, especially using the
3-cylinder mixed card. With a production rate of 20 kg/hour for a
delivered sliver count of about 7 ktex, this card has opening powers
fully comparable with conventional industrial sets of the Prato type. We
have noted that an unbeaten carbonised noil gives results which are good
from the aspect of nep content. It has been possible to demonstrate the
role of the card in the removal of oxydised (weathered) tips of fibres
which create dust in the sliver that is deposited in the rotors.

Figure 6 shows the layout of what may be regarded as the prototype
of the new card emanating from the ITF work whilst the next (Figure 7) is
the Rosique 2-cylinder mixed card, equipped with rollers and with flats,
which is being marketed commercially (Figure 8).

Depending on the fibre-length diagram of the sliver delivered by the
card, it is desirable to provide the minimum number of drawing passages,
with or without autolevelling, to improve the medium-term and long-term
evenness. If the fibre-length diagram does not allow this, an autoleveller system on the card is highly desirable.

Figure 6

Figure 7

<u>Figure 8</u>

The two cases illustrated, namely : autolevelling at the card delivery and autolevelling on the drawframe, have been studied. The evenness of a sliver may be defined by its sectional regularity like that given by the Uster apparatus and its variance-length curve or CBL which is an indication of the variations in mass between elements of equal length.

Our basic reasoning was that in the case of direct feed of card sliver to open-end machines, it would be desirable for this to have largely the same evenness characteristics as a conventional sliver from the second drawframe stage on the cotton system, in other words, an Uster CV % of 4 for a count of 5 ktex, and correct development of medium-term and long-term evenness. The autoleveller system situated at the card delivery must therefore be capable of providing these three functions (Figure 9)

In our industry we have used an RRC unit by Texcontrol. This type of autoleveller (Figure 10) at the card delivery consists of a pair of sensor rollers which record the count variations over sliver elements less than 10 cm (Texcontrol says 2.5 times the mean fibre length). The information is transmitted to a pair of drafting rollers the speed of which is varied by a low-inertia motor. As the optimum range of draft lies between 1.25 and 1.75, the linear levelling for sliver count variations lies between ± 20 %.

Determination of the CBL curve

The method involving the weighing of pre-cut lengths, although tedious, is the most appropriate for sliver. In this study, the procedure was to cut over a length of 3000 m, 30 samples measuring 1m, 2m, 5m, 10m and 50m, and weigh them as they are cut. By cumulating these results, it is then possible to calculate the CV corresponding to these lengths from a minimum of 30 measurements.

110

Figure 9

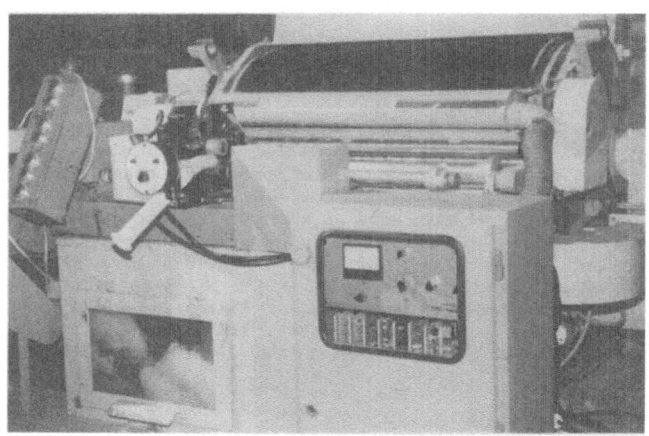

Figure 10

Two series of measurements were performed on one sliver without autolevelling and one sliver subjected to autolevelling on the RRC. The sliver consists of 50 % polyester (1.66 tex) and 50 % "open-top" wool (23 u). (The characteristics are shown in Figure 11).

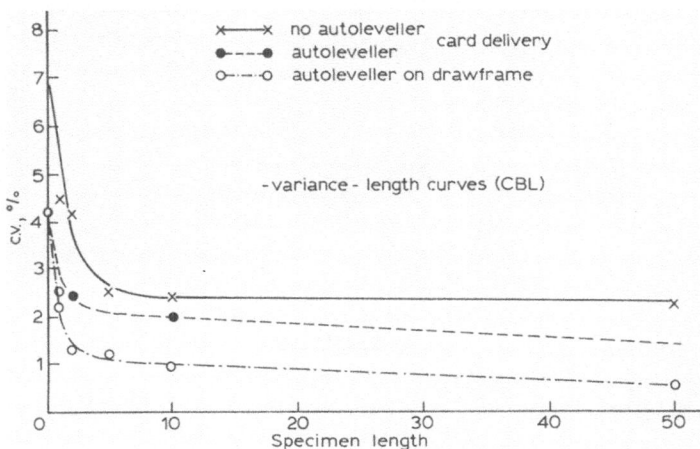

<u>Figure 11</u>

Inspection of the curves shows that levelling has taken place in the short, medium, and long terms. In the short and medium term an improvement of the order of 40 % in evenness proves that the response time, between measurements being taken by the sensor roller and the speed of the drafting rollers being modified, is very short, since the distance between the rollers is only 49 mm. In the long term the curve flattens out to approach its asymptote which is the limiting CV for sections of infinite length.

An autoleveller system at the card delivery thus allows good levelling to be attained in the short and medium term as well as in the long term : the levelled sliver is quite adequate for feeding present-day open-end machines directly with card sliver. Levelling may be performed on the drawframe for materials having sufficiently homogeneous fibre length diagrams. In fact, the drawing action brings about improved fibre parallelisation and thus suggests better fibre performance in spinning, especially at the beater stage.

In our study we used the new S.A.C.M. autoleveller system, the A.R.C., fitted on the ER 700 drawframe (Figure 12). This is a drawframe with two independant heads; its drafting system is of the three-on-four type with a steep slope, thus providing optimum fibre flow towards the coiler (Figure 13). The ARC autoleveller is designed to correct short-term and long-term variations; the count of each sliver is measured by mechanical means.

Figure 12

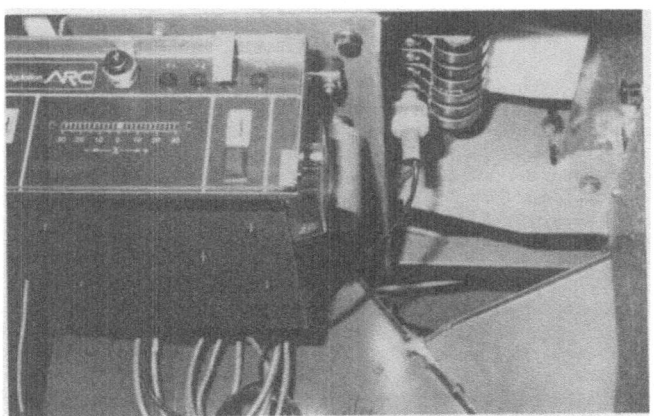

Figure 13

The information is subjected to data processing and is transmitted to a speed regulator of low inertia which acts upon the input speed, the delivery speed remaining constant. The correction range is ± 25 %.

As in the previous case of autolevelling, we obtained a CBL curve, following the autoleveller drawframe stage, from the same 50/50 wool/ polyester blend. (The curves obtained are shown in transparency 1). The main feature shown is that a drawframe stage further improves the medium-term and long-term evenness. Therefore, where permitted by the fibre length diagram, a drawframe passage should be included. This permits not only improvement of evenness but also better parallelisation of the fibres before they reach the beater unit. In fact, there will be less damage to fibres by the teeth of the beater if there is better fibre parallelisation.

Another route investigated for obtaining an open-end feed sliver was "short-staple stretch-break converting" (Figure 14).

Figure 14

This technique is currently practised on combed sliver, i.e. on fibres in almost perfectly parallel formation and having no fibre hooks at their tips. The difficulty encountered in stretch-break converting using card sliver or any other form of uncombed sliver results precisely from the presence in these slivers of a large number of fibre hooks which lower the efficiency of the stretch-break conversion process. The purpose of the "short-staple stretch-break converting" process is to facilitate spinning on open-end machines by removing fibres of length exceeding a certain threshold, equal to the diameter of the rotor. These fibres of excessive length are present not only in new wools of fairly high mean length, but also in recycled materials of very low mean length, such as noils from combing. A proportion of fibres of excessive length causes an excessive rate of end breaks and an unacceptably high occurrence of "wrapped" fibre bundles. To overcome this drawback, it is sufficient to replace a drawing stage by a stretch-break converting stage. Research attempted to determine the optimum production conditions for this new system of preparation for spinning (Figure 15).

SCHEMA CYLINDRES DE RECRAQUAGE COURT

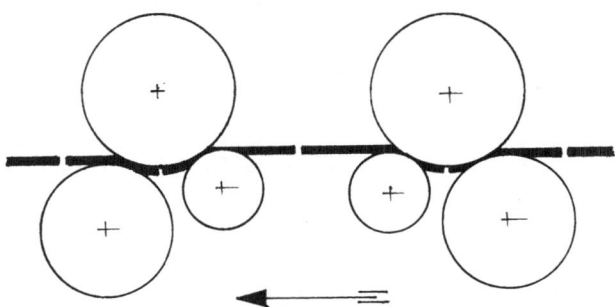

Figure 15
Diagram of rollers in short-staple stretch-break converting

For economic production of "woollen spun" type yarn, stretch-break converting must be performed on card sliver or on card sliver that has been subjected to a drawing passage. Stretch-break converting is thus carried out on sliver with fibres of fairly poor parallel formation, lacking squareness, and above all, generally having hooks at the leading or trailing ends or at both.

The theoretical investigation consisted of the development of a mathematical model of the stretch-break converting process on a sliver possessing a large number of fibre hooks in order to provide optimum machine settings for the stretch-break converter. Practical trials where then carried out with this system of preparation followed by spinning on commercially available open-end machines.

Experience has shown the impossibility of attaining adequate efficiency in stretch-break converting directly from semi-worsted card sliver even by reversing the direction of sliver feed.

The production system finally adopted comprises a high production card of the semi-worsted type (approx. 300 kg/h) of 2.5 m width across the wire (which represents a considerable advantage over the conventional "woollen" process), and then a drafting stage, followed by a short-staple stretch-break converting stage with carefully selected machine settings.

Besides this well defined production plan, an initial series of conclusions may be drawn on sliver preparation for spinning on open-end machines :

- the mixed card possessess good opening powers, the wool card has better blending action than the cotton card, and the mixed card generally gives results between the two;
- the cotton card delivers less neppy slivers than the wool card. However, the best result was obtained with the experimental mixed card of Tecnotessile;
- the card removes about 50 % of the residual vegetable matter. The results is that materials that have not been carbonised produce slivers with a relatively high vegetable matter content;

- it is confirmed that noil that has been carbonised but not beaten produces a card sliver generally containing fewer neps;

- the card effects a preferential removal of the wool component of wastes. The result is that blend composition changes (loss of 3 to 5 % wool) and this should be taken into account in processing.

These results are encouraging and they open-up interesting possibilities for subsequent development and the industrial application of the mixed card, not just for open-end spinning but equally for feeding conventional spinning frames. This could be of still greater benefit if the reductions in capital and maintenance costs are taken into account. These are made possible by a card of reduced dimensions compared with the highly complex and very expensive card sets currently used in the woollen industry. Finally, there is a need for levelling the slivers, especially where they are intended for supplying the spinning frame directly from the card.

Now we know how to produce a sliver of good quality; but how does it perform in open-end spinning and what are the properties of the products obtained ? The comparative distinguishing features of open-end spinning machines are mainly the feed system comprising the actual feed unit and the beater or fibre selector, the rotor and its geometry, and the device taking up the thread from the twist tube.

Adaptation of the open-end technique to wool is succeeding through a better knowledge of the influence of these factors (form and speed principally) and of the interactions created by the material itself (blend composition, amount and type of additive, amount and nature of dust generated). In the early stages we relied particularly on the feed unit developed by SACM-T on the ITG 300 which we compared with the more conventional devices of the machines of the AUTOCORD, BD 200, and RU II types which enable the fibre cluster problem to be resolved. These more conventional devices generally consist of a roller and a support apron with a rapidly rotating beater of small diameter. The fringe emerging from the feeds is combed out and the fibres pulled out singly or in groups of two or even three. The beater (Figure 16) provided excellent fibre separation. Numerous comparative studies with cotton ringframe drafting units have demonstrated the superiority of this system as far as yarn evenness is concerned. However, this carding type procedure has drawbacks : it causes the formation of numerous fibre hooks because the teeth of the beater are only able to withdraw the fibres after folding over one end (Figure 17).

There is also a tendency to shorten the fibres by the sudden stress imposed on them when they are released by the feed system. Their speed is increased by 1500 to 8700 times, and this causes great stress. Fibre breakeage creates the formation of debris which collects in the rotor and causes a gradual fall in yarn quality. Furthermore, the shortening of the fibres brings about a reduction in yarn strength.

These last three observations apply especially to wool. The development of the fibre selector fitted on the ITG 300 is a direct consequence of these factors. This selector is made up of an assembly of superposed discs a half millimetre thick in line with the axis of rotation of this unit (Figure 18). Each disc comprises three active portions (or plots) thirteen millimetres in length positioned at 120 °. The discs are assembled in such a way that the plots of two adjacent blades are staggered by 29°15'. This geometry was used as a standard for the remainder of the study (Figure 19).

116

COMBED SLIVER   CARDED SLIVER

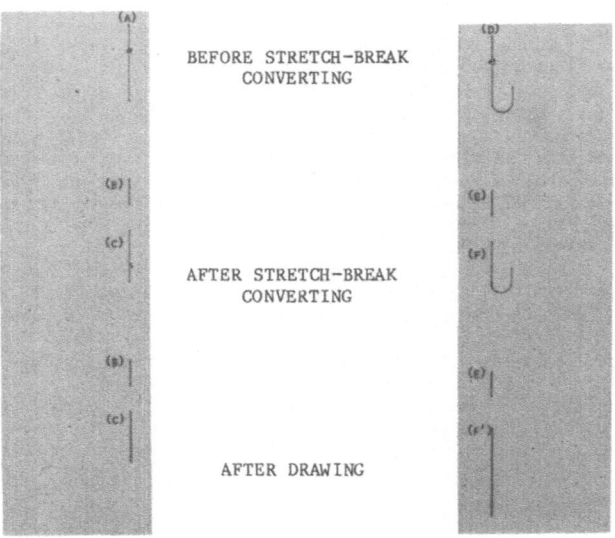

BEFORE STRETCH-BREAK
CONVERTING

AFTER STRETCH-BREAK
CONVERTING

AFTER DRAWING

Figure 16

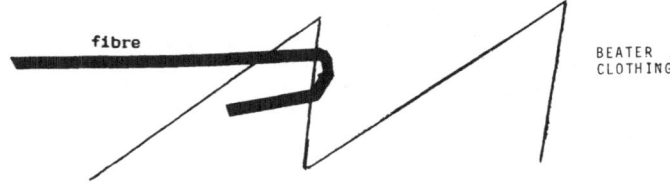

Figure 17

117

Figure 18

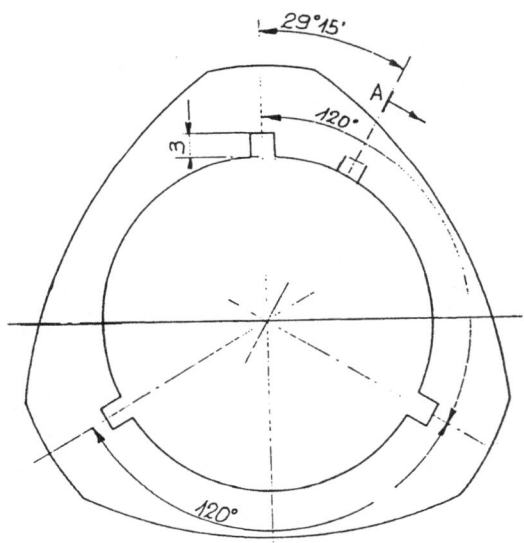

Figure 19

SPINNING ROUTINE WITH STRETCH-BREAK CONVERTING

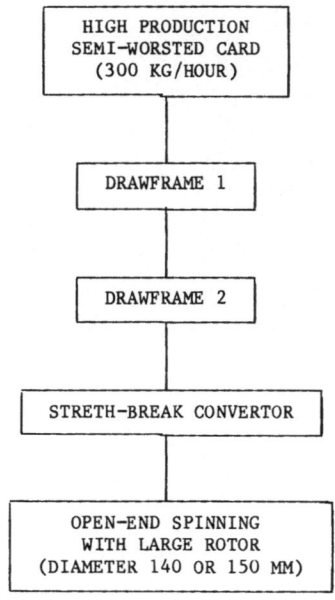

Figure 20

The mode of operation is as follows (Figure 20) : a rubber roller is pressed against the selector by means of a spring. The fibres are carried towards this assembly by a double apron which simultaneously controls their rate of progression, because as in the case of a conventional drafting system, only those fibres nipped by the drafting rollers assume the delivery speed, the others remain at the feed speed. Because of the form of the selector, the nipping of the fibres does not occur with a rectangular contact area corresponding to the common area of a rubber roller. The full width of the fringe is in theory only gripped after a 120 ° rotation of the selector. The fibre cluster formation phenomenon is greatly diminished without the fibres being damaged by hooking. The selector should thus allow the advantages of the beater to be allied to those of the drafting unit without having its faults.

This is what we have tried to prove by carrying out a comparative investigation, fitting a beater on a BD 200 and a selector on an ITG 300 (slide 18). All the trials demonstrated the superiority of the selector, provided that no fibre exceeded 82 mm in length. This restriction is dictated by the apron of the feed system (Figure 21).

Figure 21

The profile of the throat of the rotor has a significant influence on yarn characteristics. It was possible for us to study on the ITG 300 a number of profiles (Figure 22) and we draw the following conclusions :

REF.	ROTOR DIAMETER	ROROR CHARACTERISTICS	DIAGRAM FORM
A	88	$\alpha = 20$ °	
B	88	ROUNDET THROAT BASE $\alpha = 20$ °	
C	88	GROOVED THROAT BASE $\alpha = 26$ °	
D	88	$\alpha = 12$ °	
E	60	$\alpha = 20$ °	
F	66	GROOVED THROAT BASE $\alpha = 35$ °	
G	56	GROOVED THROAT BASE $\alpha = 18$ °	
H	55	GROOVED AND SUNK THROAT BASE $\alpha = 31$ °	

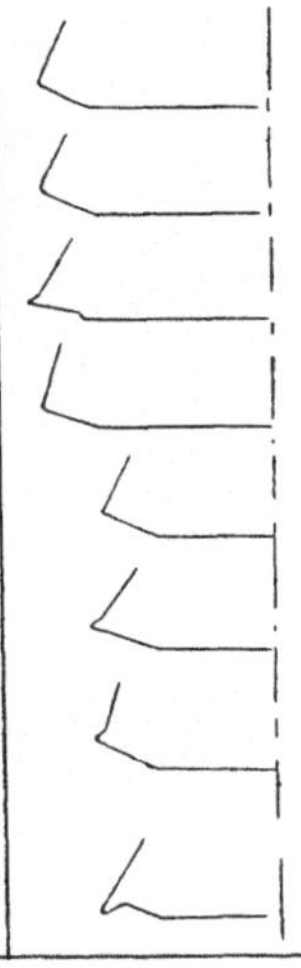

Figure 22

- With respect to evenness, the rotors with steeply sloping walls (large α) give the best results.
- With respect to mechanical properties, small rotors produce a slightly stronger yarn, but especially with a greater extension at break. It must be pointed out that rotor B with a rounded throat base and rotor D with a gently sloping wall (small α) give the worst results both in breaking strength and extension.
- Abrasion resistance tends to be improved by a steeply sloping wall and a narrow restricted throat base.
- Finally, yarn bulk is greater with large rotors, rounded throat bases or gently sloping walls.

These highly encouraging results have unfortunately been offset by the announcement of the suspension of manufacture of the ITG 300. Work has naturally been concentrated on the development of products on the other open-end machines such as :

- the RU 11 (Figure 23)
- the Autocore already mentioned (Figure 24)
- the DREF II (Figure 25).

Figure 23

Figure 24

Figure 25

It transpires that wool in pure form is very difficult to spin on these machines, except perhaps with the DREF. However, in blends with synthetic fibres in various proportions, a range of counts up to 40 tex or even 33 tex may be produced perfectly well depending on the fibre length diagrams, provided that certain precautions are taken. These include selection of beater and its speed (Figure 26 and 27), selection of coefficient of twist and of twist tube (Figure 28 and 29), and selection of rotor diameter and its profile in accordance with fibre length diagram (Figure 30 and 31).

SURVEY OF BEATERS				
FORM	DESIGNATION	CLOTHING	SPEED	APPLICATION
25° 2.1	OB 20	SAWTOOTH	6500-8000	COTTON MEDIUM COUNT
15° 3.25	OS 21	SAWTOOTH	7000-8500	COTTON, FINE COUNT
	OS 21 K	SAWTOOTH WITH DIAMOND COATING	7000-8500	POLYESTERS AND BLENDS
6°	NW 20	PINS	6000-7000	VISCOSE AND ACRYLIC

Figure 26

Figure 27

123

SURVEY OF TWIST TUBES					
FORM	TYPE	SURFACE	α m	YARN STRUCTURE	SPINNING STABILITY
	BG	SMOOTH STEEL	120		
	CG	CHROMED SMOOTH STEEL		VERY SMOOTH	GOOD
	BK 3	STEEL 3 SLOTS, BORON COATED	120		
	CK 3	CHROMED		SMOOTH	VERY GOOD
	BK 4	STEEL 4 SLOTS, BORON COATED	110	SLIGHTLY	GOOD
	CK 4	CHROMED		ROUGH	
	CK 8	STEEL 8 SLOTS , NOTCH CHROMED	100	ROUGH	GOOD
	KG	SMOOTH CERAMIC	125	VERY SMOOTH	VERY GOOD
	KG 4	CERAMIC 4 SLOTS			

Figure 28

Figure 29

SURVEY OF ROTORS				
FORM	DESIGNATION	YARN STRENGTH	YARN STRUCTURE	SELF-CLEANING
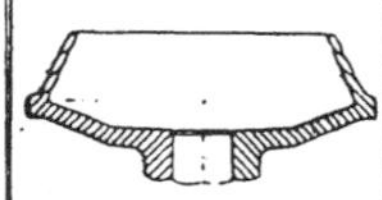  YS 46 SB YS 56		APPROX. 1cN/TEX WEAKER THAN ROTOR 6	Bulky yarn but non-slip	VERY GOOD
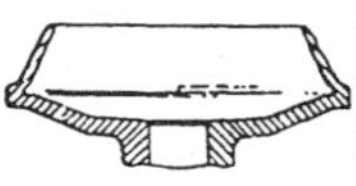 YG 40 SB YG 46 SB (YG 46 D)		VERY GOOD	SMOOTH YARN	ADEQUATE POOR
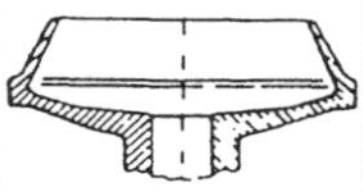 YU 46 SB (YU 46 D)		APPROX. 0.5 cN/TEX WEAKER THAN ROTOR 6	BULK YARN	GOOD ADEQUATE

Figure 30

Figure 31

A few comments may be made on the more consistent results obtained on the Autocoro.

Figure 26 : increase in speed of beater rotation improves yarn cleanliness, but reduces evenness and yarn strength, especially in the cases of beaters OB 20 and OS 21, and has an influence on the amount of dust generated. Nevertheless, it is acknowledged that the OS 21 series at 7000 rpm gives results that are industrially acceptable, as also do the pinned beaters.

Figure 28 : twist tube selection is very important, especially in the case of low coefficients of twist (α 120). Generally speaking, it is the twist tube imparting the greatest amount of false twist which gives the best spinning performance. However, its influence drops as the coefficient of twist increases. The notched twist tube is a "safety precaution".

It is very difficult to take account of all the blends which have been studied by the various laboratories and for which we have been able to determine the optimum coefficients of twist, the number of fibres to be adopted in the yarn cross-section, information has also been obtained about the physical/mechanical properties to aim for, the probable rate of end breaks in spinning, and the nature and acceptable amounts of additives.

Interested parties from industry to get in touch with the research workers for further details, but important findings include : (Figure 32)

- 55/54 blends of noils/polyester or open-top/polyester reach their spinning limit at 33 tex with coefficients of twist of the order of 130;
- with the same blends but in proportion of 80/20 their spinning limit falls to 84 tex, because of the reduction in polyester;
- wool/acrylic blends have been very difficult to process,
- pure wool remains difficult to spin by the open-end process and it is a matter of regret that the ITG 300 is no longer available.

The amount of fatty matter on the sliver demands particular attention. Between 0.5 and 0.8 % is to be regarded as the limit for avoiding lapping at the beater. Furthermore, the amount of dust deposited in the rotors and containing a high proportion of cuticle is caused especially by the mechanical action of the beater on fibre surfaces. Dr. Knott will discuss this aspect of the subject. Nevertheless, the problem posed by dust has resulted in our designing a simulator, intended eventually to allow classification of lots likely to contaminate the rotors (Figure 33), to give a numerical value to sliver cleanliness and thus to estimate the material yield.

We must also mention the work carried out on hollow-spindle spinning which, in counts of 60 to 150 tex is capable of producing yarns competing with folded yarns produced by worsted or semi-worsted spinning in 2x30 to 2x75 tex (Figure 34). This very thorough study has concentrated on investigating the effect of the density of wrapping on the yarn properties including pilling, strength and extension at break, evenness, and handle and appearance (Figure 35). Fabrics and garments have been made by knitting and large-scale wearer trials have been carried out, these have shown that it is possible to obtain products which bear comparison with those obtained by conventional spinning, provided that certain conditions in hollow spindle spinning are observed. (Figure 36)

Figure 32

Figure 33

Figure 34

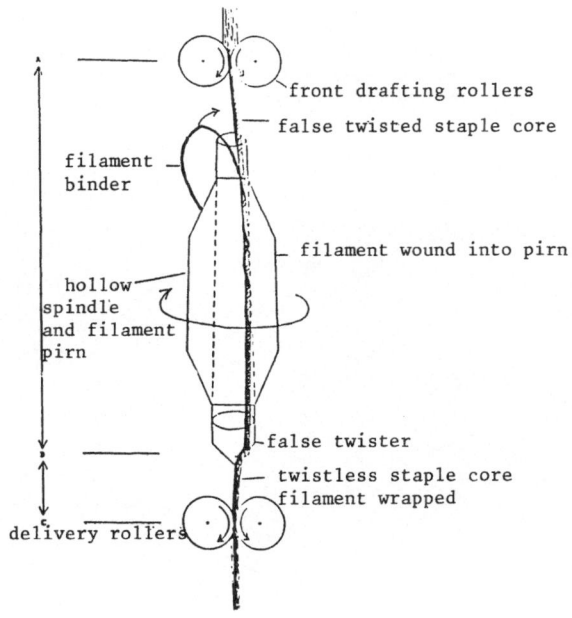

Figure 35

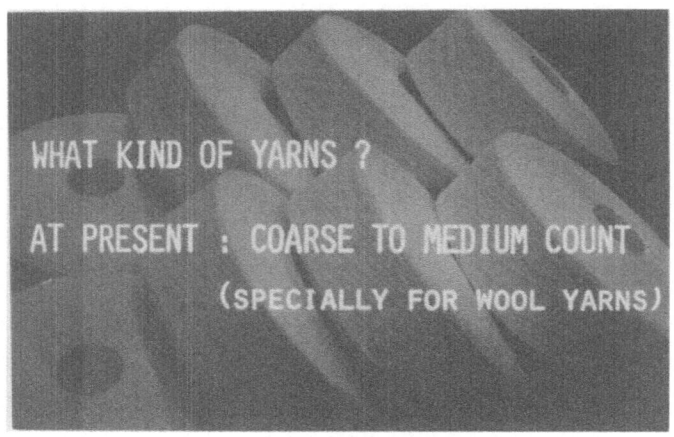

Figure 37

In the knitting sector (Figures 37/38) it appears that open-end spun yarns have generally only mediocre pilling performance because of the selection of blend components, in particular polyester, although this factor has been the subject of closer investigation. It appeared that the spinner's choice of high-tenacity polyesters was governed largely by a desire to attain adequate breaking strengths in spinning. A compromise will always have to be reached between the strength of the yarn and its pilling propensity.

Figure 38

129

In furnishing fabrics, blankets, or tufted materials including carpets, trials have proved most encouraging, although there are differencies in appearance from the same products manufactured with conventional yarns due essentially to very short-term. These deficiencies may be overcome irregularities in open-end yarns and to an occasional excessive presence of wrapped fibre clusters by using a sliver levelling system.

As far as apparel fabrics are concerned, opinions remain divided on the use of these yarns in either warp or weft, although open-end spun warp has the correct size take-up characteristics (Figure 39). however, its use as weft in conjunction with a "conventional" warp no longer appears to raise objections from users.

The results obained within the scope of this vast research programme have only been made possible by close cooperation between the various EEC laboratories who also acted as spokesmen for the industrial interests involved in the studies and whom we wish to thank once again for their cooperation.

Figure 39

PHYSICAL AND CHEMICAL ANALYSES

J. KNOTT
Centexbel

INTRODUCTION

It is well known that physical parameters such as the length and diameter of wool fibres and roving levelness are not the only factors to which consideration must be given in open-end spinning. The physico/chemical parameters related principally to the surface condition of the fibre play a significant role, especially in the generation and of deposition of dust in the rotor. The objectives of co-operative research carried out by Deutsches Wollforschungs- institut Aachen, Institut Textile de France (Nord) and Centexbel (Verviers) were as follows :

1. to propose new test methods or to adopt existing ones in order to try to predict the performance of wools in open-end spinning,

2. to study the generation of dust in the process of yarn production,

3. to study dust generation on open-end spinning machines equipped with beater or with selector opening systems.

2. – RESULTS OF EXPERIMENTS

2.1. Development of specific test methods

2.1.1. Determination of dust content in roving and in yarn

The original method for the estimation of dust in the yarn, in the roving or in the loose wool described below has been tried on dozens of specimens.

Method

The test material is cut to a length of 1cm in the case of loose wool and sliver, and to 0.5 cm in the case of yarn. A 1g specimen is placed in a 50 ml polyethylene flask with 30 ml filtered denatured ethanol. Four flasks are treated in this way simultaneously. The flasks are then placed on a 30 Hz highfrequency shaker with 5 mm amplitude. After shaking for 15 minutes, the contents of the flasks are filtered on a 100 um metallic screen. The residue collected is squeezed by a flat- ended glass rod. The specimen is then replaced in its flask, treated again in 30ml denatured ethanol and screened as before. The denatured ethanol solution that is thus collected is filtered on previously weighed fluoropore (porosity 0.5 um) and the flask that has held the solution is throughly rinsed to prevent any losses. The filter is dried and the amount of dust extracted is determined by weighing.

To determine the amount of protein the residue is treated with sodium hypochlorite solution and the amount of protein is determined by difference.

Precision of the method

The results of trials of the method carried out in the Verviers and Aachen laboratories on both roving and yarn are given in Table 1.

No statistically significant difference exists between the results obtained in Aachen and in Verviers, regardless of the condition of the specimens.

Table 1
Determination of percentage dust in roving and in yarn

% DUST	ROVING		YARN	
	Aachen	Verviers	Aachen	Verviers
	0.46	0.45	1.16	1.13
	0.49	0.49	1.21	1.16
	0.51	0.46	1.16	1.19
	0.49	0.44	1.21	1.17
	0.48	0.51	1.12	1.16
	0.50	0.47	1.15	1.21
		0.49	1.07	1.19
		0.50	1.11	1.21
X	0.4688	0.476	1.15	1.17
S	0.017	0.025	0.048	0.027
CV %	3.52	5.25	4.19	2.35

2.1.2. Kinetics of dust release

We have made a particular study of the release of dust in the roving and in the open-end spun yarn. The results in Table 1 show that the yarn generates the greater amount of dust and that it releases this dust more quickly. This suggests that mechanical degradation occurs during the spinning process and that a proportion of the damaged scales does not become fully detached from the fibre. These scales are removed progressively during shaking, which gives a rate of dust generation greater than that of intact fibre. The high dust content in the yarn could be related to the type of spinning frame; in this case a 95 mm diameter rotor intended for long fibres is used.

We have also investigated the influence of the origin of the wool. On a range of wool lots scoured in the same plant but of different types (South American, Australian, and New Zealand) the dust content of the sliver and the profile of dust release over the course of time are very similar. The dust directly removable is 0.2 to 0.3 % and the dust removable in 10 minutes is 0.4 to 0.5 %.

2.1.3. Determination of the tippy nature of wool

The development of a technique for determining the percentage of oxidised (weathered) fibre tips, which are consequently mechanically brittle, is a particularly apt in view of the high mechanical stresses to which the fibre is subjected during open-end spinning.

Two methods are investigated :

1. A method restricted solely to research laboratories uses a high-speed scanning photometric microscope. The detector records the light transmitted by the stained portions (the oxidised zones of the fibre). A detailed description of this method is given in the final report prepared by Deutsches Wollforschungsinstitut and Centexbel.

2. A simple detection method based on the fact that the oxidised fibre tips possess a higher cysteic acid content with a greater capacity for the selective absorption at pH 2 of certain chosen cationic dyestuffs. This is especially useful for loose wool as it allows a quick estimation to be made of the extent of fibre tip degradation.

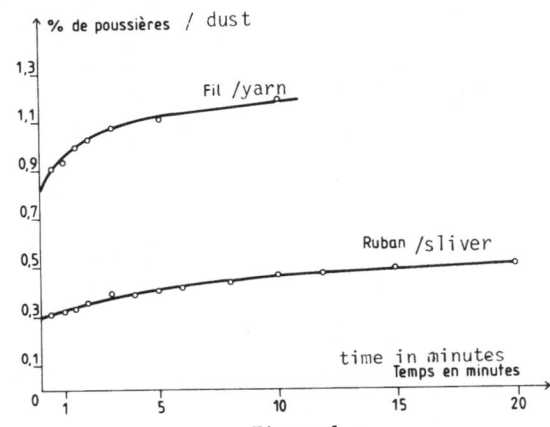

Figure 1
Comparative kinetics of dust release by sliver and by yarn during shaking

2.1.4. Analysis of dust present in the rotor

The dust was first segregated into fractions which could be and could not be sedimented. Separation of the matter which could not be sedimented is achieved by screening with various mesh sizes from 100 to 10pu. Identification of the materials involved in each fraction is performed by scanning electron microscopy or sometimes by optical microscopy. This test method permits determination of the "morphological" composition of the dusts (fibre tips and fragments, cuticle and cortex). Using these new methods along with other classical chemical methods such as determination of dichloromethane extractable fatty matter, estimation of cysteic acid by chemical means or by multiple internal infrared reflectance and estimation of amino acids by the Moore and Stein method, measurement of alkali, solubility and estimation of terminal amino groups by ninhydrin, has made it possible to study the various influences occurring in the open-end spinning process.

2.2. Evolution of damaged portions at the various stages of preparation before spinning

To gain a better knowledge of the origin of the contaminants responsible for processing faults in open-end spinning, we had to characterise the material at the various stages of production starting with loose wool. Visual detection of the partial degradation of loose wool is achieved by treating the material with a cationic dye. The materials and the residual dusts have been analysed following the operations of carding, combing, stretch-break converting and spinning. We have also compared the yarns produced on open-end spinning machines equipped with a selector and with a beater, as well as those produced by conventional spinning.

The loose wool (150 Kg) used in the tests was first treated with 2 % methylene blue in order to make visible the damaged portions of the material. The distribution of the dye on the wool reveals three distinct zones : the highly stained tips (10 % by weight of the material), the middle zone with medium staining representing (25 %), and the base portion with little staining (65 %).

To determine the distribution of dye along the fibre (figure 2), each of the portions obtained in this way was extracted with 50 % formic acid. This test enables the profile of degradation in the roving to be illustrated. Thus, a reduction in dye content in the material during processing can roughly be equated to a loss of damaged portions.

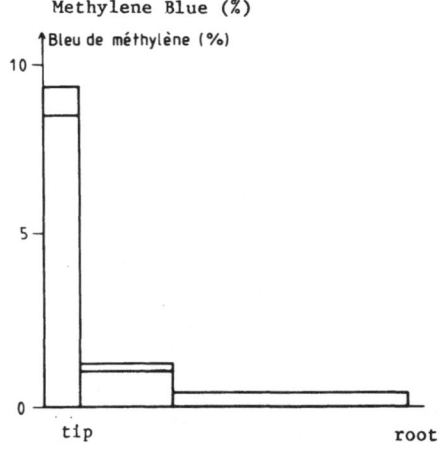

Methylene Blue (%)

Figure 2
Distribution of dye from fibre root to fibre tip

2.2.1. Estimation of the residual amount of dye at the various stages of preparation and spinning.

Figure 3 shows the dye content at these different stages.

On the basis of these results it appears that processing on the card greatly reduces the percentage of stained fibres, showing that the card plays an important role in removing the damaged portions of fibres. Also, in the course of open-end spinning, stress is imposed on the feed roving by the fibre opening system and a significant loss of stained portions might therefore be expected, but this is not the case ! The explanation is to be found in the cleaning action of the yarn in the rotor (see dust content). In the course of conventional spinning, there is a considerable loss of fibre tips, although this system is less "aggressive" than open-end spinning. In this case, the loss of dyestuff must be attributed to the "free" fibre tips which were present in the feed roving.

% dyestuff

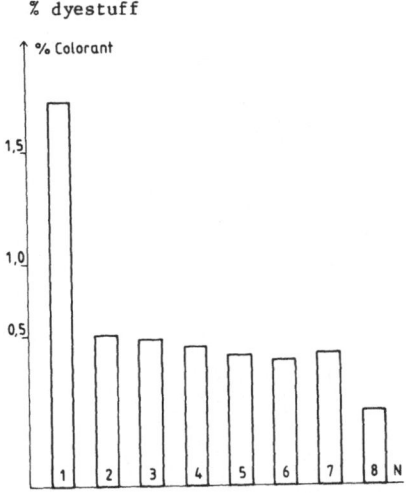

Figure 3
Estimation of dyestuff present at the various stages of preparation.

1 = loose wool
2 = card sliver
3 = combed sliver
4 = stretch-break converted sliver
5 = roving prior to spinning
6 = yarn – open-end spinning with beater
7 = yarn – open-end spinning with selector
8 = conventionally spun yarn

2.2.2. Analysis of amino acids

A comparison of the analysis of the yarn and of the loose wool does not reveal any significant difference. The dust on the card, of which the wool portion is made up of tips (determined by microscopy) is marked by a high cysteic acid content and by the combined cystine + cysteic acid being distinctly lower than that of wool. It has not yet been possible to explain the very high glycine content. However, the drop in the contents of histidine and proline provides confirmation that photo-oxidation and some degree of thermal degradation have occured (table 2)

Table 2
Amino acid composition (mol/100 mol)

AMINO ACIDS	LOOSE WOOL	O.E. YARN Machine with beater	CARD DUSTS	ROTOR DUSTS machine with beater	ROTOR DUSTS machine with selec.
CgSO3H	.158	.133	.2315	1.177	1.980
Asp	6.840	7.097	7.736	6.514	3.776
Thr	6.327	6.690	6.987	6.116	5.188
Ser	11.233	12.449	11.549	14.319	16.722
Glu	12.413	13.451	13.356	11.309	10.023
Pro	7.322	6.539	4.896	7.922	9.383
Gly	8.624	7.881	10.246	10.268	10.703
Ala	5.696	5.334	6.424	6.574	6.539
Val	5.629	5.445	6.014	6.098	7.033
(Cys)2	5.303	5.665	2.115	4.254	6.895
Met	.415	.449	.435	.259	.315
Ile	3.083	2.983	4.164	2.656	2.197
Leu	8.519	7.931	8.255	7.276	6.220
Tyr	4.095	3.938	3.395	3.434	2.729
Phe	3.196	3.053	3.251	3.005	1.945
Lys	3.050	2.937	2.656	2.520	2.643
His	.712	.767	.423	.535	.766
Art	7.384	7.259	6.783	5.765	4.942

Dusts on spinning frames equipped with beaters and with selectors consist principally of cuticle, as shown by high cystine and serine contents. Their respective amounts vary in the fibre tip content which is greater in the case of the spinning frame with beaters. Cysteic acid content, just like microscopy, sugggests that the extent of oxidative damage to the cuticle is less than that to the fibre tips. The contents of glycine and "cystine + cysteic acid" indicate that the skin-flake content does not exceed 10 %.

2.2.3. Dust content at the various stages of conversion of the material.

Extraction of dust at the various stages of conversion of the material have been performed by high-frequency shaking for 3 minutes. The results in Table 3 again show the important role of the card in the removal of dust. Streth-break conversion of the material is an additional source of dust generation which cannot be ignored.

Table 3
Evolution of rate of dust in the spinning cycle
(* = mean of 4 Measurements)

Sample	% dust (*)
Loose wool	1,06 - 0,91
Wool at card delivery	0,55 - 0,46
Wool prior to stretch-break converting	0,43 - 0,43
Wool after stretch-break converting	0,52 - 0,51
Wool prior to combing	0,40 - 0,35
Roving at open-end feed	0,50 - 0,51
Open-end yarn - beater	0,76 - 0,76
Open-end yarn - selector	0,59 - 0,61
Conventional yarn	0,53 - 0,51

The dust content in open-end spun yarns is always greater than that occuring in conventionally spun yarn. Open-end spinning machines equipped with a selector are less damaging to wool than open-end machines with a beater.

2.2.4. Spinning performance

2.2.4.1. Spinning machine with selector

After 8 hours running, there was no deposition of dust in the throat of the rotor and the number of end breaks was zero. The dusts sampled were deposited on the upper circumference of the rotor where they escape the cleaning action of the yarn. Because the principal action of the selector is by friction, the generation of dust is low and the dust is largely composed of cuticle.

2.2.4.2. Machine with beater

Four rotors of the machine were used simultaneously. After 105 minutes, three yarns had broken, this occurring during a period of ± 20 minutes. Deposition of dust by the roving and by the action of the beater exceeds the self-cleaning powers of the yarn. In fact the throat of the rotor is clogged up in a short time (105 minutes). The fibres projected into the rotor are thus no longer able to come together in the throat to form a coherent strand. The dust collected from the rotor contains significant amounts of cortex and fibre tips. These results clearly show the aggressive action of the beater. The cortex content is greatest in the throat of the beater.

2.3. Comparison of open-end spinning machines with beater and with selector

The machines differed not only in the type of fibre opener but also in the diameter of the rotor. However, the wools used were quite similar in both chemical and physical properties. The component which generates dust is obviously the fibre opener, but it is the rotor which retains the dust. In the rotor, a dynamic state of equilibrium is established which is the result on the one hand of the deposition of dust by the sliver and by the action of the fibre opener, and of the retention of this dust by centrifugal and adhesive forces (thermolysis of oils), and on the other hand is the result of the cleaning action of the yarn itself on the rotor (friction, adhesion, pick-up) or of the action of a dust extraction device (compare especially wools H_4 and R_1 which are quite similar in their physical and chemical properties).

Examination of the data on dust analysis confirms that the beater is a more aggressive device than the selector. In fact a much greater amount of roken tips is found in the first type of dust and in some cases the attack on the cortex of fibres results in a greater occurrence of cortex cells (Table 4). Fibrillar fracture at the wool stage is a feature of mechanical fatigue phenomena as is also the weakening of the cellular membrane by chemical modification (photo-oxidation, hydrolysis). However, if the totality of the fibre is damaged, the fracture appears as a clean break : this is the case with the extreme tip of fibres. Although the selector is equally capable of breaking the brittle tips and even attacking the cortex of fibres, the principal action of this device is frictional. There are two consequences of this : dust generation is generally lower and the dust is composed at least predominantly of cuticle. The extent of damage of this dust is shown by the fact that it is more easily detached from the upper third of the fibres where the Methylene Blue test on loose wool demonstrates oxidative degradation and where microscopy is able to show the partial detachment of scales. Certain oxidising treatments assist the detachment of scales.

Fine dust generally contains more fatty matter, undoubtedly because of its greater specific surface area.

table 4
Comparison of open-end spinning machines equipped with beater
and with selector

INITIAL RAW MATERIALS

P A R A M E T E R S	H_4 (beater)	R_1 (selector)
diameter (um)	22.1	21.0
hauteur (mm)	58.5	56.7
barbe (mm)	71.4	70.9
dichloromethane extract (%)	0.90	0.87
alkali solubility (%)	17.5	18.0
pH of aqueous extract	7.5	7.6
content of groups (+) NH2 (umoles/g)	222.0	225.0

Note : R_1 has greater absorption of methylene blue than wool H_4. Therefore R_1 is more greatly oxydised than H_4

DUST IN THE ROTOR

P A R A M E T E R S	H$_4$ (beater)	R$_1$ (selector)
General appearance	dusty	fine and sleek
Composition (%)		
– fibre tips	54.2	10.0
– cortex cells and coarse debris	31.9	3.3
– cuticle	5.6	61.0
– fatty matter (%)	7.9	24.0

3. – CONCLUSIONS

Various methods have been developed for the analysis of spinning dusts and fibres : staining tests for demonstrating tippy nature; quantitative determination of the tippy nature of fibres by microspectrophotometry; extraction of the dust from sliver and from yarn by high frequency shaking and weighing; separation of dusts by wet microscreening and preparation for examination by electron microscopy.

Various influences affecting the open-end spinning have been investigated with these methods and other classical chemical methods. The specimens studied have been obtained through the collaboration of the German, Belgian, French and Italian laboratories and industries.

Other points that have been demonstrated are the role of the card in the removal of fibre tips and the influence of the fibre opener surface on the type of dusts generated. The sources of dust include not only the cleanliness of the material before spinning (scouring efficiency) but also the intensity of the mechanical action of the spinning machine on apparently clean raw materials. Measurements have shown that the yarn plays an important cleaning role at the rotor stage. The dusts picked up by the yarn may be the cause of faults.

Open-end spinning machines equipped with a selector are less damaging to wool than those with a beater, as indicated by a much lower cortex : cuticle ratio in the dusts generated on this type of machine.

The use of well carbonised wools greatly diminishes contamination of the rotor, by reducing vegetable matter and dust.

The nature of the lubricant applied also influences dust generation and the presence of silicic acid in these lubricating additives has an adverse influence on satisfactory performance in open-end spinning.

Photomicrograph 1/Lot A
Electron micrograph of the cuticle fraction of rotor dust

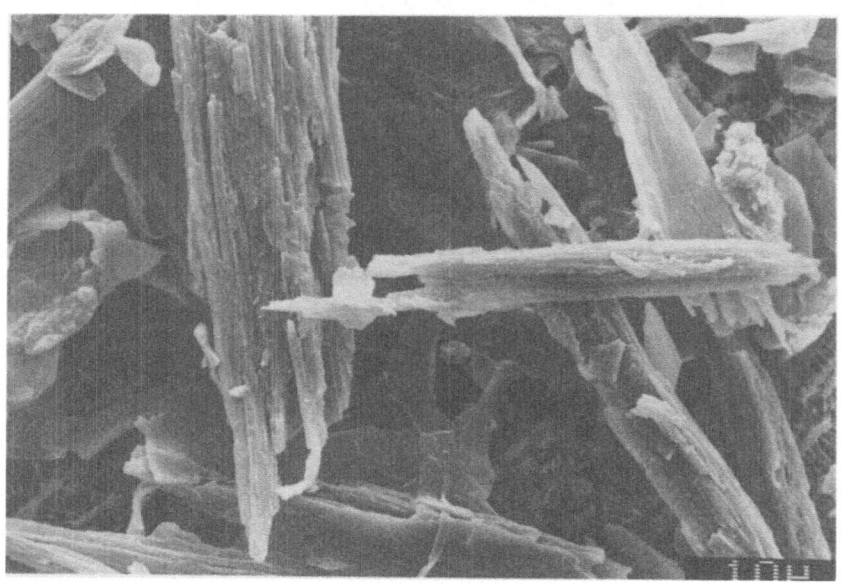

Photomicrograph 2/Lot A
Electron micrograph of the cortex fraction of rotor dust.

141

Photomicrograph 3
Dust at the card stage

Photomicrograph 4
Dust at the rotor stage (machine equipped with selector)

SUMMARY AND CONCLUSIONS BY THE CHAIRMAN OF THE SESSION

M. BONA
Città Degli Studi

The session was a great success with about 70 delegates attending.
As arranged, three reports were given, summarizing the work carried
out by eight institutes in five Member States. The paper, by Dr Artzt,
Preparation of recovered materials for non-coventional spinning was based
on worked carried out by the Institut für Textiltechnik, WIRA and CELAC.
The one by Mr Bruggeman, new carding and spinning technologies included
the results from ITF-Nord, Centexbel, CELAC, WIRA, UMIST (GB) and
Tecnotessile. Dr Knott's paper, physical and chemical analyses, reported
the research of Centexbel, Deutsches Wollforschungsinstitut, and
ITF-Nord.
The reports were followed by a lively and constructive discussion
during which specific technical aspects were explained in greater detail,
including the following :
- The choice between conventional carbonization and the new mechanical
  purifying processes was explored in detail. While mechanical processing may give satisfactory results in combing noils containing little
  vegetable matter, carbonizing, when properly carried out, still
  appears to be the best solution for waste containing a high proportion
  of vegetable matter. It has been shown that beating can be omitted in
  order to avoid felting during neutralization. After carding, materials
  which have not been beaten have fewer slubs and are virtually as clean
  as those carbonized in the traditional manner. The various procedures
  will also have to be compared further under industrial conditions,
  especially from the point of view of cost.
- The question of dust is crucial for the functioning of the turbines.
  Mechanical faults cause dust to be produced throughout the whole of
  the manufacturing process to a greater or lesser degree, depending on
  the state of the raw material and working conditions. It is as
  important to devise means of eliminating the dust as to investigate
  how it is produced. An interesting reference was made to dust of
  mineral origin which, although not present in large quantities, may
  have a significant impact due to its abrasive effect.
- This was a discussion about the use of card-wire opening rollers or
  selector rollers in the feed mechanism of rotor-type open-end spinning
  machines. The former system, which is by far the most widely used,
  subjects the fibres to rougher treatment, and many figures were quoted
  in the main report and during the discussion showed that this results
  in a higher rate of breakage of the fibres although only indirect
  proof is available, since actual measurements of length have not yet
  been made on the yarn.
It is particularly to be regretted that the only firm which has so far
produced an open-end machine with selector rollers has now withdrawn
it from the market.
This would appear to confirm the machine manufacturers' lack of
interest in the wool sector, and it is hoped that the results of this
symposium (which delegates are urged to make every effort to bring to

their notice) will help to rectify the situation.
- Attention was drawn to the properties of products manufactured with the new yarns, in comparison with traditional knitted garments and fabrics, with particular reference to covered yarns (hollow spindle). A range of samples, manufactured with the different techniques tried out by the various institutes, provide concrete evidence of these properties.
- Some information was given on future research, the most interesting being the study of additives specially designed to eliminate the problems currently encountered when spinning wool on open-end machines. Whas has to be done is to identify the weak points in the fibre (and, as a result of this research, it is now possible to do this precisely) and to take systematic remedial action. Thus, the objective is to adapt the fibre to the machine, and not merely the machine to the fibre. This is a break with tradition, which has been brought about by the higher manufacturing costs in relation to raw material costs.

Finally, to turn to more general matters, attention should be drawn to two contributions which, for different reasons, are of greater importance than purely technical matters :
- The contribution to the discussion made by a Spanish representative, attending the symposium as an observer, who confirmed some points made by the authors of the reports. This suggests that the great spirit of co-operation shown throughout the duration of this programme will be further enhanced with the enlargement of the Community.
- The remark by the Chairman of Interlaine, the highest representative in office in the Community wool industry, who commended the institutes on their work and stressed its practical value, which he intended to mention at the forthcoming general meeting of his organization.

S E S S I O N   B

NEW   SPINNING   TECHNOLOGIES   IN   THE   WOOL   INDUSTRY

*   *   *

Contract 004–TEX–D – P. ARTZT

Contract 015–TEX–F – J. EDME

Contract 018–TEX–I – T. TORELLI

Contract 026–TEX–B – J. BRACH

Contracts 002–TEX–D – G. BLANKENBURG
027–TEX–B – J. KNOTT

Contract 028–TEX–B – J. GRIGNET

Contract 033–TEX–UK – R.P. HARKER

Contract 038–TEX–UK – W.D. COOKE

CONTRACTOR : INSTITUT FÜR
            TEXTILTECHNIK
ADDRESS : Körschtalstraße 26
          D - 7306 DENKENDORF

PROJECT LEADER : Dr.P. ARTZT

CONTRACT N° :004-TEX-D

STARTING DATE : 1/9/1982

FINISHING DATE : 31/8/1985

TELEPHONE :711/34.08.0

TELEX : 725.66.65

TITLE OF PROJECT

APPLICATION OF NEW SPINNING TECHNOLOGIES IN THE
WOOL INDUSTRY

AIM OF PROJECT

The aim of the project is the recovery of wool from card waste and noils.Replacement of carbonising by mechanical preparation and further processing by unconventional spinning techniques are primary considerations.

SUMMARY OF RESULTS

Card waste

The proportion of fibre in uncleaned material lies between 30 and 35 % depending on the degree of contamination. Fibre recovery by mechanical means amounts to approx 30 %. The trials were carried out on a cleaning machine which is now available from a machine maker. It is a machine operating on the batch system in which the material to be cleaned is subjected to the action of a drum equipped with spikes for a specified length of time, to be finally removed by suction. An advantageous feature is that fibre quality is preserved as no damage occurs by fibre breakage or by chemical treatment (carbonising).

Card waste of fibre fineness 21um and 27um was studied. Comparison was made between these two raw materials and also between mechanical preparation and chemical preparation by carbonising.

The following yarns were spun :

	21 um	27 um
ring-spun	Nm 2, Nm 28, Nm 40	Nm 28
wrap-spun	Nm 14, Nm 28	Nm 14

Noils

Cape noils of fineness 17.9um were used. Cleaning was on a modified cleaning machine usually used for the preparation of cotton wastes.

There is a continuous cleaning action with saw-tooth roller, knife/grid system and finally a rapidly rotating trash roller. Recovery of fibre amounts to approximately 76 %. Modification of the special machine for cotton is technically possible. It would thus be possible to market relatively quickly a machine by means of which it is possible to further process wool noils as a blend component in rotor spinning without previous carbonising.

Yarns were spun both in blends of polyester with wool noils (50/50) and of cotton with wool noils (50/50). Yarn count was Nm 18, 20, 28 and 34 in rotor yarns and Nm 28 in ring-spun yarn.

Conclusion

The weaving performance of all the yarns was investigated. Marketable products including final finishing are now available. Mechnical cleaning represents an economic alternative to the carbonising of wool wastes. The appearance of the end-product is consistently better than is the case with carbonised wools.

CONTRACTOR : ITF-Boulogne   CONTRACT N° :015-TEX-F

ADDRESS : 35 rue des Abondances   STARTING DATE : 1/10/1982
          B.P. 79
          FR - 92105 BOULOGNE BILLANCOURT   FINISHING DATE : 31/10/1984

PROJECT LEADER : J. EDME   TELEPHONE : 1/ 48.25. 18.90

                                             TELEX : 250.940

TITLE OF PROJECT

NEW PROCESS FOR SPINNING WOOL

AIM OF PROJECT
     Application of new spinning technologies in the wool textile
industry.

SUMMARY OF RESULTS

Sliver preparation
     In view of the fibre diagrams of the qualities used in open-end
spinning, an interesting solution comprises the preparation on a
"short-staple" type card of wool sliver intended for feeding to open-end
spinning machines. Nevertheless, the need for a certain degree of
"preopening" of the material and homogenous blending favours the
application of a mixed solution (workers-deburring rollers and hoods).
This solution has been adopted by ITF-Nord and a cotton card has
therefore been modified in collaboration with the French machine maker
SACM, a card on which systematic trials have been carried out.
     In view of the interest shown by the industry in this technique,
Société ROSIQUE has installed an 2-cylinder mixed card at the Machinery
Centre of ITF-Nord. Numerous trials are being conducted and various
capital investment projects are planned.
     With regard to sliver levelling, the study has shown that a sliver
levelled at the delivery of the card allows direct feeding to open-end
spinning machines to be envisaged. However, where the fibre diagram
allows it, levelling on an auto-leveller drawframe will be advantageous
since the medium-term and long-term regularity and the parallelisation of
fibres will be improved.

Modification of open-end spinning machines for wool
     Study of rotors shows that :
1. Increase in slope of the walls and narrowness of the rotor throats
   improve the eveness as well as the frictional resistance but reduce
   the bulk.

2. Reduction in rotor diameter improves the strength and extensibility of
   the yarn.

     Study of the separator elements proves the superiority of the
selector as against the breaker : at lower twist levels there may be less
dust formation and a lower spinning limit.

148

Simulator

The simulator made from an ITG 100 open-end spinning head allows
sliver cleanliness to be very quickly evaluated, and consequently an
approximate estimation of material yield. Furthermore, initial results
lead to the suggestion that, after appropriate study, it should also be
capable of classifying lots likely to produce poor running
characteristics as a result of deposition in the rotors, and it could
possibly act as a delivery "test".

Correlation between physico-chemical characteristics of fibres and their spinning performance

Production trials carried out on the "Autocoro" have been followed
systematically on a chemical basis.

Fibre performance in spinning is closely related to the amount of
extractable matter the fibre contains as well as to the amount of dust
present in the sliver. The two features have been the subject of
particular analysis.

It is found essentially that :
- the constitution of the dust in the rotors is largely cuticle. This
  proves that in the spinning operation there is superficial attack on
  the fibres by the teeth of the breaker roller.
- the amount of dust in the yarns is always close to that of the sliver
  which reflects the proportionality based on the amounts of dust in the
  blend components.
- the amount of dust in noils is always reduced by carbonisation.

Integration of all these results allows establishment of a limit for
dust in feed sliver at about 0.12 %.
With regard to analysis of extractable matter, the result is that :
- the amount of matter extracted varies according to the nature of the
  solvent, however, a maximum limit value for this "fatty matter" is
  about 0.5 to 0.6 %.
- Spectral analysis reveals very great differences in composition of
  lubricants : the presence of silicic acid in these auxiliaries appears
  to be detrimental to open-end spinning.

CONTRACTOR : TECNOTESSILE

ADDRESS : Centro di Richerche SpA
Via Valentini 14
It - PRATO

PROJECT LEADER : Ing. T. TORELLI

CONTRACT N° : 018-TEX-I

STARTING DATE : 1/10/1982

FINISHING DATE : 31/3/1985

TELEPHONE : 35.741

TELEX

TITLE OF PROJECT

STUDY OF POSSIBLE VARIATIONS OF PREPARING PROCESS AND
WORKING MATERIAL MACHINES, INCLUDED RECOVERY TECHNOLOGIES
BEFORE TO FEED OPEN END SPINNING.

AIM OF PROJECT
The current methods of preparing and carding re-used raw materials
require very expensive and large machinery, which is not suitable for
non-conventional spinning such as open end. The purpose of this research
has been therefore to design and construct a preliminary prototype of new
machine, derived by the concept of both wool (cylinder) and cotton (flat)
cards.

SUMMARY OF RESULTS
The experiments carried out with the new card during the research
have demonstrated the high carding effect and the high mixting capability
with the kind of regenerated materials worked at Prato.
These are interesting results because they indicate possible uses of
the system for both traditional and open-end spinning.
Installation of the new machine will require reducec capital
expenditure and will lead to reduced running costs compared with the
present carding system.

CONTRACTOR : LABORATOIRE CELAC

ADDRESS : Avenue du parc 69 H
          B - 4655 CHAINEAUX

PROJECT LEADER : J. BRACH

CONTRACT N° : 026-TEX-B

STARTING DATE : 1/10/1982

FINISHING DATE : 31/3/1985

TELEPHONE : 87/ 22.13.29

TELEX : 49.289

TITLE OF PROJECT

THE USE OF WOOL NOILS IN OPEN-END SPINNING

AIM OF PROJECT
    Study of cleaning, opening and carding processes for wool noils with a view to open-end spinning in blends with polyester or acrylic fibres.

SUMMARY OF RESULTS
1. The elimination of vegetable matter by chemical means is more effective than mechanical treatment. The effect of preliminary treatment remains significant as far as the yarn stage.
2. Careful carbonising treatment with omission of beating permits a well opened product of adequate cleanliness to be obtained, the reduction in felting having a marked influence on the nep content of the card web.
3. In the case of raw materials with a moderate vegetable matter content, it is possible to eliminate 50 % of the latter by preliminary opening treatment and by carding. Mechanical treatment of uncarbonised noils allows the raw material costs to be reduced. Nevertheless, the spinner will have to provide an extra opening machine requring additional capital costs.
4. For blends of wool noils with polyester and wool noils with acrylics, the mixed system of carding enables the advantages of wool carding (better homogenisation and better cleaning) and of cotton carding (lower nep content) to be combined.
5. Spinning trials on unconventional systems (open-end or friction) have shown, that spinning on DREF 2 of blends based on wool noils does not appear to cause any great problems in the case of coarse count yarns (count   200 tex). In the case of fine yarns (count   100 tex) a continuous filament core must be incorporated.
A more systematic study of conditions for rotor spinning, is necessary. However, the use of the selector appears to be better than to using a breaker.

151

CONTRACTORS : 1.CENTEXBEL
             2.DEUTSCHES WOLLFORSCHUNGSINSTITUT
ADDRESS :1. Rue Montoyer 24
        B - 1040 BRUXELLES
        2. Weltmanplatz 8
        D - 5100 AACHEN
PROJECT LEADERS :J. KNOTT
            G. BLANKENBURG

CONTRACTS N° : 002-TEX D
             027-TEX B
STARTING DATE : 1/10/1982

FINISHING DATE :31/10/84

TELEPHONE : 1. 87/22.45.38
           2. 24/13.99.21
TELEX

TITLE OF PROJECT

NEW TECHNIQUES FOR WOOL SPINNING

AIM OF PROJECT

Chemical and physical tests on yarns and auxiliary agents to determine the origin of impurities responsible for faults in the manufacture of wool yarns.

SUMMARY OF RESULTS

Various methods of analysis for spinning dusts and fibres have been developed :
- staining methods for demonstrating the tippy nature of wool fibres, based on the presence of sulphonic groups capable of fixing basic dyes such as Astrazon G Blue or Methylene Blue at pH 2;
- quantitative determination of the tippy nature of the fibres by micro-spectrophotometry after staining with the above dyes;
- extraction of the dust from the sliver and from the yarn by high frequency shaking and weighing;
- fractionation of the dusts by wet microscreening and preparation for study by electron microscopy.

Using these techniques and other conventional chemical techniques, it has been possible to investigate various factors influencing the open-end spinning process. The samples investigated have been obtained thanks to the cooperation of German, Belgian, French and Italian laboratories and mills.

Amongst the points demonstrated are the following :
- the role of the card, in removing fibre tips has been demonstrated, by following through spinning a lot stained with Methylene Blue to show up the fibre tips,
- the influence of the type of clothing of the opening machine and the effect of other spinning variables on the kinds of dusts products have been studied by morphological and quantitative analysis.

It has been observed that sources of dust are not only the impurity content of the raw material before spinning (influence of scouring) but also the severity of the mechanical action of the spinning machine processing apparently clean raw materials. Estimations have shown that the yarn plays the role of a cleaner at the rotor stage. These dusts picked up by the yarn may obviously cause faults. Spinning machines using selectors are less severe in their action as indicated by the much lower cortex to cuticle ratio in the dusts created on this type of machine.

CONTRACTOR : CENTEXBEL

ADDRESS : Division Verviers
Avenue du Parc 69 H
B - 4655 CHAINEAUX

PROJECT LEADER : J. GRIGNET

CONTRACT N° : 028-TEX-B

STARTING DATE : 1/10/1982

FINISHING DATE : 31/3/1985

TELEPHONE : 087/ 33.21.46

TELEX

TITLE OF PROJECT

IMPROVEMENT OF ROTOR SPUN YARNS BY SHORT-STAPLE
STRETCH-BREAK CONVERTING OF CARD SLIVER

AIM OF PROJECT

The aim of the programme was to produce yarn of the woollen type of satisfactory quality on an open-end spinning machine, with a drawing routine similar to the semi-worsted process, but including a stretch-break converting stage in order to resolve problems relating to the presence of excessively long fibres.

PROGRAMME

The research programme comprised two points :
1. Development of a mathematical model of stretch-break converting on card sliver or on card sliver that has been subjected to one passage of drawing.
2. Development of a method of sliver preparation with stretch-break converting of card sliver for the production of woollen type open-end spun yarns from virgin wools of fibre diameter greater than 25um and from recycled raw materials of 20 to 30um.

RESULTS AND CONCLUSIONS

The "short-staple stretch-break converting" of wool for the purpose of open-end spinning is currently practised on combed sliver, i.e. on fibres made almost perfectly parallel and which do not have any hooks on the extremities. The difficulty of stretch-break converting card sliver or any other type of sliver that has not been combed arises from the presence in these slivers of a large number of fibre hooks which reduces the effectiveness of the stretch-break converting operation.

This difficulty has been approached simultaneously from the theoretical and practical aspects. A method has been developed for measuring fibre length in three phases (comprising sampling in the two sliver directions and sampling of "straightened fibres") enabling all the necessary parameters of length and hook description to be obtained. A method has been established for calculating the joint distribution of the lengths of fibre segments having one or two hooks from these three measurements of length distribution.

Next, a method has been developed for calculating bundle diagrams (upstream and downstream) of fibres nipped by the stretch-break roller, taking into consideration the partial straigthening of fibre hooks in the converting zone. These bundle diagrams are then used for determining the optimum settings of the stretch-break convertor.

It has been shown that stretch-break converting could not be applied directly to card sliver, even with the suppression of sliver direction reversal between the card and the stretch-break convertor.

The process used finally comprises three passages : a card of the high production semi-worsted type (approx. 300 kg/hour in 2.50 m effective width, which constitutes a significant advantage over the conventional "woollen" process), one passage of drawing and stretch-break converting at carefully selected settings.

It has been possible to produce yarns of appearance similar to woollen-spun yarns (especially the high bulk) with eveness of yarn linear density as good as that obtained on ringframes, but with lower yet adequate strength.

It appears that this type of yarn could be used in numerous applications of "woollen spun" yarn; for carpet yarns, however, the occurrence of "belly bands" still appears to be too frequent at present and creates problems at the stage of "bursting" the tufts.

154

CONTRACTOR : W.I.R.A.
            Technology Group
ADDRESS : West Park Ring road
          GB - LEEDS LS16 6QL

PROJECT LEADER : Dr.R.P. HARKER

CONTRACT N° : 033-TEX-UK

STARTING DATE : 1/10/1982

FINISHING DATE : 31/3/1985

TELEPHONE : 532/ 78.13.81

TELEX : 557.189

TITLE OF PROJECT

APPLICATION OF NEW SPINNING TECHNOLOGIES IN THE
WOOL INDUSTRY.

AIM OF PROJECT
A survey of textile waste and an evaluation of carded wool in open-end
spinning.

SUMMARY OF RESULTS

Waste survey
      The objective of the survey was to determine the position of that
part of the textile industry concerned with re-cycling textile wastes.
The original conception was concerned mainly with wool, with some
references to open-end spinning. This has been extended within the terms
of reference to include all types of fibres traded internationally as
waste.
      Details of individual import-export figures for EEC countries
trading with other major users have been obtained. For brevity, the
import-export figures for all categories of waste are presented in the
final report and detailed for each EEC country, with EEC and "Other"
breakdown.
      The figures show which countries are nett importers or nett
exporters in each type of waste. The amounts involved show the importance
of either the waste reclamation or utilisation in recycling for each
member state. Due to the very involved nature of the statistics and the
large "across the border" transactions in textile materials, these
figures have to be interpreted in terms of the textile usage in each
country. Only a very short discussion on the activity in EEC countries is
given, but an indication of the current position has been attempted in a
general manner.
      World fibre production figures are detailed, plus EEC figures for
certain fibre usage. Waste production is indicated so far as it was
possible to ascertain. Per capita figures are given for major categories
and from individual annual usage and EEC populations, an estimate of
total waste can be approximated. These calculations indicate the
difference between recovered waste (actual) and the potential for the
re-cycling industry.

Spinning
      Three different applications for Open End spun yarns produced from
carded only slivers were selected, namely woven blankets, flat woven
upholstery fabrics and woven cut pile carpets. In all three trials only
wool fibre were used which had not been reprocessed or reclaimed.

155

The carded slivers were produced from a modified woollen card and spinning was carried out on an SACM ITG 300 machine.

The problems encountered within the spinning of certain slivers were generally caused by a build–up of deposit within the yarn formation area. Fabrics produced from each of the three trials were examined by both physical testing and visual assessment.

The products indicated that fabrics requiring a milled and raised finish can be satisfactorily produced from rotor spun yarns, but fabrics requiring a clear cut finish do not have sufficient surface clarity when compared with fabrics produced from ring spun yarns. Carpets can be produced, using folded yarns, but plain carpets tend to have problems due to random tuft definition caused by wrapper fibres.

CONTRACTOR : U.M.I.S.T.   CONTRACT N° : 038-TEX UK

ADDRESS : P.O. Box 88,   STARTING DATE : 1/7/1983
          Dept. of Textiles, Sackville Street
          GB - MANCHESTER M60 1QD   FINISHING DATE : 30/6/1984

PROJECT LEADER : Dr.W.D. COOKE   TELEPHONE : 61/236.33.11
                                                      (ext.2853/2719)
                                           TELEX : 666.094

TITLE OF PROJECT

THE WRAP-SPINNING OF WOOL

AIM OF PROJECT

To advance the understanding of the mechanism of wrap-spinning, and to examine the effect of the main variables on the properties of wrap-spun yarns. To compare the performance of comparable wrap-spun and ring spun yarns in fabric and garment by laboratory tests and wear trials.

SUMMARY OF RESULTS

Spinning results

1. The tenacity of wrap spun yarn increases linearly with increasing wraps per metre (w.p.m.), within the range of 150-375, and filament dtex. The filament modulus appears to play a very important role in achieving strong yarns particularly when a low filament dtex is used. The higher the filament modulus the stronger the resultant yarn. Thus, for a given yarn tenacity lower wraps per metre could be used for a higher modulus filament with the advantage of a higher production speed. The yarn elongation increases with increasing filament dtex. As the w.p.m. increases the elongation decreases to a minimum value and then increases with further w.p.m. The minimum occurs at w.p.m. value of 270.

2. The presence of a false twister for wrapping would appear to be of more importance to the spinning performance than to any improvement in the yarn properties. The use of a false twister significantly reduced fly and roller laps. Although, spinning with a false twister did not result in any significant differences in the tensile and evenness properties, a noticeably hairier yarn was produced, irrespective of threading direction. Pirn length would also appear to have affected yarn hairiness in that the longer pirn gave a less hairy yarn. The longer pirn also gave an improvement in the regularity of the filament wraps. An improvement in wrap regularity was also seen when the false twister was threaded to cause wrapping inside the hollow spindle.

3. The main contributing factor to the spinning tension was the speed ratio of the front drafting rollers to the yarn delivery rollers. This may be referred to as the wrapping draft tension. Increasing this did not affect the yarn tensile properties, but the yarn evenness was observed to decrease.

Knitting results

1. A range of yarns were spun with w.p.m.of 150, 200, 250, 300 and 375
   with 22f1 and 22f5 polyamide wrappers and knitted to tightness factors
   $\dfrac{\text{tex}}{\text{e}}$ of 12, 14, 15, 16.

   The 32 fabrics samples were assessed for appearance and tested for
   pilling and abrasion resistance.

2. Pilling improved with increases in tightness factor and w.p.m. The
   22f1 wrapper produced better results than the 22f5. Commercially
   acceptable fabrics were achieved with w.p.m. 300 in fabrics with
   tightness factors 15.

3. Abrasion resistance increased with increases in tightness factor and
   w.p.m. There was no clear difference between the 22f1 and 22f5
   wrappers.

4. The mens hose wear trial showed no conclusive difference between a
   2/28's ring spun yarn and a 1/14's wrap spun yarn with 265 w.p.m. and
   a 22f1 polyamide wrapper.

5. The full fashioned garment wear trial showed the wrap-spun yarn
   performed as well as the ring spun yarn in terms of pilling, fuzzing
   and abrasion.

THE QUALITY OF KNITTED FABRICS AND ARTICLES

J. STRYCKMAN
Centexbel

INTRODUCTION

The knitwear industry occupies a significant place within the European textile industry. Employing 410,000 people in some 25,000 companies, four-fifths of them in italy, this sector contributes one eight of European textile production. In 1984, it consumed approximately 695,000 tonnes of yarn, while turnover currently stands at nearly 16,000 million ECU. Although Europe exports 2.5 times the amount of knitted fabric, by value, that it imports, there is a clear deficit in the balance of trade in finished articles. This loss outweighs the suplus on fabrics by 5 to 1. Consequently, all companies in the sector will need to observe the principles of creativity, productivity and quality.

Knitted articles must be functional, properly made and attractive. Points to be marketed are quality, imaginative design and tastefulness at an affordable price. If the sector is to succeed, it must be through rapid adaptation to market conditions, especially in the demanding fashion market. There must be a permanent drive for improvement and renewal of its techniques and its products, paralleling the greater emphasis placed by consumers on hard-wearing and high-quality products.

The small and medium-sized enterprises in which the majority of knitted articles are produced can only meet this demand if they are provided with external technological support from specialized institutes and research centres. With the aim of improving this support, and increasing their fund of knowledge eight research institutes in seven Community countries signed a total of ten contracts with the Commission, pooling and coordinating their efforts in a research programme which is intended to give producers the necessary information to maintain product quality and persuade consumers to buy European goods.

The production of an attractive, functional and reliable article requires a mastery of a considerable number of parameters in the fields of raw materials, manufacture, packaging, storage and freight.

Quality control for knitwear articles is, however, a complex problem involving a knowledge of the properties the article must possess, the availability of objective tests of these properties, and a knowledge of standard values for judging quality levels. In addition, it is necessary to establish which parameters affect quality and how quality will be influenced by any variation in these parameters.

The basic aim of the programme is to establish how the parameters of raw materials and manufacturing conditions affect the quality of articles in terms of the market for that knitwear and the use to which the article will be put.

All manufacturers, regardless of the kind of knitted articles they produce, have to satisfy the major fundamental criterion that the article should keep its shape during both wear and cleaning. Particular attention has been paid to this within the programme.

This characteristic is complemented by other quality criteria to form a group of properties to be assessed in any planned appraisal of the quality of a given article and determining its quality level in terms of the use to be made of it.

The research programme has made it possible to draw up a number of guidelines, regarding raw materials and their characteristics, knitting conditions and subsequent treatment, which manufacturers must observe to guarantee that their products will be of the best possible quality. The results are presented under the following three, linked headings :

- quality criteria for knitwear as assessed by the consumer,

- quality aspects for the knitting process and for knitted articles,

- objective quality assessment.

Taken together, the results presented in these three reports comprise an invaluable guide for the knitwear manufacturer intent on controlling, enhancing and stabilizing the quality of his products.

THE QUALITY OF KNITTED FABRICS AND ARTICLES

H.J. SUURMEIJER
Fibre Research Institute T.N.O.

INTRODUCTION

The joint activities of the textile research institutes in the EEC, coordinated by Comitextil, were focussed on the study of the quality of knitted products. Special attention was paid to the good and comfortable fitting of knitted garments. The institutes, participating in this project, were Hatra, ITF Maille, Stazione Sperimentale, Fibre Research Institute TNO.

This aspect of quality was chosen, since the good and comfortable fitting of knitted garments, close fitting garments in particular, is essential, because, in spite of carefully applied manufacturing methods and thoughtfully chosen raw materials badly and uncomfortably fitting garment fails to fulfil the expectations of the consumers.

For the majority of consumers the comfortable fitting of knitted garments in the current limited number of sizes should be ensured, since to-day many of these garments are purchased in stores and supermarkets providing self-service systems. In these information about the performance and the fitting of garment cannot always be obtained from shopkeepers. In many cases it is not possible or usual to try on the garments before buying.

Knitted garments, like underwear garments, have to be a recommandation by themselves.

PERCEPTIONS OF QUALITY

Manufacturers', retailers' and consumers' perceptions of quality do not always agree :
- knitters' specifications for garments stress constructional details,
- retailers' specifications vary widely in extent but stress performance and quality of make-up,
- consumers stress fit and appearance.

There is a common care of agreed quality concepts, but, beyond these few basic concepts, lack of agreement and differences in perception and emphasis are a source of dissatisfaction.

Over the last thirty-five years research has identified the basic parameters that determine the characteristics of knitted fabrics. The importance of loop length and its relations to many other less but nevertheless significant fabric and garment properties, is much more fully understood. However, research has not produced agreement as to what tests are essential to specify the performance of different types of knitted goods, nor has it resulted in full agreement as to the parameters, essential to factory quality control.

At one extreme, there is a very good agreement and understanding on requirements such as colour fastness and, at the other extreme, there is a marked lack of agreement on factors such as yarn variables and the quality of make-up. Between these, there is only limited agreement on the realization of important factors such as fit and dimensional stability.

The aim of this particular project was to examine the way in which different groups of people, concerned with the production, distribution and use of knitted garments regard "quality'", how it is defined, how adequate and satisfactory such definitions are and how more agreement and a better understanding may be achieved.

Although knitted fabrics are used in a great variety of commodity and high fashion garments, it was considered most appropriate for this project to examine a more limited range of volume-produced goods. The garment types covered were mens' trunks, womens' briefs, classic cut-and-sew outerwear, chunky knits, fully fashioned sweaters, T-shirts and sweatshirts.

A detailed questionnaire was designed to collect information from knitters, garment makers and retailers relating to the above classes of garments. After completion of questionnaires and visits to appropriate organizations, sample garments were collected and subsequently subjected to laboratory analyses. Similarly, a questionnaire was designed to collect information from consumers on the knitted garments they currently used, their sizes, ages, origins, etc., including on the assessment of their performances.

MANUFACTURERS' AND RETAILERS' INQUIRY

The required information and the representative sample garments were collected during visits to the co-operating organizations. A questionnaire was prepared as a guideline which covered all possible aspects of a garment specification. The sections covered by the questionnaire were, shape, sketch, size, performance, wet processing/finishing, yarn, fabric information and parameters, garment assembly, stitching/seam checks, labels, and trims

In all, 24 organizations co-operated directly : 17 fabric/garment makers and 7 retailers. Totally, the specifications of two sizes of each 25 different garments were examined.

GARMENT INSPECTION

The procedure for analyzing the representative garment samples in the laboratory was as follows :
- Each garment was conditioned and measured.
- The fabric construction and fabric parameters were determined.
- Seam and stitch constructions were defined using standard methods. Seams were examined for security, distortion and balance : make-up quality was also assessed (the inspections were carried out in accordance with special guidelines for this project).
   The types of yarn and numbers of stitches/unit length were also determined.
- The dimensional stability was tested using standard washing and drying procedures.
- Fabric, seam and trim extensibility was assessed using hand-stretch techniques. For samples cut from specified parts of a garment a Fryma Extensiometer was used.
- Selected outerwear garments were washed and subjected to tests to determine the resistance to pilling using the method described in BS 5811.
- Colour fastness tests were selected from those mentioned in ISO standard 105.

ORGANIZATION OF SURVEYS

An attempt was made to assess the extent to which knitted articles were included in the wardrobes of people of different background and the perception of the quality of their knitted clothes. To provide some dataa from a number of countries within the EEC, research institutes which intended to take part in the EEC-programme were invited to carry out small-scale surveys in their own countries. England, Belgium, Denmark, France and the Netherlands took part in the survey.

The wearers in the different countries were asked to give information about, their body-measurement, the garment style, the garment measurements, the age, the use, the origin of their garments, and the type of knitted structure, their view about the fitting and the performance of a garment, and the labelling in the garments.

In this way wearers' views, measurements and their assessment of the fit of the knitted garments they owned were obtained, covering some 1,400 garments.

SUPPLIERS' AND PURCHASERS' SPECIFICATIONS

The specificiations were obtained from purchasers and manufacturers. Some of them also supplied the purchasers' specifications they normally use. The forms were not selected to provide a valid statistical sample of the companies and their customers but rather to illustrate the overall range of current practice, from major manufacturers and purchasing groups to small single product manufacturers and retailers, all within one locality.

CONCLUSIONS

The manufacturers' and retailers' specifications and standards for the garments covered in the inquiry varied largely in form and content, ranging from an integrated system containing input from retailers, sampling, laboratory tests, and procedure manuals updated by feedback from production, to complete reliance on "trade practice" and single sample garments. The specifications of the major retailers and the larger own-brand producers share much in common. Manufacturers' specifications, in particular, tend to be compartmentalized : information on required trimmings may well be best stated in a purchasing specification, whereas details on the required size- and colour-mix may have to be extracted from a contract document since it may not be included as part of the product specification. There is no merit in centralization for its own sake, but the difficulties experienced in putting together complete specifications for the survey were brought about by the reliance placed on informally-kept and verbally-communicated information necessary for the production of satisfactory goods. This situation, where the complete picture has to be assembled from separate records, all subject to periodic change and updating, is an ideal one for computerization. Though evidence was seen of computer costing, purchasing and production control systems, no company visited had yet attempted to use the computer to control and coordinate the development of its technical specifications for different costumers. This could prove to be a useful line of development, facilitating better communication and reducing the possibility of serious errors and omissions.

Specific points to emerge from the comparison of various specifications were :

- recognition that idential "performance" figures often conceal marked differences in the severity of the test applied.
  Illustration :
  * A universal agreement on a minimum pilling performance requirement of grade 4 for acrylic sweaters cannot be obtained; some retailers require this after 5,000 rubs and others after 18,000 rubs.
  * All retailers'performance specifications give for acrylic sweaters a maximum shrinkage of either + 5 % or + 3 % in length and width. In some cases this refers to hand wash, in other cases to an HLCC 6 machine wash or to extended washing treatments to simulate the effect of repeated washing.

- demonstration of the problem resulting from differences in measuring and grading practice.
  Illustration :
  * In several European countries, there are no national standards that specify garment measuring points in detail or give guidance on size grading. There is also lack of an up-to-date anthropometric survey on which such standards could be based. In the absence of such standards there is a great lack of uniformity in the relation of the references points on size charts. Many of the size charts include references to stretched dimensions and in most cases, imply "hand-stretch testing", a procedure which is common practice but demonstrably unreliable.

- reinforcement of the need for aids to rapid garment inspection, particularly of dimensions, of seam quality and of the extensibility of key items such as waistband and openings.

- the recognition of the need for wider application of the concept of stitch length and tigthness factor as key control parameters.

WEARERS' SURVEYS

The results from the wearers' surveys illustrate the need for the purchaser to be able to decide, from information on the label, what size of garment to buy so as to obtain a slack, average or tight fit. Considering the T-shirts of the wearers the difference between garment underarm width and chest/breast girth shows that females generally wear more tightly fitting T-shirts than males do. The mean differences between garment and body sizes for females are - 8.1 cm and for males + 0.9 cm.

The garments considered too slack or too tight by females show that whilst all garment considered too tight were smaller in sizes than the wearers, the garments considered too slack covered the complete range of garment fit. It appears that most but not all females prefer T-shirts with a tight to very tight fit, and most, but not all males prefer T-shirts with an average or slack fit. The wearers' views on the quality of their garments show, that most garments were regarded as having a satisfactory fit when they are bought. If the garment was not satisfactory the only comment was that in most cases it was too slack.

The condition of garments when reported on "in use" gave rise to more comments, as would be expected. The most significant areas were garment fit, where sweatshirts and T-shirts gave rise to most comments (26 % of garments) whilst cardigans and waistcoats gave least comments (11 %). Other significant areas were pilling (9 to 16 %) and colour fading observed with sweatshirts and T-shirts (11 %each).

The extent to which cleaning instructions were presented as symbols varied significantly from process to process, as could be expected. About 95 % of all garments did not give any drying information, whereas washing instructions were lacking in 37 % to 54 % of the garments, depending on style. A similar proportion of each style of garment contained information on dry-cleaning processes. The very wide range of washing instructions for T-shirts suggests the possibility of some underlabelling, where the degree of severity of the washing process might be as high as was feasible for the garment, leading to lower quality, but achieving, removal of soil.

GARMENT DESIGN AND RESOURCES
The research institutes participating in the project also examined what aids and appliances are available for trade and industry to provide a majority of consumers with well and comfortably fitting knitted garments.

The feedback in the distribution chain from consumers, via retailers to manufacturers is evidently an important source of information. This information, however, has not been arranged and its importance will sometimes depend on the potential size of orders to be obtained.

Statistical information about the body-measurements of population groups in many regions of the European markets is not available.

In the absence of testing methods to judge the quality of knitted garments the assessment of different aspects of quality is often done by visual inspection whilst garments are measured when they are lying in a flat position before and after washing tests, without being exposed to tension.

PROPOSED TESTING METHODS
Some years ago the Swedish Textile Research Institute (TEFO) at Goteborg introduced systematic measuring methods for designing and testing the sizes and the comfortable fit of close fitting knitted garments (e.g. underwear). To perform these testing methods sizing tables for knitted garments have been drawn up, based on a population search of large groups of adults and children in Sweden in the seventies. These sizing tables also contain the body-measurements which are of importance for close fitting garments, including arm girth, vertical trunk girth, crotch length, and groin girth.

In the Swedish sizing tables the current European size designations and intervals between sizes are maintained using half the chest girth for men's sizes, the body length for children's sizes, and the current size designation for women's wear 32, 34, 36, 38 etc. (in most countries not related to measurements).

For knitted garment the largest and the smallest size of each body-measurement occurring within one size are presented (coverage 96 %) in the successive sizes stated in the Swedish tables. To be able to consider these tables and measuring methods for testing the comfortable fit of knitted garments for international use it is stated that the growth and development of body sizes result from natural processes. For this reason a Gaussian distribution for the frequencies of body sizes in each body length or chest girth limit is assumed. This means that the Swedish sizing tables are less effective for the different regional markets in Europe.

From recommended national sizing schemes existing in some European countries, also based on a population search, which are used to serve regional or home markets optimally, it appears that the relative body proportions of population groups in Europe significantly differ from each other. The average buttock girth of adult men occurring in each size in the Swedish tables is much smaller (6 to 12 cm) than the average buttock girth of adult men in Germany and Spain. The adult men in France, having an average chest girth of 92 cm have a much smaller average waist girth (6 to 8 cm) and a much smaller average buttock girth (10 to 18 cm) than the men having the same average chest girth in Germany, Sweden and Spain.

Many other examples can be found in the national sizing schemes of European countries for men, women and children, showing differences in body proportions and figure types in population groups of various nationalities. Though the knitted structures of garments can fit different body sizes, these differences in relative body proportions of the population in different regions cannot be neglected in drawing up effective sizing schemes.

In the absence of appropriate "good-quality" anthropometric data of all European population groups, test results, comments on sizing and comfortable fit should be regarded to a certain extent as tentative.

A systematic testing procedure of critical garment measurements has been established by TEFO. Garment measurements are critical if they influence the user's judgement of fit. The critical garment measurements are compared to the corresponding critical body measurements.

A simple measuring device is used to test all kinds of garment openings, such as the leg opening of slips and briefs, the arm-hole openings of T-shirts, the waistband of slips and briefs,and the neck openings of T-shirts and singlets.

With a simple device the comfortable fitting of the critical garment measurements is tested under minimum required and maximum acceptable loads (see Figure 1).

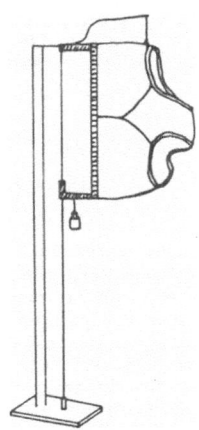

Width gauger

Figure 1

After wear trials carried out at TEFO the load have levels been established as being acceptable for the majority of users.
- The arm-hole openings are tested under a maximum load of 1 N, since pressure against the arm-pit soon becomes irritating (Figure 2).
- The leg-openings of a slip are tested under a mimum load of 1 N to judge the correct fit to the smallest groin girth occurring in one size as well as under a load of 10 N corresponding with the highest acceptable load for the largest groin girth occuring within the sizes.
- The performance and comfort of the waistband of slips and briefs are tested under a minimum load of 3 N and a maximum load of 15 N (adult sizes).

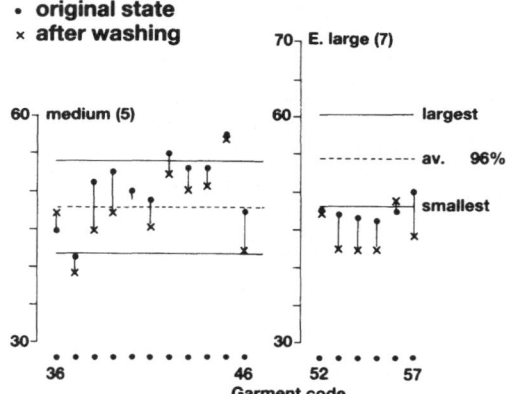

Armscye girth - shoulder width
Men's wear T-shirts

- **original state**
× **after washing**

Figure 2

The measurements of the garments in the original state and after washing, obtained in accordance with the instructions mentioned on labels and standard test procedures, can be compared with the body measurements occurring within the sizes of the garments.

The length of T-shirts and singlets is measured by means of an instruments, which can be adjusted to a width corresponding with the smallest, the average and the largest chest girth occurring within the sizes (see Figure 3). The garments have to be stretched in the width direction when they are placed on the instrument. In doing so, the garments are loaded to some degree in the length direction as well, since to measure the length, the garments should be placed on the instrument in a flat position without folds.

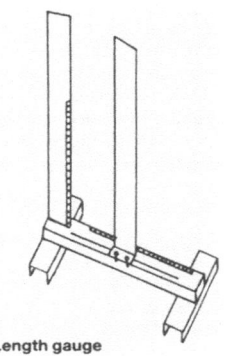

Length gauge

Figure 3

The length measurements carried out on the deformed garment are realistic because the wearer also judges the length after having put on the garment, as a result of which the garment shows additional deformation. The proportional length change, established after washing and determined according to the measuring method mentioned, is often much smaller than the length change established after performing standardized washing tests, based on measuring the dimensions of knitted fabrics and garments in a flat position, without tension.

The standard test procedures normally used to judge the dimensional stability of knitted garments do not give any information about the fit of garments. These test methods reveal the effects of the finishing processes applied in order to reduce the potential shrinkage of machine state knitted fabrics.

To test the fit of close fitting garments the research institutes participating in this project agreed to mease the length of garments preferably when they are placed on a fitting form maintaining a deformation in the length and directions within limits corresponding most nearly to practical conditions.

Although the reproducibility of this method is less than perfect, there have been no objections to it during the trials. Consumer organizations in the Netherlands also appreciate this more realistic measurement.

In the Swedish testing procedure the required length for T-shirts and singlets in the different sizes is calculated from the vertical trunk girth. The calculation is based on wear trials. The results give minimum and maximum limits of acceptable lengths for 67 % of the persons falling within one size.

With the assumed Gaussian distribution for the frequencies of body-measurements the calculated limits for an acceptable garment length in larger sizes are inadequate, since people having larger or extra large body lengths can never find a well-fitting garment in the larger sizes. To provide greatest number with well-fitting garments the addition of a certain extra length to the garments of the larger sizes is desirable. The extent to which this should be done can be derived from the anthropometric data of population groups in regional or local markets.

From the tests performed with the aid of the TEFO equipment on commercially available underwear it appears that the lengths of most of the T-shirts and singlets in small sizes fall within the calculated limits stated in the Swedish sizing tables. The lengths of the garments in the larger sizes exceed the calculated limits (see Figure 4).The results also show that the differences between garments falling within the same size designation are very large.

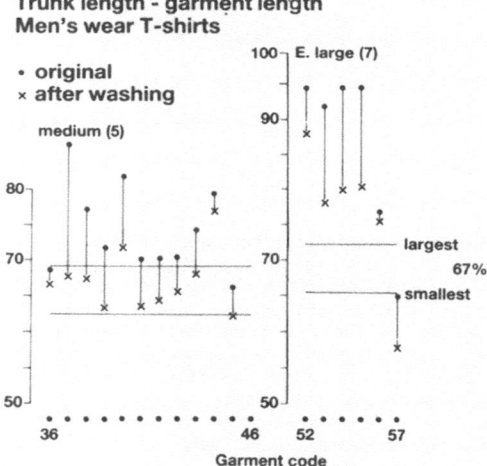

Figure 4

Instead of establishing minimum and maximum garment lengths it is also proposed, as in France, to determine only the minimum length in each size. Overlong garments cannot simply be judged as being too long, since the dressing habits, individual preferences, and the conditions for use should also be taken into account.

Considering these differences within one size it can be noticed, that the current size designation is rather limited. because of this and as reflected in their views it is difficult for consumers to select comfortably fitting garments.

In the Swedish test procedure the width of singlets and T-shirts is measured when the garments are lying tensionless in a flat position. The results are compared with the waist girth occurring within the size of the garment.

To ensure comfortable fit to the chest girth, the width of the garments is tested with a device which stretches them in the width direction (see Figure 5). The load applied corresponds with a tension of 0.25 N/cm in the stretched garment. The results obtained are compared with the smallest and the largest chest girth for the size of the garment.

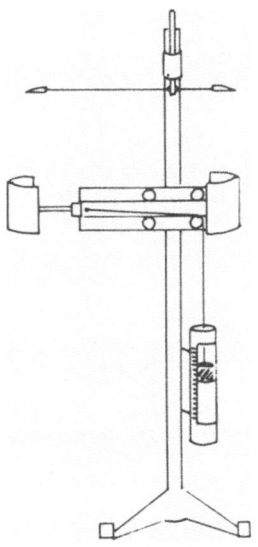

Width extension guage

Figure 5

The load of 0.25 N/cm, which was judged to be acceptable by the majority of wearers, involved in the Swedish wear trials, should probably be reconsidered, since the acceptability of tightly or loosely fitting garments is different between males and females.

The width measurements of the garments also show large differences within one size. With respect to the comfortable fitting to the waist and the chest, the current size designation can easily lead to disappointments (Figure 6).

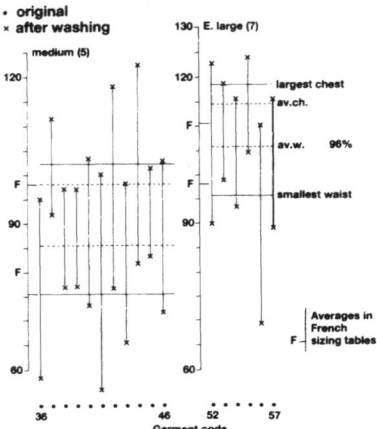

Figure 6

If adequate data about body measurements and systematic testing methods had been available for the design of the garments tested, the differences between garments of the same size would presumably have been smaller.

The fit of pants, slips, briefs, etc. is judged on flat fitting forms (Figure 7). For the TEFO testing procedure, flat fitting forms for lower torso garments of each size mentioned in the Swedish tables are available. The length of the garments is measured on the forms, whilst the results are compared with the crotch lengths occuring within the different sizes. The style of these garments (e.g. pants or bikini slips) should be considered when judging comfortable fitting. In the Netherlands the Dutch Association of Housewives decress that a waistband, worn approximately 12 cm below the waistline, is acceptable for the majority of wearers.

The results obtained with the TEFO method, compared with body measurements within successive sizes allow the comfortable fit and the fitting capacity of underwear garment to be judged.

For T-shirts the testing of the comfort and the fit of sizes is still insufficient. The width and the extended width measured do not give sufficient information about good and comfortable fitting of the sizes.

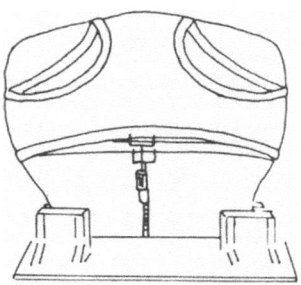

Template

Figure 7

The shoulder width of T-shirts and the extensibility of shoulder seams, coupled with the size of the neck opening, largely determine of the good and comfortable fit. These garment measurements and their properties should be compared with the shoulder breadth of wearers.

ADDITIONAL TEST PERFORMED WITH T-SHIRTS

ITF-Maille has paid particular attention to the judgement of the fit and the fitting capacity of T-shirts.

Based on experiences with flat forms for feet to judge the comfortable fitting of socks and stockings, 26 two-dimensional shapes of the upper torso in different sizes were made, starting from a French standard containing 250 sizes of the adult male population in France (see Figure 8).

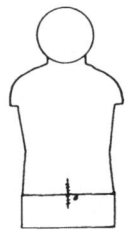

Chest 96 cm   Chest 96 cm   Chest 96 cm
Waist   Wast   Waist 96 cm
Differene 10m   Difference 6cm   Difference 0

Figure 8
PLAT FORMS

The 26 forms cover 95 % of adult men in France. The forms include the head in a flat form to judge the neck-opening of shirts as well as the upper part of the upper arm, to give the total shoulder-breadth. At the bottom of the form the minimum required length, detetermined experimentally for different sizes and figure types, is indicated.

Using new and washed T-shirts, the garments are placed on several forms, starting with one which is too small and visually judging the fit, until the smallest and largest providing good fit are found. The garments are considered not to be of a good fit if they do not reach the defined length marks, or if they are too tight or too slack across the chest (see Figures 9 and 10). Assessment of fit, measured with the use of the French forms, provides an adequate guide to the range of body sizes, with varying waist measurements, which a garment can be expected to fit, without its being too slack or exerting significant pressure on the body. The disadvantage of the French forms is that in judging the garment sizes which will fit, there is no was of estimating the tension imposed in the width direction on the form, and thus of judging whether or not it is an appropriate fit.

Results of wear tests indicating that, for at least some end-uses and wearers the load of 0.25 N/cm, used in the TEFO system, is too large. Garments which wearers considered too small for them obtained maximum size designations on the French forms that were smaller than the labelled sizes. Consequently, with regard to the chest/waist width assessments the French forms provide a better guide than does the TEFO system.

This project emphasizes that one of the major problems in providing the majority of consumers with well fitting knitted garments is the lack of reliable/anthropometric data detailed, showing the effects of regional, age and other variables on different body measurements. To carry out quality assurance programmes the availability of these data is essential. Recently an anthropometric study was initiated in the Netherlands.

Provided that the manufacturers and retailers have these data about population groups at their disposal, the following test advantages will result the design of well and comfortably fitting knitted garments, will be the size designation of knitted garments will be based on test improved results, improved communication between manufacturers, retailers and consumers will occur product specifications for quality assurance schemes can be drawn up.

Additional studies have been started in France and England for the development of improved testing methods.

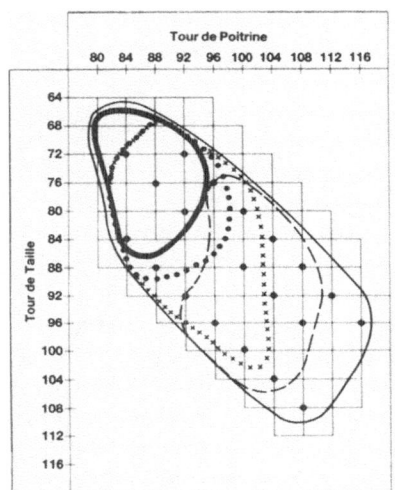

Graphique III
stature 171 cm

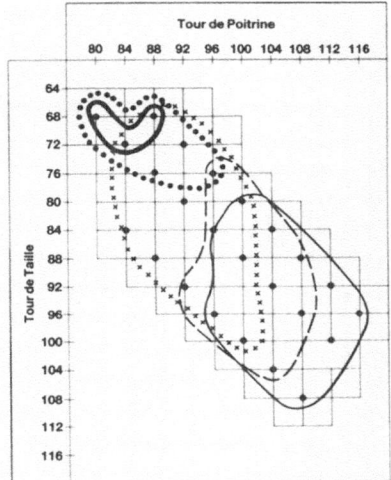

Graphique IV
stature 175 cm

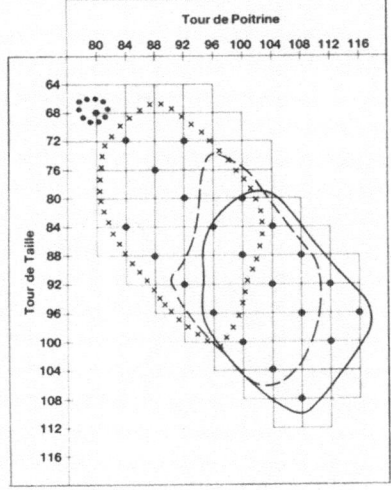

Graphique V
stature 181 cm

Tailles
━━━━━━ 1
•••••• 2
xxxxxxxxx 3
– – – – – 4
———— 5

<u>Figure 9</u>
<u>T-shirts, original state</u>

176

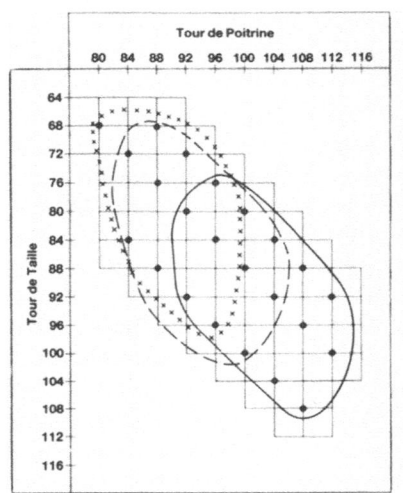
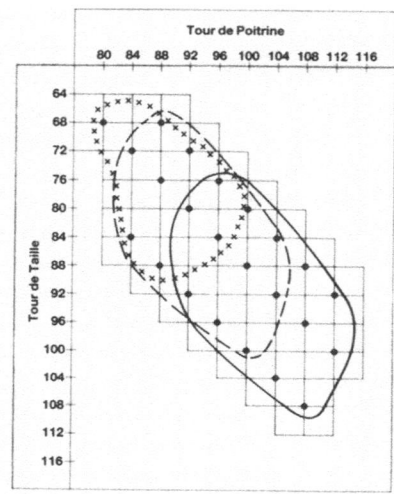
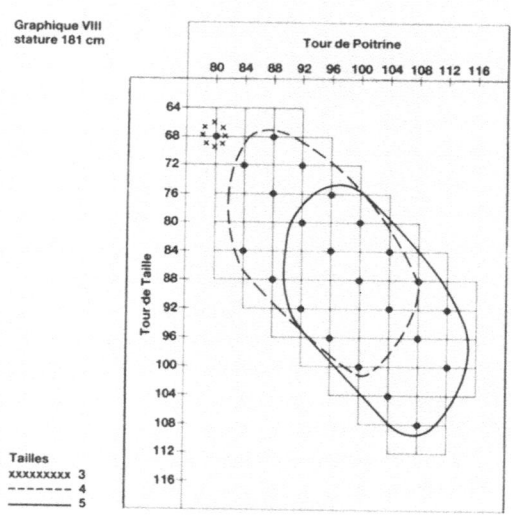

Graphique VI
stature 171 cm

Graphique VII
stature 175 cm

Graphique VIII
stature 181 cm

Tour de Poitrine

Tour de Taille

Tailles
XXXXXXXXX 3
------- 4
———— 5

Figure 10
T-shirts after washing

QUALITY ASPECTS IN THE PRODUCTION OF KNITTED
FABRICS AND GARMENTS

M. BALLAND
I.T.F. - MAILLE

The results presented summarise the projects carried out in the laboratories of Centexbel, the Danish Textile Institute, the Institut für Textiltechnik and the Institut Textile de France.

1. - COSTS OF FAULTS

In order to put into perspective the industrial importance of the cost of sub-standard goods, we shall first examine the fabric cost structure and secondly the cost of faults. Table I gives a distribution of costs as a percentage in flat knitting and in circular knitting. This shows the high percentage of material costs and the extent of the cost of faults, amounting to 6 %. It is thus easier to appreciate the significance of action taken in industry both for preventing material losses and for preventing the occurence of faults. Table II gives the costs distribution for faults as a function of their causes. It shows that in flat knitting, 65 % of fault costs are due to the yarn whilst in circular knitting this is 70 %. It appears that useful progress could be made by using yarn splicing techniques and by introducing methods to make yarn friction properties consistent.

TABLE I
COST DISTRIBUTION (%)

	FLAT KNITTING	CIRCULAR KNITTING
Faults	6	6
Material	56	64
Knitting	25	11
Finishing		13
General costs	13	6

TABLE II
COSTS OF FAULTS

FLAT KNITTING		CIRCULAR KNITTING	
Thick places	5	thick places & fly	11
thin places	9	holes	14
foreign fibres	5	wastes	19
slipped stitches	20	needle stripes	18
holes	5	needle wear	3
fraying	24	seconds	31
various	14	dyeing faults	4
needle wear	7		
seconds	11		

II. - INFLUENCE OF THE DYEING OF WOOL FIBRES

In order to demonstrate the influence of dyeing on the quality of wool yarns, 1000 Kg of worsted spun wool yarn were divided into three lots and each lot was dyed by a different method, namely using metal complex dyes, chrome dyes and reactive dyes. The yarns were then knitted in 1 x 1 rib construction and under identical conditions. The most significant differences appeared in the mechanical properties of the yarn.

The extensibility of the undyed yarn was in the range 11.4 to 15.8 %. The dyeing processes reduced this by as much as 65 %. The least bad results were obtained with the reactive dyes. In the same way tenacity (cN/tex) was reduced by 15 %. More faults were found in dyed yarn and good correlation existed between the number of faults and machine stoppages. On the other hand, no correlation was found between the yarn analysis results and either the presence of fly on the machine or pilling of the knitted fabric.

III. - INFLUENCE OF FIBRE DIAMETER IN PURE WOOL AND IN WOOL/ACRYLIC BLENDS

The influence of fibre selection on handle and appearance has been studied with wool fibres of 19.5μ and 22μ with and without shrink-resist treatment, and with standard and pilling-resistant acrylic fibres of 2.2. and 5.6 dtex. The tests were performed by five persons on knitted fabrics, first decatised and then washed three times and tumble dried. The handle criteria chosen were flexibility, surface characteristics and compressibility. For appearance, the factors considered were the clarity of stitch structure and the general aspect.

1. Appearance

1.1. circular knitting

The yarns used were 25 tex in the qualities shown in Table III.

TABLE III
YARNS USED

W O O L	A C R Y L I C
100 % 19.4 μ untreated and shrink-resist	
100 % 22.0 μ untreated and shrink-resist	
50 % 19.5 μ +) untreated and 50 % 22.0 μ) shrink-resist.	
60 % 22.0 μ untreated	40 % 2.2 dtex pill-resist
60 % 22.0 μ untreated	40 % 3.6 dtex standard
60 % 22.0 μ untreated	40 % 5.6 dtex standard
60 % 22.0 μ shrink-test	40 % 5.6 dtex standard

179

The appearance of the decatised knitted fabrics both in microstructure and macrostructure shows that the 19.5 ɥ wool to be better (Figure 1). However, as far as microstructure is concerned, the influence of structure can be seen. In flatknit jersey, the 22 ɥ wool is judged to be best whilst in the more compact structures the finer wools are preferred. The influence of shrink-resist treatment on appearance is only apparent in the interlock fabric where a better grading is given to the shrink-resist treated wools.

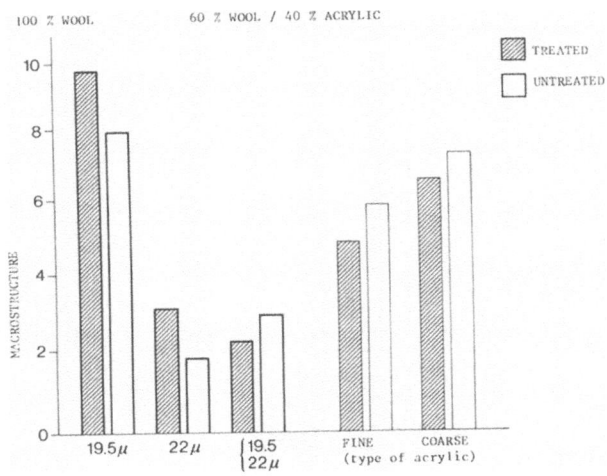

Figure 1

"Aspect" punto-di-Roma decatised circular

Effect of washing

The knitted wool fabrics without shrink-resist treatment were felted and unusable. In the shrink-resist treated wools, the finest wools were the best. No difference was found between the blends. By comparison with the pure wools, the blend of 22 ɥ wool with 3.3. dtex standard acrylic gave the best grading in 1 x 1 rib. In jersey and punto-di-Roma, no differences could be detected.

1.2. Flat knitting

In flat knitting, yarns of 35tex x 2 were used, the range being extended by including a 26 ɥ wool. The appearance of the decatised knitted fabrics shows the advantage of the 19.5 ɥ wool. In practical terms, the shrink-resist treatment has only a slight influence and only in 1 x 1 rib and jersey (Figure 2).

In blends, the best, which is superior to pure wools, is that containing the 22 ɥ shrink-resist wool and 5.6 dtex standard acrylic, particularly in 1 x 1 rib. In jersey and punto-di-Roma no differences could be detected between the wools. After washing the untreated wools had felted greatly and could not be used and of the rest, only the blend of 22 ɥ wool with 5.6.dtex standard acrylic graded better than the wools.

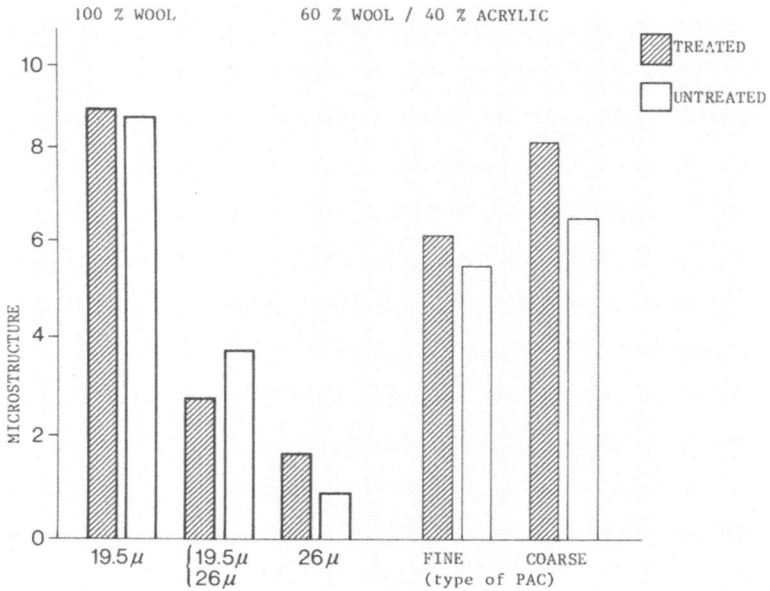

Figure 2

2. Handle

As far as handle is concerned, the fineness of the fibres exerts a significant influence.

The results of the decatised fabrics from flat and circular knitting may be described together. The finest 19.5μm wool was graded best for the three criteria in all structures (figure 3), and the shrink-resist treated wool possessed advantages in all criteria. No differences are revealed in comparing the blends with one another but when comparing the 100 % wools and the blends, the 19.5μm pure wool is well placed in all criteria.

After washing, the fine wool was best for flexibility and smooth surface characteristics, whereas in compressibility, the blend of 19.5μm and 26μm wool performed best. Comparisons of the blends did not reveal any differences except in compressibility, where the best was the blend of shrink-resist 22μm wool with 5.6 dtex acrylic.

SUMMARY

In handle and appearance, the best fabrics were made from 100 % wool yarns with the finest fibres. Additionally, the wools without shrink-resist treatment were the best.

In the wool/acrylic blends, no differences were found to be caused either by differences in fibre fineness or by differences in treatment.

After washing, there is no difference between the best wool and the blends. The blends were graded better than the other types of wool.

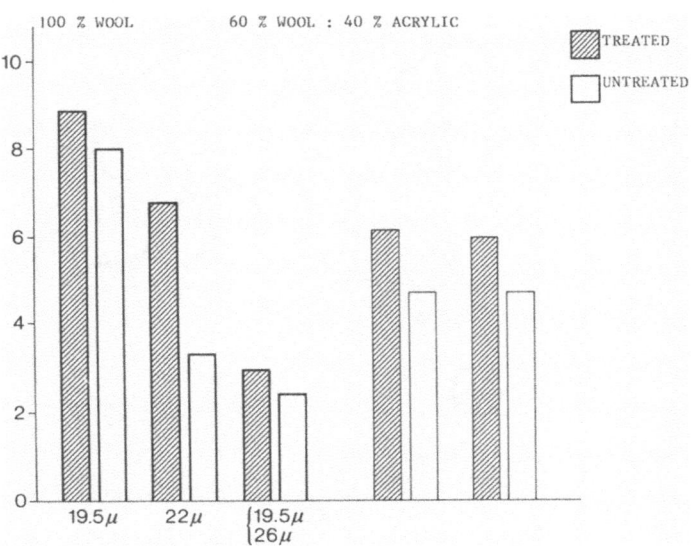

Figure 3
Handle (decatised circular knitted jersey)

3. Stitch distortion : cockling

Stitch distortion is principally a phenomenon of jersey knitted fabrics produced with wool yarns or containing a high proportion of wool. Either the phenomenon involves one or more stitches which have one longer leg which appears on the face of the fabric whilst the other disappears into the body of the fabric (distorsion), or it involves a whole course of stitches and a cockling fault occurs.

An important result has been obtained by studying cross-sections (figure 4). Where the fault occurs, cross-sections show that there are very coarse fibres at these points, a factor which considerably increases the bending resistance. This finding has been confirmed by trials using different blends. The origin of the wool is of no significance and the greater the proportion of acrylic contained in the blends the lower is the incidence of the fault.

4. Pilling

Comparative tests were performed with the Random Pilling Tester. The knitted fabrics produced with fine wools pill the most, this being greatest in the case of flat-knitted fabric. A difference between wools with and without shrink-resist treatment is only apparent in flat-knitted fabrics where the results for treated wools are about 10 % better. In the wool/acrylic blends, the best results are obtained with the 5.6 dtex fibres. Comparing the blends with the pure wool knitted fabrics, the blends only are better when compared with fine wool fabrics without a shrink-resist treatment. Good results are obtained with the 100 % wool ,ntreated fabrics with high stitch density.

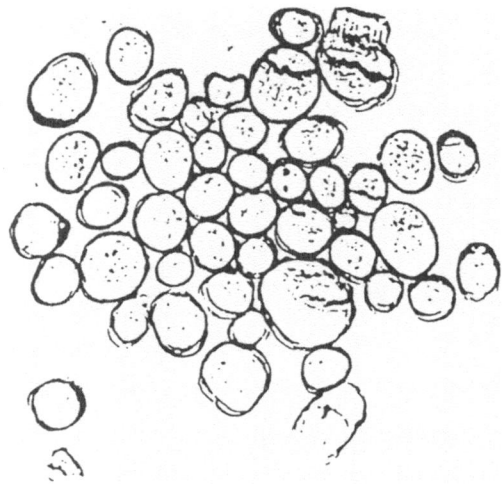

Figure 4
Cross-section at the site of a distortion

IV. EFFECT OF PRODUCTION PARAMETERS ON THE DIMENSIONS AND STABILITY OF COTTON KNITTED FABRICS

Another important problem is the dimensional stability of knitted cotton fabrics.

In an investigation of the effects of repeated washing, in processing, and in the wear and wash cycle, an attempt is made to find a method for predicting the final dimensions of knitted fabrics.

1. Determination of washing shrinkage

The dimensional variations caused by successive washes and drying either on a line or in a tumble-drier have been studied. Tumble drying to different levels of residual moisture gives practically the same amounts of shrinkage down to 6 % moisture content, but continuation of drying to about 0.5 % moisture content causes additional shrinkage. Consequently, the method given in British Standard 4923 (1980), procedure I A, has been adopted as the general washing method, but with the procedure controlled to give a residual moisture content of 5 to 6 %. Figure 5 is one example of what is generally found.

The longitudinal shrinkage obtained after the first wash continues to increase during subsequent washes although at a diminishing rate until a practically constant level of shrinkage is attained after about ten washes. Typical longitudinal shrinkage values for unfinished 1 x 1 rib knit fabric are 23 % after 5 washes and 24 % after 10 washes.

Laterally, the interlock and the 1 x 1 rib have the greatest shrinkage in the first wash whilst subsequent washes cause a gradual increase in width, sometimes to such a degree that the initial shrinkage of the first wash is little by little converted into an extension.

Longitudinal and lateral shrinkages in jersey are similar to those obtained longitudinally for interlock and rib knit.

Drying on a washing line gives much lower shrinkage values than those obtained by tumble drying.

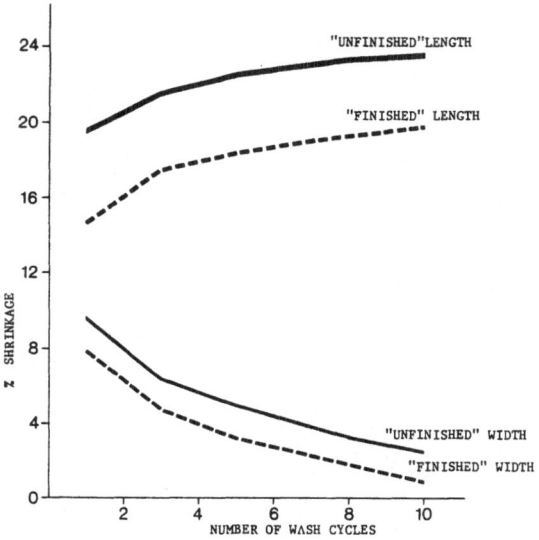

Figure 5
Shrinkage in repeated washes : 1 x 1 rib knit

2. Influence of finishing deformations on washing shrinkages.

Washing shrinkages have been determined on samples taken at different stages in the finishing cycle. Comparisons of the shrinkage results with the corresponding dimensions of the knitted fabrics at various stages show how these data may be correlated. Taking the dimensions of the unfinished knitted fabrics as the reference point, it has been found that extension of the knitted fabric during a specific process generally causes an increase in washing shrinkage whilst contraction in the process generally reduces it. This is clearly illustrated by showing on the same graph, the respective dimensions following different stages of processing and the washing shrinkages of the samples taken.

Figures 6 shows an example of a process causing longitudinal extension and lateral contraction during wet finishing and especially during drying which is performed using a suction drum. This has caused an increase in longitudinal shrinkage and a decrease in lateral shrinkage of the sample taken immediately following drying (in fact a transformation of the lateral shrinkage into an extension "in the wash"). During calendering there is lateral extension and longitudinal shrinkage and the consequence is that the washing shrinkage values do not differ greatly from those found for unfinished fabrics.

Figures 7 shows that treatment of the tubular fabric on a more modern dryer caused considerable lateral contraction and consequently reduces the longitudinal washing shrinkage. The length is maintained almost without change during the final process, resulting in a reduction in longitudinal shrinkage of the unfinished fabric from approximately 20 % to about 10 %.

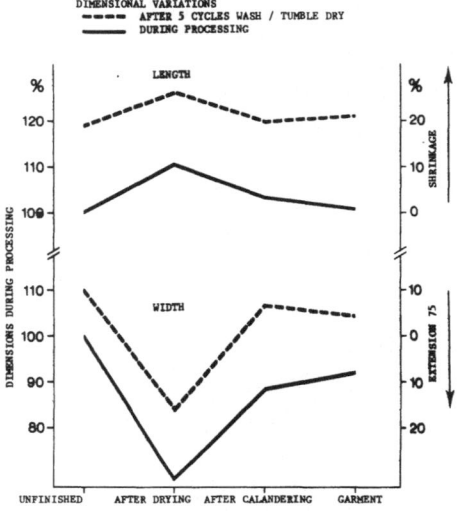

Figure 6
Dimensional variations

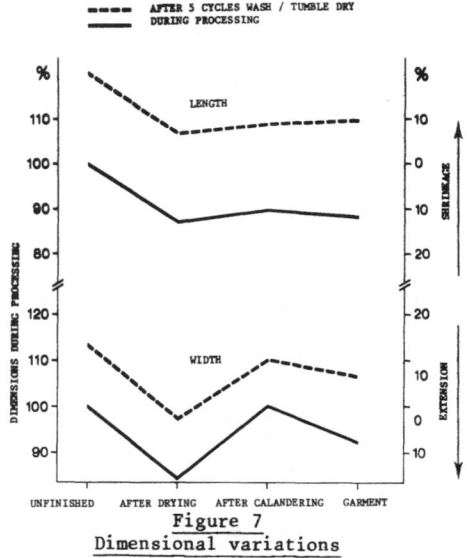

Figure 7
Dimensional variations

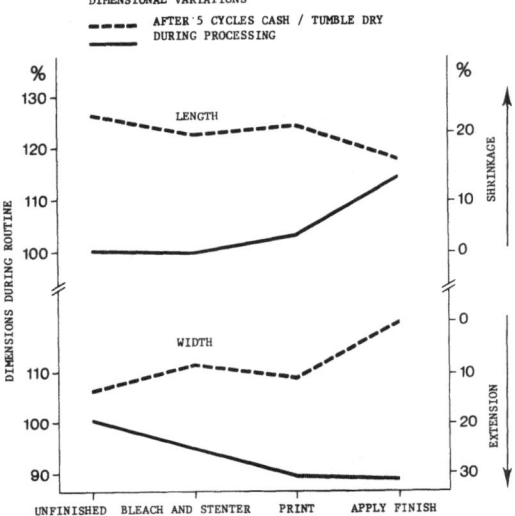

DIMENSIONAL VARIATIONS

---- AFTER 5 CYCLES CASH / TUMBLE DRY

—— DURING PROCESSING

Figure 8
Dimensional variations

Crosslinking finishing treatment reduces the shrinkage that would otherwise be expected. Figure 8 illustrates this effect on a knitted fabric with high inherent longitudinal shrinkage. The first stages have not caused any length variation and the printed knitted fabric consequently has practically the same longitudinal washing shrinkage as the unfinished fabric. The subsequent application of finish is accompanied by appreciable extension which should be expected to cause a distinct washing shrinkage, but because of the finish applied, the shrinkage is in fact reduced.

3. Wear trials

Men's vests in interlock and in 1 x 1 rib knit either not worn or worn for 24 hours by the same person were washed 25 times under standard conditions and dried either on a line or by tumble dryer. In practice, constant dimensions are reached after 10 consecutive washes (figure 9). Drying on a washing line generally gives considerably less longitudinal shrinkage than tumble drying, whilst the effect on lateral shrinkage is insignificant. Wearing results in a reduction in lateral shrinkage. Thickness of the knitted fabric, which was not affected by wearing, was generally considerably greater after the first wash and then ceased to change. Increase in thickness was greater in tumble drying than in drying on the line. Lateral extension under loading together with recovery after loading were reduced relatively more during the first two washes, but only very little during subsequent washes up to the 25th.

The weights of the vests shown in Figure 10 decrease during the first washes but then show an increase from wash to wash up to 105 to 108 % of the initial weight after the 25th wash. These weight gains are caused by deposits due to washing in hard water (20 to 24 degress of hardness on the German scale) and are reduced by loss of fibre during drying and/or wear.

186

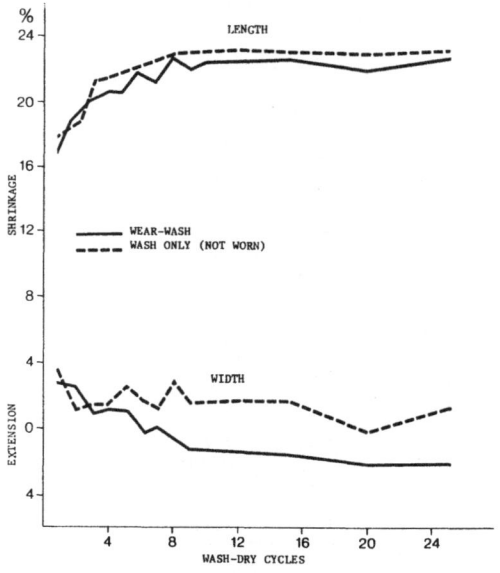

Figure 9
Dimensional variations of garments in washes

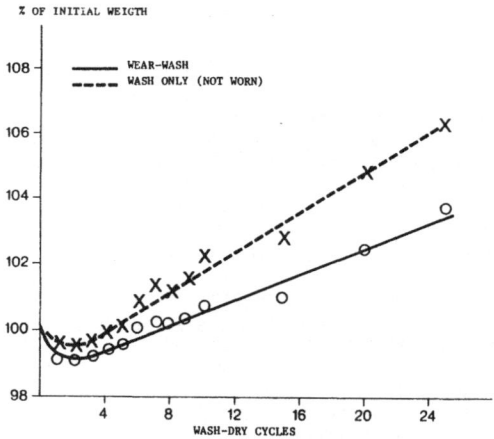

Figure 10
Weight variation of garments in washing

4. Discussion

The principal factors affecting the washing shrinkage of cotton knitted fabrics treated by conventional methods are :

- the shrinkage of the unfinished knitted fabric which is governed by machine settings,

- the dimensional variations appearing during finishing processes; these depend on the equipment used.
  For tubular knitted fabrics, the dryer is a crucial machine; use of the continuous tumble dryer should enable considerable progress to be made.
  For open-width finishing, the stenter is the most important machine as it allows considerable degrees of over-feed.
  The fact that finished knitted fabrics have relaxed dimensions differing considerably from those of the corresponding unfinished fabrics complicates calculations for predicting final dimensions with specific shrinkage values. The differences in values between unfinished and finished in this study may, however, be used to correct predictions made by stitch geometry of unfinished knitted fabrics.

V. STUDY OF THE PERFORMANCE OF ACRYLIC SWEATERS

The performance of acrylic flat knitted fabrics will now be reviewed.

1. Stitch geometry

Any study of dimensional stability demands previous knowledge of the relaxed state of the knitted fabric concerned. After analysing the various relaxation processes, the following treatment was selected :

- immersion in a water bath for 1 hour at 40°C containing wetting agent (1g/1),

- tumble drying for 1 hour at 70°C with low load.

We have found that the acrylic fabrics obey Munden's laws and Table IV gives the coefficients for jersey, 1 x 1 rib and 2 x 2 rib.

TABLE IV
Acrylic fabrics

STRUCTURE	Munden coefficients		
	K 1	K 2	K 3
Jersey	3.7	4.9	18.1
1 x 1 rib	5.8	4.8	27.8
2 x 2 rib	6.7	4.9	32.8

188

2. Study of the dimensional stability of acrylic sweaters in steam pressing

The purpose of the tests is to check whether steam pressing treatment enables the dimensions of acrylic pullovers to be set a long way away from their relaxed state. Figure 11 summarises the tests carried out on the garment panels attached to a frame permitting extensions of 0 %, 5 % and 10 % in both directions compared with the initial state. They were then steamed and relaxed and the extension set after steaming and relaxing was compared with the extension imposed from the relaxed state.

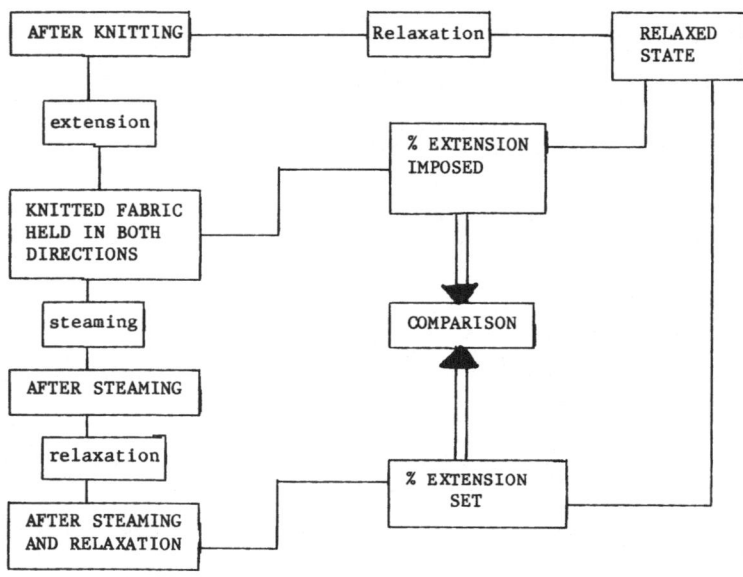

Figure 11
Acrlyic sweaters – trials on the press

Figure 12 gives an example of the percentages of variation of set by comparison with the percentage of area variation imposed in the case of the 1 x 1 rib. It is found that a quarter of the deformation is set at 70°C and two thirds at 95°C. Looking at the results more closely in the wale direction, there is found to be an influence of stitch length – the knitted fabrics of slackest construction can be set further away from their relaxed state. In applying press treatment it is therefore imperative to be aware of the relaxed state. Furthermore, it appears difficult to rely on pressing treatment for establishing the dimensions of a garment. These should be established during the previous processing stages; the press can only be used to make small adjustments.

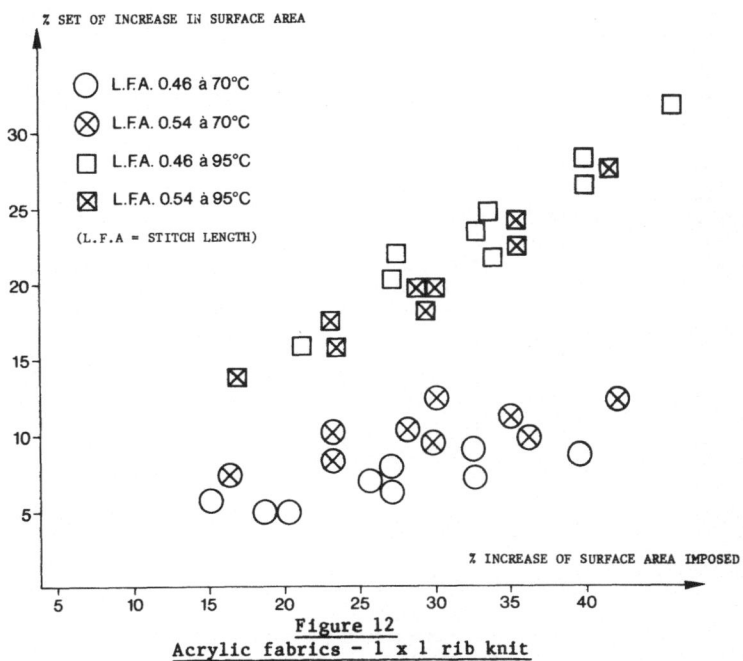

Figure 12
Acrylic fabrics - 1 x 1 rib knit

3. Study of setting on the press

To study the set of yarn on the press, 100 stitches were unravelled from the fabric and reference marks applied (figure 13) The yarn is subjected to successive loadings of 0.2.cN and then of 20 cN. The corresponding lengths are called L_1 and L_0 and the "crimp" is expressed as a percentage by the equation :

$$F \% : \frac{L_0 - L_1}{L_1} \times 100$$

The results on the press are given in the same diagram. It is found that the crimp is greater for short stitch lengths, and that it increases with the severity of treatment, in the order 100°C without steam, 100°C with steam, 120°C without steam and 120°C with steam.

4. Study of setting on the press with a mixture of air and steam

To study the influence of temperature, on old Hoffman press was converted as shown in figure 14. A three-way valve was fitted at the old outlet. The first part is linked to the press, the second to a fan and the third to a vacuum system. A steam inlet governed by an electric valve allows steam to be injected. This system enables a mixture of air and steam to be injected in a first cycle and the vacuum to be operated in a second cycle.

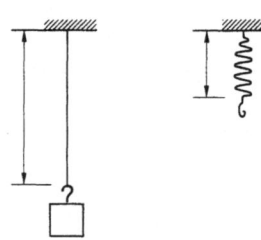

STITCH LENGTH cm/STITCH	WITHOUT TREATMENT	100°C		120°C	
		WITHOUT STEAM	WITH STEAM	WITHOUT STEAM	WITH STEAM
0.44	32.0	42.0	50.0	53.0	53.0
0.52	27.0	41.0	45.0	49.0	51.0
0.59	24.0	33.0	34.0	43.0	48.0

Effect on the crimp of treatment on the press

Figure 13
Acrylic fabric (2 x 2 rib knit)

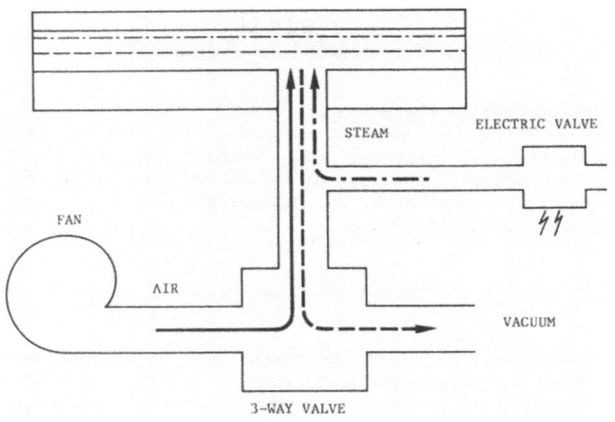

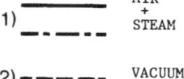

1) ———— AIR + STEAM
——— ———

2) —— —— —— VACUUM

Figure 14
Modification of the press

Four types of acrylic were treated at between 65 and 100°C. Table V gives the results for percentage crimp. Taking the temperature which gives 95 % of the set obtained at 100°C, the following treatment temperatures are deduced :

- Leacryl at 75°C
- Courtelle between 75°C and 80°C
- Dralon at 80°C
- Crylor at 85°C

TABLE V
Effect of temperature on different acrylics
% crimp obtained with different mixtures of air and steam

	Temperature (°C)							
	65	70	75	80	85	90	95	100
LEACRYL	36	42	46	**48**	**48**	**48**	**48**	**48**
COURTELLE	40	47	49	51	**52**	**52**	**52**	52
DRALON	27	35	38	40	**40**	**40**	41	41
CRYLOR	38	42	43	46	47	**47**	48	49

VI. ACTUAL WASH-WEAR TRAILS

To confirm the results obtained, we performed actual wash-wear trials. The sweaters were in jersey and 1 x 1 rib, each with two stitch lengths. The panels were subjected to four types of finishing treatment before making-up.

Treatment (1):
the panels were used after one week's storage in the same condition as they left the knitting machine.

Treatment (2):
the panels were soaked in a water bath for 1 hour and dried at 40°C.

Treatment (3):
the panels were treated on the steam press at 100°C by the method used in industry.

Treatment (4):
the panels were relaxed for 1 hour in a water bath then tumble dried for one hour at 70°C.

Three sweaters were allocated to each of 16 wearers.
After being measured flat on the table in the "new state", the sweaters were subjected to 10 wear-and-wash cycles. Analysis of the results shows that after the 6th cycle, the subsequent cycles cause no

further significant variation of dimensions. The shape of the graphs for the three wearer subjects is always of the same type. As far as treatment is concerned, the date of Table VI show that it is treatment (4) with relaxation and setting that achieves the best results. The direction of the dimensions has some influence, as in all cases there is shrinkage in the lengthway direction, slackness appearing in the lateral direction and very good stability in sleeve length.

It is therefore possible to produce acrylic sweaters with good wear performance if care is taken not to knit too loosely and if the finishing process is carried out not too far from their relaxed dimensions.

TABLE VI
Wash-wear trials - Acrylic sweaters
Dimensions of sweaters at the 10th wash as % of initial state

STRUCTURE	DIMENSIONS	TREATMENTS			
		1	2	3	4
TIGHT 1 X 1 RIB	Length	92	97	97	98
	Width	98	100	104	104
	Sleeve	92	97	98	100
LOOSE 1 X 1 RIB	Length	89	95	96	99
	Width	106	106	106	101
	Sleeve	91	97	98	99
TIGHT JERSEY	Length	94	96	96	97
	Width	105	110	101	106
	Sleeve	97	99	99	99
LOOSE JERSEY	Length	93	95	97	97
	Width	108	109	104	106
	Sleeve	97	100	99	101

VII. PRODUCTION OF PANELS OF EQUAL LENGTH ON FLAT KNITTING MACHINES.

1. Study of waxes

It is accepted that stitch length is a function of the frictional properties of the yarn. It is therefore of interest to study the effect of different lubricants.

Six paraffin waxes were studied having the properties shown in Table VII. Chemical analysis of these six waxes reveals that the hydrocarbon molecules contain from 21 to 38 carbons.

A distinguishing feature of wax PB 91 is that it contains a greater proportion of short-chain hydrocarbons causing it to have a low melting point. Relation of the wax application and the coefficient of friction, as shown in figure 15, gives the general trend of the results obtained (wax SI 6 and various wool, acrylic and acrylic/wool yarns) for which a wide spread is found due to yarn differences.

TABLE VII
Paraffin waxes

WAX	MELTING POINT	PENETRATION
PB 91	50 / 52	12,0
PB 71	54 / 56	13,5
PB 43	54 / 56	11,5
PB 46	54 / 56	12,0
OC 75	54 / 56	16,0
SI 6	54 / 56	15,0

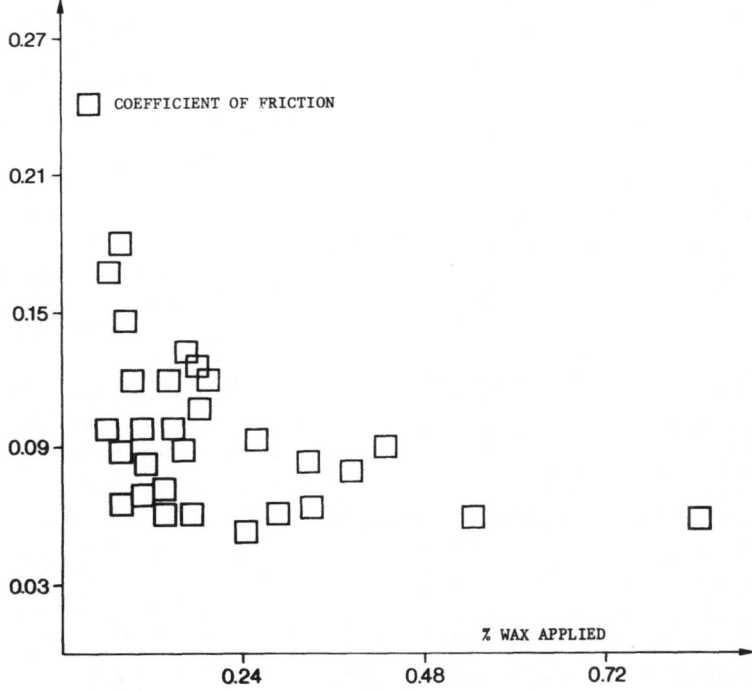

Figure 15
Relations between waxing and coefficient of friction

From the practical aspect, this means that the waxing rate should be set separately for each type of yarn. The same problem of widely spread results arises when an attempt is made to correlate the stitch length and the waxing rate, as shown in Figure 16.

further measurements have been made of friction coefficient and stitch length, firstly with new wax and secondly with stored wax. The results obtained were very similar and there has been little deterioration of the waxes.

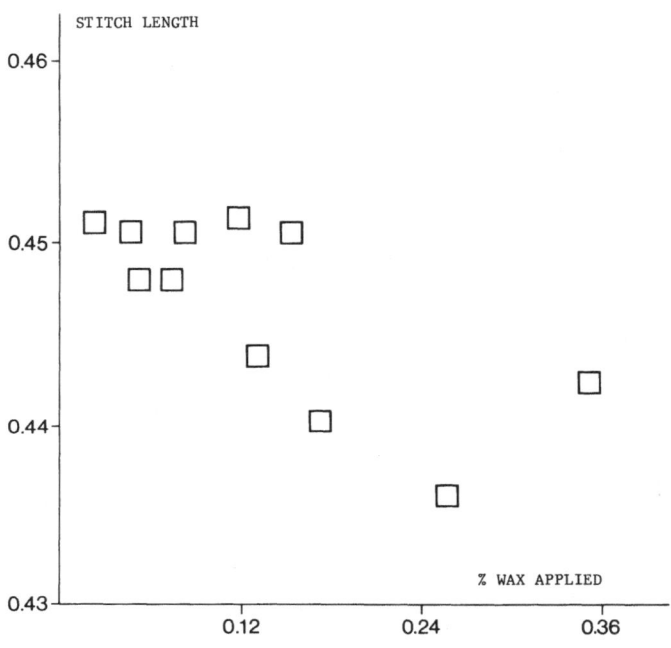

Figure 16
Relation between waxing and stitch length

Conclusions

- the chemical composition of the waxes influences the amount of wax picked up by the yarn,

- the material composition of the yarn influences the waxing rate,

- the rate of wax application must be between 0.4 and 1 % according to the type of yarn and the wax used,

- at the waxing rates studies, ageing causes only slight deterioration.

2. Monitoring the yarn length in a knitted loop (stitch length)

All these variables have an influence on the stitch length and it appears necessary for it to be monitored, especially on flat knitting machines.

Two methods have been investigated :

(A)

In this method, shown in figure 17, the apparatus consists of the parts 1 - 6 and ψ.

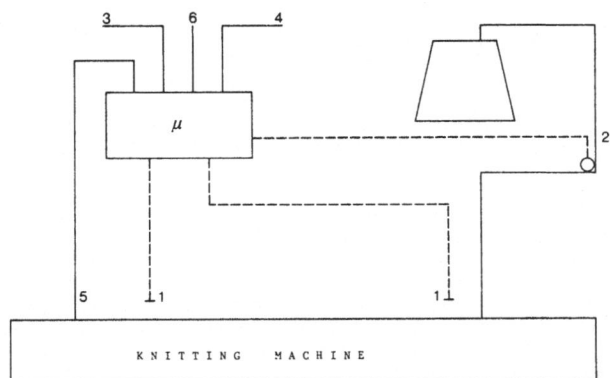

1. carriage position detectors; 2. yarn travel sensors;
3. knitter warning device; 4. machine stop device;
5. correction of machine setting; 6. interface with a
central computer; μ microprocessor based system.

Figure 17
Stitch length controller - method (A)

Carriage position detectors (1)
These are small components which provide a very clear pulse free from surge when activitated by the field of a megnet located on the carriage of the machine. The pulse from the first detector encountered in knitting a course of stitches gives the start of measurement whilst the pulse produced by the second detector defines the end of measurement. These pulses from the two carriage position detectors are fed into the microcomputer which is described below.

Yarn travel detectors (2)
These are composed of small wheels which, when set in motion by the yarn, produce one pulse for each centimetre of yarn delivered. Up to twelve such wheels may be fitted on each machine. The pulses from them are received by the microcomputer and processed simultaneously.

Device for warning the knitter (3)
This system is a binary output from the microcomputer and is used when it is not possible for the microcomputer to regulate the stitch length within the programmed limits. A warning light is generally used to perform this function.

Machine stop device (4)
This device is a binary output of the same type and is activated under the same circumstances as (3). it is connected to a safety device in such a way as to provide an identifiable cause of machine stoppage.

Device for correcting the machine setting (5)

This system is an analog output (256 levels) which drives a solenoid governing the yarn tension. Twelve systems of this type can be driven in pairs by the microcomputer. In the install cycle, the voltage fed to the solenoids is a value in the centre of the range (128) to permit maximum correction in both directions during production.

Interface to a central computer (6)

This is a serial interface of the RS422 type which allows a single line to be used in common by a number of feeds (normally 256) and makes it possible to exchange information between the minicomputer and a computer acting as a large memory for it, storing the individual stitch lengths without needing to reconstitute a reference panel every time.

Central processor (μ)

This is an 8K RAM memory which permits storage of the individual stitch lengths of the different courses in a panel. These memories may possibly be supplied with measurements made during the install cycle or by the central computer via the serial interface when the item has already been produced. A 4K EPROM memory contains the programme which runs on a 6809 microprocessor. The connections with the keyboard and the various inputs/outputs are performed by VIAs which allow parallel inputs and outputs.

In producing the "reference" panel, the various measurements are taken of stitch length and stored, using a mean delivery tension for the yarn. When reproducing the panels, all the measured lengths are compared with those in the memory and the variations are corrected by varying the tension applied to the yarn. During the knitting cycle, other information may be obtained, such as the mean production time for a panel, the total length per colour knitted in a pannel, the running time, and the downtime. Numerous other statistical data can be entered in the system and these may be centralised on the main computer which may be linked to all the microcomputer in a workroom.

Trials with different types of yarns revealed that in most cases it is possible to stabilise the stitch length and to limit its variations to values below 1 %. From the economic aspect, the study is being continued to investigate whether it is economically feasible to fit a relatively costly device on every machine. If this is not the case, with technological development it would be possible to design a more centralised system requiring less expensive measuring apparatus and correction devices on each machine.

(B)

The principle of operation of the second method is shown in Figure 18.

From a bobbin 1, the yarn passes through an unwinding eyelet 2, into an adjustable tensioner 3, into a fixed guide 4, then to the tip 5 of a lever 6, before being directed by the fixed guide 7 to the knitting machine 8.

The lever 6 diverts the yarn from the straight path joining guides 4 and 7. A counterweight 9 and a pointer 10 are located at the end of lever 6. According to the yarn feed conditions, the lever 6 assumes a position of balance which depends on the tensions T_1 and T_2 exerted by the yarn on both sides of eyelet 5 and on the influence of the constraint T_3 imposed by lever 6. This balance is registered by the position of the pointer 10 on a graduated scale 11. The system itself for a given yarn at a given

197

setting. If the frictional characteristics change, for example if they increase, foreces T_1 and T_2 vary and cause displacement of the lever.

If action on the yarn tensioner causes a return to the initial configuration, there is re-establishment of equilibrium of the initial forces and consequently the effect of the tensions exerted on the yarn. This is the principle of the stitch length controller.

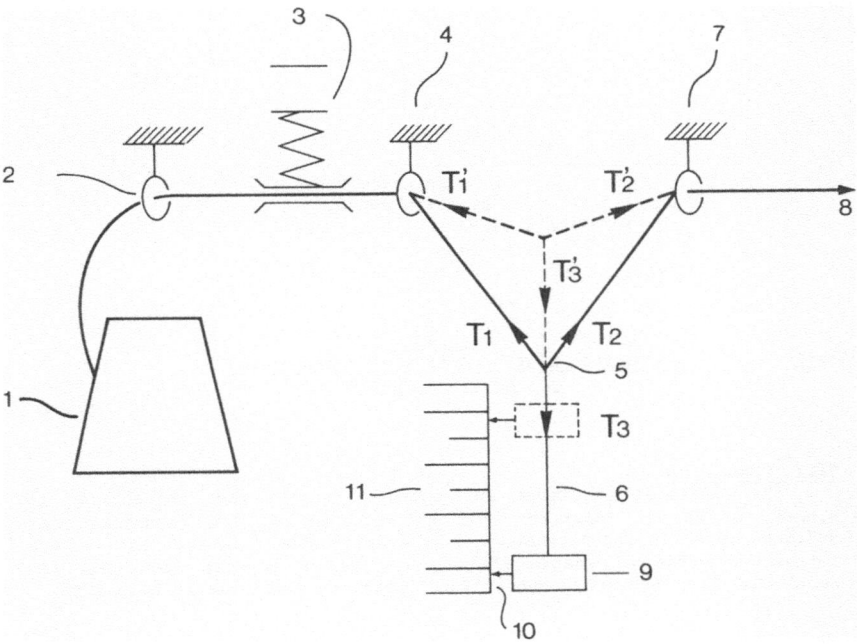

Figure 18
Stitch length controller (method (B)

It is in fact made up of two parts (Figure 19):

- a position sensor located at the end of the needle bed which detects variations in yarn friction and,

- a motor-driven tensioner 3 controlled by the sensor and which corrects the variations detected.

This stitch length controller is already available in industry and stitch length variations of the order of $\pm$ 1 % are obtained.

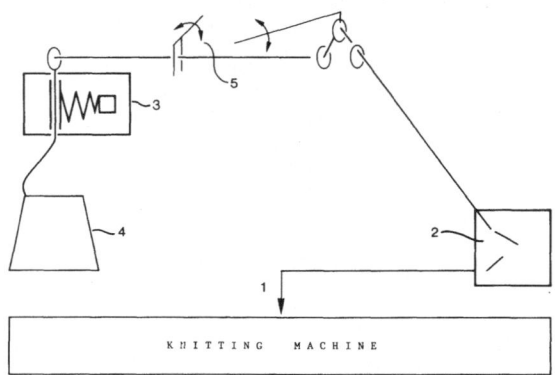

1. yarn guide
2. position sensor
3. motor-driven tensioner
4. yarn package
5. knot catcher

Figure 19
Stitch length controller - method (B)
Knitting machine

VIII. CONCLUSION

The extremely important work carried out by the laboratories is clearly of use to industry. This quick survey of the results cannot suffice. Those in industry who are interested should contact the various research centres to enable their companies to benefit from this new knowledge. The laboratory specialists there are available to help them.

OBJECTIVE QUALITY ASSESSMENT

E. FINNIMORE
Deutsches Wollforschungsinstitut

1. - INTRODUCTION

The general theme of the papers presented in this section of the Symposium is the quality of knitted fabrics and articles. The European knitting industry will only be able to remain competitive compared to low-wage countries if it can supply the customer with a better-quality product at a reasonable price. This means that quality standards have not only to be maintained but constantly improved, in order to keep ahead of developments elsewhere. When setting up quality standards it is important to have objective methods of measurement, since these facilitate both quality control on the production line and the setting up of international standards. Furthermore, there will be an increasing tendency in the future to "engineer" fabrics and garments to meet specific requirements. Constant improvement in quality also means that new processes and treatments will be developed and their effect on fabric quality has to be assessed. For these applications it is also important to have objective assesments of quality which are reliable and reproducible. In this paper several aspects of objective quality assessment of knitted fabrics, based on work carried out in three European Institutes, will be discussed.

2. - OBJECTIVE ASSESSMENT OF FIT OF HALF-HOSE ARTICLES

For the consumer an important aspect of quality when purchasing socks is a good fit, which also leads to greater comfort in wear. At the ITF Maille in France a study was carried out to develop a means of objective control of the size and fit of socks (half-hose). When socks are worn they are in a state of extension. A suitable objective measurement of fit has to take this into account. Ideally the article should be placed on a form where it is subjected to the same deformation as in wear. The forms used for boarding are not appropriate for this purpose. For practical reasons in the industry it is preferable to have flat forms.

An initial set of forms (1) which was developed for sock sizes 36 to 44 was found to have the following shortcomings : the angle between the axis of the leg and foot sections of the form was too small, thus causing puckering of the sock at the base of the leg. The calf region was symmetrical about the axis of the leg, but this did not conform to reality. It was thought better for the forms to include the knee, especially for judging the fit of socks with long legs.

In order to obtain a better set of forms, a study was made to obtain data on standard leg and foot dimensions. This was based partly on information obtained by the Centre Technique du Cuir (2), with regard to foot dimensions, and the centre Technique des Industries de l'Habillement for leg dimensions (3). After supplementing this data with their own measurements, ITF Maille was able to set up a table of standard leg and foot dimensions for sock sizes 24 to 46 (men's, women's and children's sizes). It was found that the leg size was generally related to the foot size.

In order to make the appropriate forms, it was necessary to develop a method for allowing a flat surface to be made equivalent in area to the external surface of a three-dimensional volume (foot and leg). The steps taken to obtain the forms were :

- conversion of a fitted sock to a rigid form, using paraffin wax. The sock was then removed from the leg by cutting it into two half-shells,
- insertion of incisions at appropriate points, in order to obtain a flat shape (Figure 1a),
- conversion of this shape to one of equivalent area having a continuous contour (Figure 1b).

The resultant form represents the mean of the two half-shells (Figure 1c). The proportions were also corrected for deviations in the dimensions of the foot and leg of the test persons from those in the standard table.

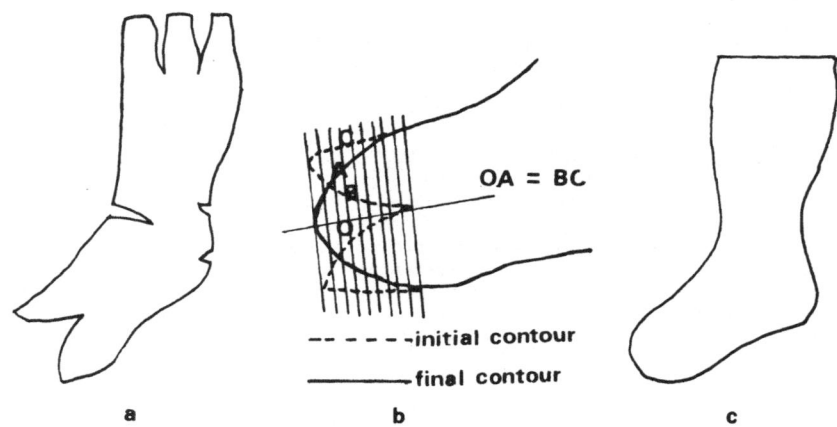

OA = BC

-- -- -- -- -- --initial contour
————————final contour

a   b   c

Figure 1
Steps in development of flat forms from three-dimensional
fitted socks

These forms represented an improvement on the original forms mentioned above in that the angle between the axis of the leg and foot sections was now 120° and the proportion of calf behind the axis of the leg was greater than the proportion in front of the leg (Figure 2).

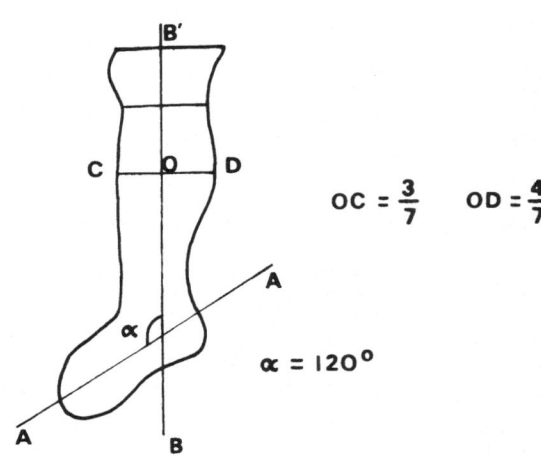

$$OC = \frac{3}{7} \qquad OD = \frac{4}{7}$$

$$\alpha = 120°$$

Figure 2
Proportions of flat form

    A set of 12 forms was prepared corresponding to the even numbers of French hosiery sizes 24 to 46. A total of 90 articles covering various types of socks was tested both on these forms and by 23 wearers. The criteria for judgement of fit on the forms are shown in figure 3. Of the 90 samples tested only three showed lack of agreement between the objective test and the wearer trial. In these three cases it could be shown that the foot and leg dimensions of these wearers deviated from the standard values for the same foot size.

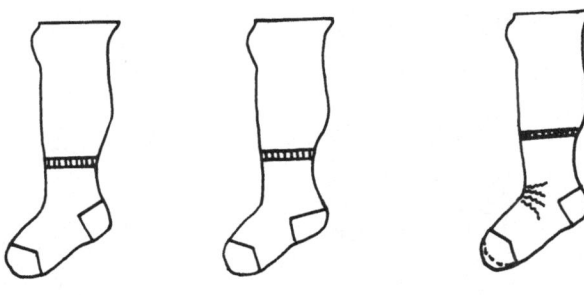

Good fit   Sock too small.   Sock too large :
                         Heel displaced   with correctly placed
                         towards sole   heel too much material
                                             at toe.

Figure 3
Criteria for objective judgement of good fit of socks

The forms have been successfully tried out in hosiery mills and are now being produced commercially (4). The method has been introduced as a French standards (NF G 30-101). Although the forms have been constructed on the basis of standard measurements on the French population the method of obtaining flat forms from three-dimensional socks is generally applicable and the forms can be made to comply with any chosen standard. In the long term it is hoped to extend the French norm to a European standard.

3. - OBJECTIVE ASSESSMENT OF SEWABILITY OF KNITTED FABRICS

An important source of faults in the knitted garment industry is the poor sewability of some knitted fabrics, which can lead to loop damage and the formation of holes during sewing, thus impairing the quality of the finished articles. If the fault is only detected during sewing it is usually too late to overcome it since it is difficult to treat cut pieces with finishing agents. At the Institut für Textiltechnik in Denkendorf, West Germany, an objective test has been developed for testing the sewability of knitted fabrics under production conditions before the fabric piece is cut.

Preliminary studies showed that there is a very good correlation between the needle penetration force and loop damage. Figure 4 shows the needle penetration force (lower curve) and the reaction force on the sewing plate (middle curve) during the penetration of the sewing needle into the material. The signals for the two forces are seen to be identical. These signals are digitalized and read into a computer.

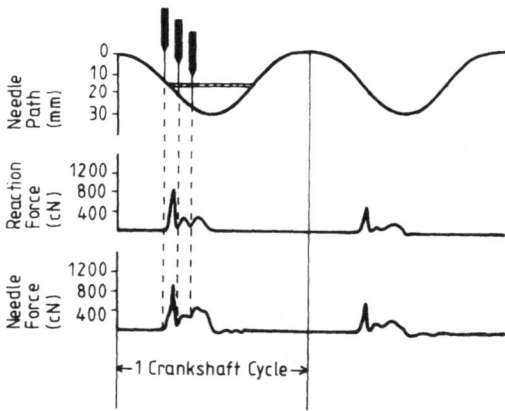

Figure 4
Needle penetration force as function of needle position

Figure 5 shows the modular construction of the measuring apparatus for penetration force. The data read into the computer can be used to plot the penetration force curve and to calculate statistical data. The results can be presented on the monitor or printed out. Details of material, finishes, needle used, etc. can be typed into the computer so that a complete documentation for the sample is printed.

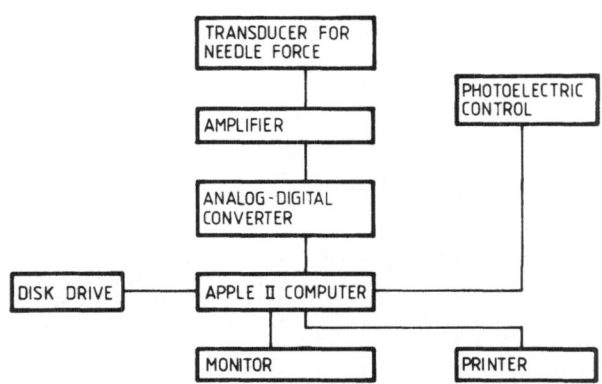

Figure 5
Schematic diagramm of test apparatus.

The results can be presented either as a curve of penetration force v. needle motion for one stitch or in the form of a bar chart showing the maximum penetration force for 50 stitches, with a printout of mean, standard deviation, etc. for the maximum penetration force.

The advantages of this test method are that it can be carried out on production machines, under production conditions (relevant needle type, speed, type of stitch, pretreatment of material). The test can be made as a precautionary measure on representative samples before cutting, thus allowing for the possibility of treating the whole piece with sewing auxiliary agents. EDP allows rapid accumulation of data and delivery of complete sample documentation.

The advantages of proper finishing of knitted goods to give improved sewability can be readily demonstrated by this method. Figure 6 illustrates the reduction in needle penetration force after treating a cotton fine rib knit with an auxiliary agent for improved sewability. The maximum penetration force with the finishing agent is reduced by about 60 %. With fabrics made of 100 % synthetic fibres the needle penetration force can be reduced by appropriate finishing agents by up to 93 %. The method thus allows objective studies to be made to optimize finishing treatments and hence has a direct influence on increasing product quality. The method also allows objective evaluation of other factors influencing sewability, such as needle size, tightness factor of knit and sewing speed.

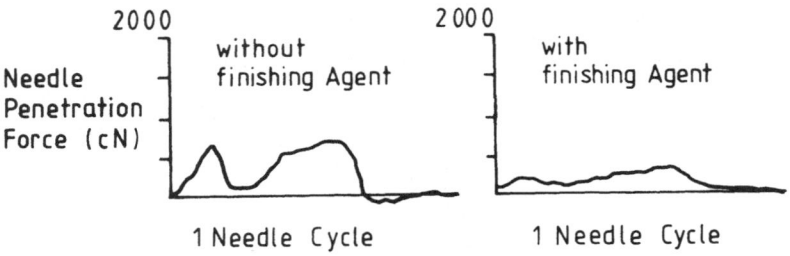

Figure 6
Objective assessment of improved sewability after
treatment with finishing agent

4. - OBJECTIVE ASSESSMENT OF HANDLE OF KNITTED FABRICS

The handle of fabrics is one of the most important quality aspects and, along with the visual impact, can be the determining factor for the consumer in a decision whether or not to purchase. However, handle is a property which is especially difficult to define, quantify and measure. By definition (5), handle is the "subjective assessment of a textile material obtained from the sense of touch". This definition would seem to preclude any hope of objective assessment of handle. However, it has been known for some time (6) that handle is related to the physical and mechanical properties of the fabric. This is indicated also by the use of words such as "thin", "stiff", "lively" etc. when describing handle. These words show a fairly clear relation to measurable fabric properties.

The advantages of objective handle assessment are obvious. Without objective measurements, it is difficult to set up quality standards for handle and the dialogue between fabric manufacturer, finisher and garment maker is made more complicated by the lack of numerical specifications which can be exchanged and compared with measured fabric properties. Systematic studies on the effect of knitting parameters and finishing treatments on handle can be more readily carried out with quantitative measurements. The mechanical properties related to handle have also been shown to be important for other aspects of quality such as tailorability, drape and comfort of fabrics (7).

The approach to objective handle assessment which has attracted the most attention world-wide is the KES-F system of Kawabata from Japan (8). This approach may be summarised as follows :

- textile industry experts in Japan agreed on the expressions to be used to describe primary handle impressions (e.g. "stiffness", "smoothness") for a particular class of fabrics e.g. men's winter suiting.
- a collection of representative fabrics was assessed with regard to these handle characteristics. For each primary handle expression the fabric was assigned a value between 0 (meaning this characteristic was not present) and 10 (strongly present). Sample cards were prepared covering the range of handle ratings. By comparing fabrics with these standards it became possible to give numerical handle values to any fabric in this class.

- A set of instruments, known as the KES-F system, was designed to measure the mechanical properties of fabrics related to those properties assessed by the experts when assigning primary handle values (Table 1). A total of 16 mechanical and physical parameters are assessed in these measurements (Table 2). The properties of the original fabric collection mentioned above were measured on this system.
- A "stepwise-block" regression was carried out on the mechanical properties and the subjective hand values. On the basis of this regression, transformation equations were set up to convert the mechanical properties into primary hand values. In these equations the mechanical properties are weighted according to their contribution to the primary handle impression.

Table 1
Instruments in KES-F system

Instrument	Properties testes
KES-F 1	Tensile and shearing products
KES-F 2	Pure bending properties
KES-F 3	Compression and thickness
KES-F 4	Surface characteristics

Table 2
Parameters describing fabric mechnical properties
related to handle

BLOCK	PARAMETERS
Tensile	Linearity of Load-Extension Curve Tensile Energy Tensile resilience
Shear	Shear rigidity Hysteresis of shear force at 0.5° Hysteresis of shear force at 5°
Bending	Bending rigidity Hysteresis of bending moment
Compression	Linearity of pressure-thickness curve Compressional Energy Compressional Resilience
Surface Characteristics	Coefficient of Friction Mean deviation of Coeff. of friction Geometrical Roughness (contour)
Fabric Construction	Fabric weight per Unit Area Fabric thickness

Although these measurements are applicable to all textile materials, there has been less work done on knitted fabrics. Adjustments have to be made to some of the conditions of measurement to take accout of the greater extensibility of knitted fabrics compared to wovens. At the Deutsches Wollforschungsinstitut in Aachen, West Germany, the application of the KES-F system to the measurement of the handle of knitted fabrics has been studied. These studies were carried out in co-operation with the Institut für Textiltechnik in Denkendorf, which supplied the fabrics and carried out subjective assessments.

In the first part of these studies the importance of various parameters such as tightness factor, type of knit, fibre material, fineness and finish on the objective handle parameters of a series of wool and wool/acrylic knits was investigated. Table 3 summarises the results of this investigation. Some of the most important points will be briefly discussed here.

The type of knit influences all the handle parameters and is mainly responsible for the "character" of the handle. The interlock structure gives high values for the handle value "smoothness", whereas French Wevenit and Punto-di-Roma have a rougher surface (Figure 7a). Punto-di-Roma has high values for "stiffness", whereas the rib construction is particularly flexible for its weight (Figure 7b).

Table 3

Effect of various factors on the handle properties of a
series of wool and wool/acrylic knitted fabrics

Factor:	Knitting Parameters		Fibre Parameters			Subsequent Treatments		
Handle Properties	Type of Knit	Tightness Factor	Fibre Material	"Superwash" Finish	Fibre Fineness	Steaming	Washing	Fabric Softeners
Surface	+	0	0	0	0	0(+)	0	0
Compression	+	0(+)	0(+)	0(+)	0(+)	+	0(+)	0(+)
Bending	+	++	+	0(+)	+	++	+	+
Shear	+	++	+	0(+)	+	++	+	+
Tensile	+	0(+)	+	0(+)	0(+)	+	0(+)	0(+)
Construction	+	+	+	0	+	+	+	0

0   = no influence
0(+) = influences some parameters in this block
+   = influences all parameters in this block
++   = strong influence

Handle Value for "Smoothness"

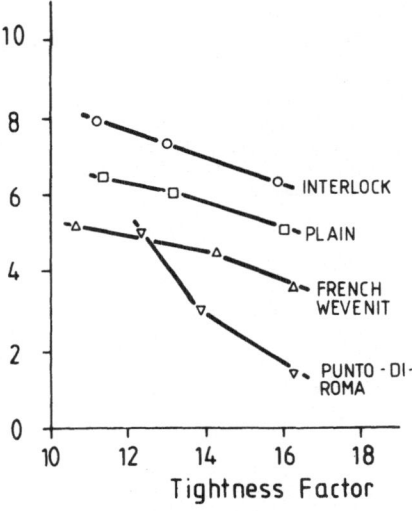

Figure 7a

Handle Value for "Stiffness"

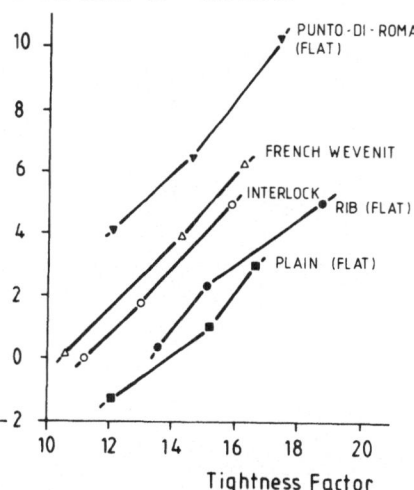

Figure 7b

Handle values for "Smoothness" and "Stiffness" for different
types of knit (wool/acrylic , 60:40).

The tightness factor, defined as :

$\overline{\text{tex}}$ /(average length of yarn in loop in cm)

also has a marked influence on many handle properties, particularly the stiffness parameters (Figure 7b).

Steaming has a large effect especially on the stiffness and resilience parameters. Figure 8 illustrates the reduction in bending hysteresis from the machine state to the steamed fabric. The effect is more marked on the wool/acrylic fabric than on the wool fabric. Steaming was found to be very important for developing a soft, flexible and elastic handle.

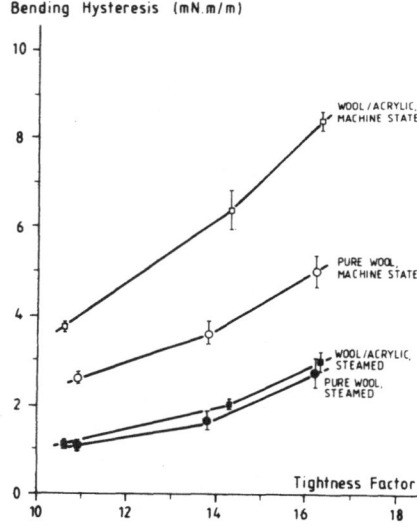

Figure 8
Influence of steaming on the bending Hysteresis of
Wool and Wool/Acrylic fabrics (French Wevenit)

Figure 9 shows a typical example for the greater compressional resilience of the untreated wool fabrics compared to the "Superwash" treated wool and the wool/acrylic blend.

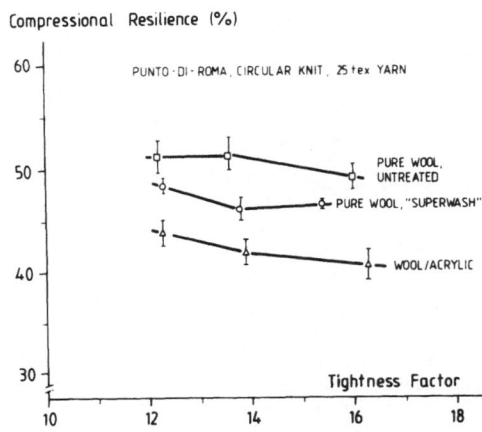

Figure 9
Effect of fibre material and finish on compressional resilience
(punto-di-Roma)

Increases in fibre diamter resulted in increased values for the stiffness parameters (Figure 10).

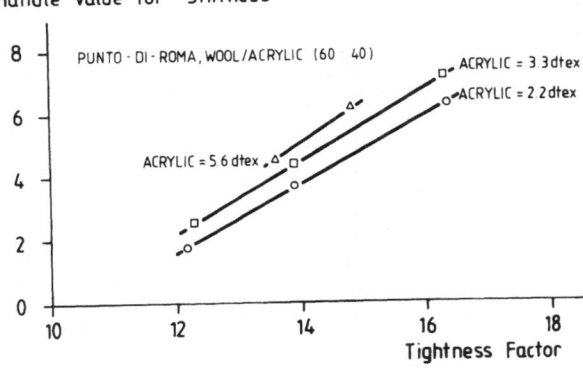

Figure 10
Effect of fibre fineness on stiffness

The second aspect of these studies was concerned with the quantitative assessment of the effect of softening agents on knitted fabrics made of wool and wool blends. In the machine state, knitted fabrics normally have an unacceptably hard, stiff handle (cf. Figure 8). This is due to residual tension in the fibre when the yarn is bent to form knitted loops. Some form of relaxation process after knitting is therefore unavoidable. Steaming of wool knits leads two stress relaxation in the fibres and thus to a reduction in stiffness. An additional treatment with softening agents can reduce the stiffness parameters even further.

The KES-F system allows the stiffness parameters of knitted fabrics to be quantified so that the effect of treatments such as the application of softening agents can be followed quantitatively. The bending and shear tests are particularly useful in this respect. The hysteresis width is closely related to frictional forces between fibres and yarns and these frictional forces can be reduced by reducing the pressure between the fibres, caused by residual tensions. This is the effect of steaming. Treatment of the fibre surface with softening agents can reduce the frictional coefficient of the fibre surface and thus give a further reduction in bending and shear hysteresis.

Softening agents based on cationic surfactants have a marked effect, especially on cotton fabrics. Previous work on wool (9) showed that on untreated wool softening effects were slight. However, on the "Superwash"-treated samples studied here a definite softening effect was found. Figure 11 shows the effect of increasing concentrations of cationic softener applied by padding to a French Wevenit fabric made of "Superwash"-wool. The stiffness parameters pass through a minimum at relatively low concentrations of softener. Higher application levels are not advantageous. This minimum in stiffness was also noted in subjective evaluations (see Table 4). One can see that there is good agreement between the objective and subjective rankings.

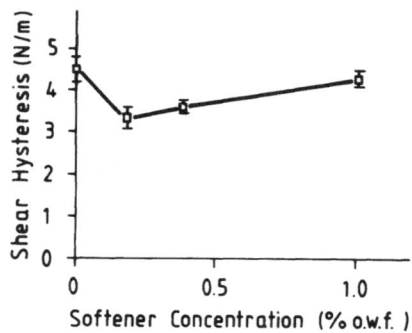

Figure 11
Effect of Cationic Softener on a stiffness parameter of
knitted fabric ("Superwash"-wool)

Table 4
Comparison of objective and subjective assessment of effect
of Cationic Softener on "Superwash"-wool

| TREATMENT | Objective Assessment | | Subjective ranking |
	Shear Hysteresis	Handle Value "Stiffness"	soft *
Blank	4.5	6.1	1.3
0.2 % Softener	3.3.	5.5	3.0
0.4 % Softener	3.6	5.5	3.0
1.0 % Softener	4.3	6.1	1.6

* average rank , 1 = hardest, 4 = softest

A comparison was also made between objective and subjective handle assessments on samples of the same fabric which had been treated in different ways (e.g. steaming, washing, treatment with different cationic softener using different application methods). Table 5 lists the Spearman rank correlation coefficients (10) between the subjective rankings (mean of several judges) and the objective parameters for seven treatments on a French Wevenit fabric and seven treatments on an Interlock fabric, both made of "Superwash"-treated wool.

Table 5
Spearman rank correlation coefficients between subjective and
objective ratings for fabrics made of 100 % "Superwash"-wool

| FRENCH WEVENIT | | | INTERLOCK | | |
Subject. Parameter	Objective Parameter	R_s*	Subject. Parameter	Objective Parameter	R_s*
soft	shear hysteresis	0.88	soft	shear hysteresis	0.81
soft	"stiffness"	0.68	soft	"stiffness"	0.66
soft	"smoothness"	0.88	soft	"smoothness"	0.96
smooth	"smoothness"	0.93	smooth	"smoothness"	0.87

* R_s (7,90 %)= 0.71, R_s (7,95 %) = 0,79

Table 6 shows similar results for five treatments on a Punto-di-Roma and four treatments on an interlock fabric, both made of wool/acrylic (60:40).

Table 6

Spearman rank correlation coefficients between subjective and objective ratings for fabrics made of Wool/acrylic (60:40)

PUNTO-DI-ROMA			INTERLOCK		
Subject. Parameter	Objective Parameter	R_s*	Subject. Parameter	Objective Parameter	R_s*
soft	shear hysteresis	0.94	soft	shear hysteresis	0.76
soft	"stiffness"	0.70	soft	"stiffness"	0.76
soft	"smoothness"	0.03	soft	"smoothness"	0.98
smooth	"smoothness"	0.76	smooth	"smoothness"	0.98

* R_s (5,90 %) : 0,90

On the relatively stiff, rough French Wevenit and Punto-di-Roma fabrics there is a significant agreement between the shear hysteresis, i.e. the objective handle parameter which responded most markedly to treatments with softeners, and the subjective ranking according to softness. On the interlock fabrics, which had a very smooth surface irrespective of their stiffness, the correlation bertween the objective stiffness parameters and the subjective softness rankings is not as good as beteen objective "smoothness" and subjective "softness". This indicates that the subjective concept of softness, as understood by these judges, has elements of "lack of stiffness" combined with "smoothness". The objectively measured effects of the softening treatments were the same on all fabrics, irrespective of type of knit, whereas the judges obviously responded to the effects differently, depending on the base fabric.

Obviously for such a complex phenomenon as handle more work will be necessary to investigate the exact relation between subjective impressions and mechanical properties. However, it can already be seen that the quantitative data obtained in this type of study can be useful for the manufacturer in choosing the best materials and finishing conditions for obtaining the desired end-product.

213

5. - REFERENCES

(1) "Mise au point de formes pour le controlle de la taille des chaussettes", ITF Maille, Feb. 1978

(2) "Etudes statistique du pied de la population française d'age scolaire", A. Hüber, R. Rigal, J. Leveque, C.T.C. Lyon, Technicuir no. 9 (1980); N° 3,5,7 (1981); N° 1 (1982).

(3) NF G 03-001, NF G 03-002, NF G 03-003

(4) Sodemat S.A. 270, rue du Faubourg Cronçels, F - 10042 Troyes

(5) "Textile Terms and Definitions", The Textile Institute, Manchestger, 6th Ed. 1970

(6) W.S. Howorth and P.H. Oliver, J. Text. Inst. 49, T 540-553, (1958)

(7) "The objective specification of fabric quality, Mechanical properties and performance", S. Kawabata, R. Postle and M. Niwa, Eds. Textile Mach. Soc. Japan, Osaka (1982).

(8) "The standardization and analysis of hand evaluation", S. Kawabata, Textile Mach. Sock. Japan, Osaka (1975).

(9) A.G. de Boos and E. Finnimore, J. Text. Inst.,75, 81-91 (1984).

(10) See U. Graf, H. Henning and P. Wilrich, "Statistische Methoden bei Textilen Untersuchungen", Springer-Verlag, Berlin, Heidelberg, New York, 1974, p. 529.

SUMMARY AND CONCLUSIONS BY THE CHAIRMAN OF THE SESSION

J. STRYCKMAN
Centexbel

Two important issues concerning the quality of knitted fabrics were highlighted in the questions raised by the presentations : on the one hand dimensions and dimensional stability and on the other objective measurements to determine the comfort of articles. The discussion is outlined briefly below.

As regards the question of whether the stabilized dimensions of a finished knitted article can be determined beforehand on the basis of the parameters of the yarn and the knitting structure, a correlation does exist between these parameters depending on the type of stitch and finishing. With the Starfish data bank, developed by the International Institute for Cotton, the dimensions of a finished article can be calculated from the characteristics of the unbleached fabric. Equations are available for cotton in three different types of stitch: jersey, 1:1 rib and interlock.

The dimensions of an article depend very largely on the length of thread used in each stitch. It is therefore very important to monitor and control this parameter. On the circular machine this is done by the positive feeders. There was previously no device for the flat machines, but two methods have now been developed and tested. In one, the aim is to maintain a constant tension at a given point as the yarn is fed into the machine. In the other, the length of thread used in a certain section on each course of stitches is compared by microprocessor with a predetermined desired length for the course. If the two lengths are different, a signal corrects the yarn feed into the machine. Both systems enable fabrics of identical lengths to be produced with a tolerance of 1 % on the length of yarn used.

The speed of modern flat machines does not hamper the operation of the system. The acceleration of the captor could be a problem in one of the systems but measurement takes place within the needle bed.

In the other system, the sensor reacts immediately to differences in tension. The effect of the acceleration of the yarn at the beginning of the course is cancelled out by the simultaneous deactivation of the tension devices.

Waxing the yarn prior to knitting is a less sophisticated means of regulating the length of thread used, but for certain "one-off" articles is the only possibility. However, all waxes do not react in the same way. Their composition, in other words the different lengths of the hydrocarbon chains exerts an influence. It affects the melting point, which also depends on the amount of wax used: this should be beteen 0.4 and 1 % depending on the nature of the yarn. It is therefore necessary to know how the wax used behaves and to apply it according to the nature of the yarn.

The question of whether wax affects the feel of knitted fabrics brings us to another issue : "comfort" and "feel" and their objective measurement. Firstly, waxing changes the feel of the fabric, but the project did not examine this aspect in depth. The objective measurement of this property is not yet very advanced and much still has to be done in this field.

The only parameters covered by the programme were type of stitch and fibre, and a large number of samples have already been analyzed. Two types of thread were examined : pure wool and a wool-acrylic mixture. The objective measurements were carried out on the Japanese Kawabata instruments, specially designed to measure tactile quality. The results of the objective measurements were compared with the subjective assessments carried out by experts selected for their long experience with woven or knitted materials. The correlation was fairly good on the whole, although the experts occasionally made mistakes when the difference between the samples was not great.

Kawabata's concept of "softness" is a combination of two elements : surface softness and firmness. Suitable terminology is required for these two measurements in Europe.

The Kawabata instruments have been much more extensively used for woven than for knitted fabrics. The parameters therefore need to be adapted to knitted fabrics. Nonetheless, subjective assessment should gradually make way for objective measurement of articles, since the experts are not always consistent, regularly contradicting themselves even within a short space of time. For knitted fabrics, parameters such as the hysteresis of cutting, thickness or weight, and the regularity of the surface would be more appropriate. This should become clear with experience.

The importance of Kawabata and objective measurement also lies in the fact that it makes it easy to assess the effect of the parameters of material and the manufacturing process on the feel of a fabric, which for practical reasons, is more difficult, if not impossible, by subjective methods. As to whether the Kawabata system could be used in Europe, the equipment is relatively expensive and small or medium-sized enterprises would not be able to justify its purchase.

In Japan, 150 machines are in use, many of them in research institutes or purchasing agencies working in collaboration with the industry. It is therefore useful now, and will be essential in the near future, for the clothing and textile industries to make use of this technique. With advantage the industries could rely on central service laboratories which could be acquiring and using the equipment and by the experience they would gain in objective measurements, offer a valuable service to industry.

The subject "Quality of knitted fabrics and articles" was examined from three different angles. The first concerned quality criteria for knitted articles, and there the work, carried out by four institutes, showed that quality, fit, comfort and consumer satisfaction needed to be improved. In order to do this, a standard European method for measuring the dimensions of the human body was needed, taking note that even within each country, regional differences in body shapes were liable to occur data banks of these measurements need to be compiled and updated as required using the standard method adopted.

The second aspect, also dealt with by four institutes, concerned quality in the manufacture of knitted fabrics and articles.
The main issue was one crucial to the quality of knitted goods : dimensional stability both in the manufacture and in wear and care by the consumer. The results of the work will enable better control of this aspect for wool, wool-acrylic, and cotton articles. One particularly interesting result was the development of devices automatically controlling the consistency of the sizes of the blanks on flat machines.

Nonetheless, much still remains to be done in the field of dimensional stability.

The final aspect was one which will in the future be of increasing importance in assessing the quality of knitted fabrics and articles, i.e. objective quality assessment. There are three important considerations : feel, fit and "sewability". This aspect was examined by three institutes. A recent methodology has produced some initial results for knitted fabrics in an area where experience with woven fabrics is already extensive. These properties are affected by a particularly large number of parameters. The programme therefore had to be limited to those which were felt to be crucial and to a limited range of raw materials, particularly spinning methods, knitting structures and finishing processes.

The work still to be done in this field may, for many years to come, continue to provide topics for many different research programmes.

As Chairman of this session, I would like to finish by expressing two hopes.

The first is addressed to the institutes involved in this research project. They have accumulated a vast quantity of results and these should be put to use. This can only be done if all these results are actually applied by the industry. It is therefore essential that each institute in its own sphere should see to it that its findings are passed on to industry and that it provides technical back-up to ensure that the information is correctly applied in practice.

In the interests of this application the research, the Commission of the European Communities will make every effort to stimulate the transfer of information and ensure that full use is made of the results.

An enormous volume of information has been amassed, but the number of parameters involved is immeasurable and the diversity of the products leaves plenty of scope for future investigations. We have seen through co-operation that the results of our efforts can be multiplied.

Secondly, I hope that this European collaboration between the institutes, which have formed such an excellent team over the past three years, will continue. This would be encouraged if the European Commission were to grant any request for subsidies from the large European knitting and hosiery industry to enable research projects to be carried out.

The fulfilment of these two hopes would undoubtedly be a crucial step forward in safeguarding the future of the European textile industry.

```
S E S S I O N   C

QUALITY  OF  KNITTED  FABRICS  AND  ARTICLES

*   *   *

Contract 003-TEX-D - E. FINNIMORE

Contract 005-TEX-D - Dr. G. BÜHLER

Contract 012-TEX-F - E. BUCHER

Contract 013-TEX-F - M. BALLAND

Contract 014-TEX-F - E. BUCHER

Contract 019-TEX-I - Prof. G. PRATI

Contract 021-TEX-NL - H.J. SUURMEIJER

Contract 024-TEX-B - J. STRYCKMAN

Contract 032-TEX-UK - J.A. SMIRFITT and D.C. WARD

Contract 035-TEX-DK - T. SØRENSEN
```

CONTRACTOR : DEUTSCHES
             WOLLFORSCHUNGSINSTITUT
ADDRESS : Veltmanplatz 8
         D - 5100 AACHEN

PROJECT LEADER : E. FINNIMORE

CONTRACT N° : 003-TEX-D

STARTING DATE : 1/10/1982

FINISHING DATE : 31/3/1985

TELEPHONE : 241/ 39.931

TELEX : 832.829

TITLE OF PROJECT

THE QUALITY OF KNITWEAR MADE OF WOOL AND WOOL / ACRYLIC BLENDS

AIM OF PROJECT

To investigate the quality of knitwear made of wool and wool/acrylic blends with regard to :
a) objective assessment and improvement of handle
b) improvement of surface appearance and spirality.

SUMMARY OF RESULTS

With the help of the Kawabata KES-F instruments it is possible to measure objectively the handle properties of knitted fabrics. Measurements on knitted fabrics where knitting parameters were systematically varied showed what contribution the different knitting parameters make to handle. The "character" of the handle is determined mainly by the tightness factor and the type of knit. In addition, steaming of the fabric is very important for developing a soft, elastic handle, especially for wool/acrylic blends. Pure wool fabrics have superior softness and elastic properties to the blends. The fibre fineness influences mainly the stiffness and the compressional properties. The addition of coarser fibres has an adverse effect on the handle.

The Superwash-shrink resist treatment has an unfavourable effect on the softness and the elastic properties. However, it could be shown that certain cationic softening agents give an improvement in the handle properties of Superwash-treated wool and wool/acrylic blends. This improvement could be ascertained both objectively and subjectively. On the other hand, none of the commercial products tested had a noticeable effect on pure wool without Superwash finish. Application of softeners by foam treatments is feasible and offers several advantages.

Laboratory experiments and industrial trials showed that the surface appearance of plain-knit wool fabrics after washing can be stabilized by chemical treatments with reducing agents using kiss-roll, spray and foam treatments or gassing. The application of the chemicals necessary to achieve setting of the fabrics is chiefly a problem of technology and requires the help of the textile machinery industry. Apart from the surface appearance the handle, pilling tendency and residual bagginess are positively influenced by the setting process. Spirality in socks can also be avoided by using chemical setting with reducing agents.

219

CONTRACTOR : INSTITUT FÜR
TEXTILTECHNIK
ADDRESS : Körschtalstraße 26
D - 7306 DENKENDORF

PROJECT LEADER : Dr. G. BÜHLER

CONTRACT N° : 005-TEX-D

STARTING DATE : 1/10/1982

FINISHING DATE : 30/4/1985

TELEPHONE : 711/ 34.080

TELEX : 725.65.54

TITLE OF PROJECT

QUALITY KNITWEAR - MANUFACTURED FROM WOOL AND COMPOSITE YARNS

AIM OF PROJECT
To improve the competitiveness of the European knitwear manufacturers by increasing the product quality and reducing the production costs.

SUMMARY OF RESULTS
The European knitwear manufacturers in the garment sector must improve their competitive ability through modern high quality products. It also could be improved by optimising the production costs, which could be achieved by reducing the knitting defects. Simultaneously, the image of the product would be improved.

The aim of the project was to reduce the knitting defects through the determination of economically important defects, by analysing the reasons for their occurence and by recommending suitable measures for their remedy.

The main aim was laid on the visual and tactile characteristics. The expensive source of defects, i.e. the sewability of knitted fabrics, was investigated as well as the behaviour of yarns during their processing. The set problem was solved through :
1. industrial analysis,
2. planned small scale experiments.

In the industrial analysis, the types of defects, their frequency and their costs were obtained for various garment products for a large volume over a long production period. The costs due to defects were related to other components of production cost. The quality influenced by yarns and machinery was examined for possibilities of improvements. The dealt small-scale experiments dealt with the influence of fibre count, the felt-free treatment of wool and the type of acrylic component in the fabric, especially where the problem of cockling was concerned.

The results of the investigation provide important information to improve the product quality and to reduce the costs due to faults. It also shows how the knitting process could be optimised.

CONTRACTOR : ITF-MAILLE

ADDRESS : 270 rue du Faubourg
Cronçels
FR - 10042 TROYES CEDEX

PROJECT LEADER : E. BUCHER

CONTRACT N° : 012-TEX-F

STARTING DATE : 1/10/1982

FINISHING DATE : 15/10/1984

TELEPHONE : 25/ 82.92.41

TELEX : 840.744

TITLE OF PROJECT

CONTROL OF THE FIT OF LEGWEAR

AIM OF PROJECT

Establishment of an objective method of controlling the size and fit of legwear.

SUMMARY OF RESULTS

The measurements of dimensional variations conventionally performed for controlling the fit and care performance of textile garments are totally unsuited to legwear because the latter are garments worn in the extended state.

Control of this type may be performed by positioning the item on a former in such a way that during the test it is subjected to the same deformations as during wear.

For practical reasons of industrial use it is preferable for the formers to be flat in shape.

The study consisted of specifying and making a series of 12 formers for the foot and the lower leg, including the knee, covering sizes 24 to 46 (French system).

It was necessary to :   .

- develop a method allowing a flat area to be obtained substantially equal to the outer envelope of a volume incapable of being opened out flat (the foot and the leg),
- to define the evaluation criteria for fit on the formers produced,
- to carry out trials with these formers under industrial conditions,

The tangible results of this work are :

- standardisation with standard AFNOR G 30-101 : control of size and fit of legwear on a flat former.
- the industrial production and marketing of the formers (Société SODEMAT).

CONTRACTOR : I.T.F.Maille

ADDRESS : 270 rue du Faubourg
Croncels
FR - 1002 TROYES CEDEX

PROJECT LEADER : M. BALLAND

CONTRACT N° : 013-TEX-F

STARTING DATE : 1/10/1982

FINISHING DATE : 15/10/1984

TELEPHONE : 25/ 82.92.41

TELEX : 840.744

TITLE OF PROJECT

STUDY OF THE PERFORMANCE OF ACRYLIC SWEATERS

AIM OF PROJECT
To improve the stability in wearing of acrylic sweaters.

SUMMARY OF RESULTS
The relaxed state enables a prediction to be made of the dimensional stability of knitted fabrics. For acrylics it is characterised by the coefficients of Munden's laws given in the table below :

	K 1	K 2	K
Jersey	3.7	4.9	18.1
1 : 1 rib	5.8	4.8	27.8
2 : 2 rib	6.7	4.9	32.8

If steam pressing is performed on knitted fabrics deformed by comparison with the relaxed state, it is found impossible to set more than a quarter of this deformation if treatment is carried out at 70°C and about two thirds if the temperature is 95°C.

To be more precise, we have developed a method for measuring the percentage of "set" of the "crimp" of a yarn in a knitted fabric treated in an air-steam mixture. It may be deduced that it is desirable to treat LEACRYL at 75°C, COURTELLE between 75 and 80°C, DRALON at 80°C and CRYLOR at 85°C.

Actual wear and wash tests prove that whatever the previous treatment on the press, sweaters have a tendency to shrink in the length direction and sag in the width direction;

However, if the knitted fabrics are not too slack and are treated at a carefully selected temperature near to the relaxed state, it is possible to produce acrylic sweaters having good wear performance.

222

CONTRACTOR : ITF-MAILLE

ADDRESS : 270 rue du Faubourg
Cronçels
FR - 10042 TROYES CEDEX

PROJECT LEADER : E. BUCHER

CONTRACT N° : 014-TEX-F

STARTING DATE : 1/10/1982

FINISHING DATE : 15/10/1984

TELEPHONE : 25/ 82.92.41

TELEX : 840.744

TITLE OF PROJECT

FIT OF MEN'S UNDERWEAR

AIM OF PROJECT
     Development of flat formers for the control of the size and fit of
T-shirts.
     Manufacture of knitted jersey fabric with performance appropriate
for wear purposes.

SUMMARY OF RESULTS
     T-shirts, whether rib-knit or jersey, may be worn in the extended
state. Control methods which comprise making flat measurements of these
garments are not always adequate for determining the fit. Flat templates,
corresponding to persons of different stature, would facilitate this
control. This was the objective of the first part of our study.
     Basing our work on the experience gained in developing formers for
legwear, 26 formers representing the upper body (head, bust, shoulders
and upper arms) were produced. These formers cover 95 % of the adult male
population.
     Evaluation criteria for fit are proposed for T-shirts :
- in width : the tightness, the good fit or the slackness of the garment
  are studied,
- in length : an experimentally defined length is given.
     The ease of passage for the head in putting on the garment is
checked.
     Experiments carried out in the laboratory and in industry have
enabled these criteria to be confirmed.
     In a second study, the dimensional performance of jersey knitted
fabrics made with different absorbed lengths of yarn has been studied by
simulation of deformations in wear of the garments. The shortest absorbed
lengths of yarn produce the most stable garments.
     The flat formers used for studying the fit of these garments and its
permanence are an auxiliary tool in maintaining product quality by
assisting the design of garments well adapted to size.

CONTRACTOR : STAZIONE SPERIMENTALE PER   CONTRACT N° : 019-TEX-I
LA CELLULOSA, CARTA E FIBRE
ADDRESS : Piazza L. da Vinci 26   STARTING DATE : 1/10/1982
I - 20133 MILANO

FINISHING DATE : 30/4/1985

PROJECT LEADER : Prof. G. PRATI   TELEPHONE : 2/ 29.29.57/60

TELEX : 320.579

TITLE OF PROJECT

RESEARCH ON PARAMETERS AND OPTIMISATION OF CONDITIONS FOR
THE QUALITY IMPROVEMENT OF SPORTS AND LEISURE KNITWEAR

AIM OF PROJECT

To define the qualitative parameters which determine the value of
knitwear and to correlate these qualitative parameters with the nature of
the fibre used and with the operating conditions employed during the
manufacture of the end-product. To propose, as far as possible, constructional standards optimising the product.

SUMMARY OF RESULTS

An initial important result of the research has been the success of
involving all the constituents of the textile cycle from fibre producer
to commercial distribution (with particular involvement of wholesale
distributors). The various constituents have thus been able to conduct a
dialogue in a spirit of cooperation with the opportunity of giving
substance to the respective requirements and problems; this in a field
where working has almost always been carried out in closed sectors.

The aim given to the research, although useful in the sense stated
above, has created general organisational difficulties which have allowed
only partial attainment of operational phases envisaged; in particular it
has not been possible to include stockings in the study and to extend the
study of alternative solutions to improve certain features.

For the four classes of garments considered, it has been possible to
define which are the knitwear structures and fibre compositions most
commonly used, and also the criteria judged to be the most important in
the selection of an item of sports or leisure clothing. It must be stated
in advance that the decision cannot exclude from consideration the
relationship between quality and cost and that the investigation has been
aimed at clothing of average quality. The most important criteria are :
garment fit, dimensional stability, dye-fastness, durability and care
performance. The identification of these quality criteria and their
relative importance is a result of significant relevance within the scope
of the research project.

The analytical tests have therefore been defined according to the
criteria governing selection, with a wide range established so as to
identify the quality of the raw material used and the fabric structure.

From all the samples the result is that there are substantially no
problems in guaranteeing adequate dye fastness to washing, to perspiration, to rubbing, to light etc. The results obtained are largely higher
than the tolerance limits, grade 3/4 for solid shades and 4/5 for mixed
colours (wet fastness); light fastness was generally higher than 4/5.

A different situation occured for dimensional stability; for the garments considered within the project, the tolerance limits were given as maximum 4 - 6 % variation, the length dimension being the most critical. For some constructions no substantial problems exist in ensuring these levels, but on the other hand some concern is aroused by slippas made in plush fabric, polo shirts in piqué and sweaters in pure wool. Problems have been identified in these garments, either fabric finishing problems, or problems of the correct design of the fabric in terms of compatibility between the adopted values of mass per unit area and cloth width (also the length of the panels in the case of large pattern repeats), taking into account the selection of yarn count and diameter and gauge of the machine.

Very different behaviour was found amongst parameters relating to garment durability; rather low results were found for abrasion resistance of some plush fabrics, for some jersey fabrics, for polo shirts and for some sweaters, and also low values for pilling on other fabrics. It is considered that this situation is not serious in a market which still has a fairly fast rate of replacement of items of clothing.

In some cases a lack of homogeneity in performance was found between items belonging to the same batch and greater attention to this question is recommended which may have its origin in low efficiency of quality control of raw material and at various stages of processing.

Finally, with regard to fit of garments, discrepancies have been observed between dimensions of garments of the same size, and in some cases, a lack of balance in dimensions, with consequent risks of dissatisfaction on the part of consumers. Rationalisation in this field is recommended with better definition of size and dimensions better suited to the comfort of users.

CONTRACTOR : T.N.O.

ADDRESS : Fibre Research Inst.
P.O. box 110
NL - 2600 AC DELFT

PROJECT LEADER : H.J. SUURMEIJER

CONTRACT N° : 021-TEX-NL

STARTING DATE : 1/12/1983

FINISHING DATE : 31/10/1984

TELEPHONE : 15/ 56.93.30

TELEX : 38.071

TITLE OF PROJECT

THE QUALITY OF KNITTED GARMENTS AND KNITTED FABRICS

AIM OF PROJECT

The comfortable fitting of close fitting knitted garments.

SUMMARY OF RESULTS

Published sizing tables and methods for size designation in the
E.E.C. countries have not been set up for the design and production of
close and comfortably fitting knitted garments. Directives are not available for the design of these garments in the current limited number of
sizes utilizing the characteristic properties of knitted garments for a
large majority of consumers.

Sizing tables for close fitting knitted garments in current sizes
containing the largest and the smallest body measurements occuring within
each size have been derived from anthropometric studies of the Swedish
population in the seventies. These Swedish tables provide more data
necessary for the design of close fitting knitted garments than existing
sizing tables for clothing in the E.E.C.

Because of the differences between the relative body proportions of
population groups of different nationalities in Europe, the Swedish
sizing tables cannot be regarded as representative for all the population
groups in the E.E.C.

To provide a large majority of consumers in local or home markets in
the EEC with good and comfortable close fitting knitted garments, sizing
tables derived from anthropometric studies of the population in these
markets should be available.

The design and production of good and comfortable close fitting
knitted garments can be based on :
- the available knowledge about the properties of knitted structures,
- the production methods based on this knowledge,
- the application of systematic measuring methods to test the good and
  comfortable fit of garments as proposed by TEFO Textile Research
  Institute in Göteborg,
- the comparison of measured critical garment properties with sizing
  tables derived from the anthropometric data of population groups
  containing the largest, the average and the smallest critical body
  measurements occuring within sizes.
- A product information system added to the size designation based on
  measured garment properties.

Measurements obtained with the TEFO test methods in a survey of the
fit of different commercially available underwear garment allow the fit
capacity of the garments to be judged.

CONTRACTOR : CENTEXBEL

ADDRESS : St.Pietersnieuwstraat nr 41
B - 9000 GENT

PROJECT LEADER : J. STRYCKMAN

CONTRACT N° :024-TEX-B

STARTING DATE : 1/10/1982

FINISHING DATE : 31/10/1984

TELEPHONE : 91/ 23.28.21
(ext. 2430)
TELEX : 20.183

TITLE OF PROJECT

QUALITY OF KNITTED FABRICS AND GARMENTS

AIM OF PROJECT

Quality aspects in the manufacture of knitted fabrics and garments. The aim of the research is to achieve a better control of knitting parameters, especially the length of yarn absorbed on knitting machines not equipped with positive yarn feeds. The research work comprises control of the length of yarn absorbed and the waxing of yarns.

SUMMARY OF RESULTS

Waxing of yarns

Six types of wax were studied and applied to six types of yarns. It appears that the waxed studies are composed of hydrocarbon chains with 21 to 38 carbon atoms. Good correlation has been found between the length of the chains and the physical properties : basic melting point and penetration of the different waxes. The amount of paraffin wax picked up by the yarn is also related to the chemical characteristics. Waxed yarns suffer only slight degradation during storage according to results of ageing tests. Generally speaking, it was shown that the rates of wax application, in order to be situated in the flat zone of the waxing curve, must lie between 0.4 and 1 % according to the type of yarn and the paraffin wax used.

Monitoring and control of the length of yarn absorbed on machine.

A prototype has been produced and installed on a flax knitting machine. The following components are included :
- a yarn delivery sensor consisting of coded small wheels actuated by the passage of the various yarns knitted.
- a carriage position sensor consisting of two hall-effect electronic register markers located in the centre of the active needlebed.
- an impulse on the knitting machine releasing the cut-out of the machine or by the instigation of the knitter.
- a monitoring unit consisting of a "single board computer" designed by Centexbel.

Machine trials have demonstrated the positive benefit of the system and have enabled the level of variance of the length of yarn absorbed to be reduced to plus or minus 1 %, although at the cost of numerous intervention actions by the knitter. A second prototype has been developed with an automatic control which reacts to yarn tension. This second device has undergone successful trials on an industrial scale.

The economics of the installation of the device on every machine must, however, be analysed.

CONTRACTOR : H.A.T.R.A.

CONTRACT N° : 032-TEX-UK

ADDRESS : 7 Gregory Boulevard
GB - NOTTINGHAM NG7 6LD

STARTING DATE : 1/10/1982

FINISHING DATE : 31/1/1985

PROJECT LEADERS : J.A. SMIRFITT
C.D. WARD

TELEPHONE : 602/ 62.33.11

TELEX : 378.230

TITLE OF PROJECT

QUALITY OF KNITTED FABRICS AND KNITTED ARTICLES :
UNDERWEAR, T-SHIRTS AND KNITTED OUTERWEAR

AIM OF PROJECT

The aim of the work described was to examine the perception of the "quality" of knitted garments by manufacturers, retailers and consumers.

SUMMARY OF RESULTS

Several hundred samples of knitwear, knitted underwear and leisurewear were obtained and tested for compliance with the corresponding U.K. manufacturers' or retailers' specifications. Wearers' views, measurements, and an assessment of the fit of the knitted garments they owned were also obtained in a U.K. survey which was subsequently extended to wearers in Belgium, Denmark, France and the Netherlands, covering some 1400 garments.

Computer programs were devised to handle the large amounts of data resulting from the surveys and laboratory measurements. Early in the work, T-shirts were identified as one type of garment that warranted more detailed study.

Examination of the specifications to which garments are produced demonstrates that the larger U.K. retailers and own-brand manufacturers show a large measure of agreement in their minimum performance standards, but that, since their test methods differ, except for colour fastness, much of this agreement is more apparent than real. There are considerable differences in sizing practice and demonstrable needs for improved methods of stretch testing and seam assessment. The purchasing specifications of the smaller retailers covered are totally inadequate and important parts of manufacturers' specifications are often too informal and communicated orally. Apart from manufacturers of fully-fashioned garments, few included stitch length in their specifications, though this is the key parameter in controlling production.

The laboratory tests and inspection of garments confirm that, manufacturers find it difficult to remain within currently specified size tolerances. There are considerable differences in sizing practice, relating body and garment measurements, both "as sold" and, after garment washing. With garments such as T-shirts some consumers, usually males, are concerned primarily with the relaxed chest measurement : others are more concerned with the maximum "comfortable" extended measurement.

Consumers cannot select garments giving the tightness of fit they require unless there is greater consistency in size marking. In the U.K., there is no generally available set of anthropometric data giving all necessary parameters and showing regional, age and sociological class variations.

Firms who have the financial and technical resources to assemble their own information are thus able to produce more reliably-sized garments.

There is a need for improvement in specifications, and for instruments for measuring neck stretch, so as to avoid high tensions on the head when putting on garments. The work also demonstrates that, though special garment forms and measuring devices can make a contribution to improving fit, there is a need for a quicker and more reliable way of making measurements on flat garments. Subsequent to the project, work has begun to develop such aids.

229

| CONTRACTOR : DANSK TEXTIL INSTITUT | CONTRACTS N° :035-TEX-DK |

CONTRACTOR : DANSK TEXTIL
INSTITUT
ADDRESS : Gregersensvej 5
DK - TÅSTRUP

PROJECT LEADER : T. SØRENSEN

CONTRACTS N° :035-TEX-DK

STARTING DATE : 1/10/1982

FINISHING DATE : 30/4/1985

TELEPHONE : 2/99.88.22

TELEX : 33.754

TITLE OF PROJECT

EFFECT OF PRODUCTION PARAMETERS AND PRACTICAL USE ON
DIMENSIONS AND STABILITY OF COTTON KNITWEAR

AIM OF PROJECT
To improve the possibilities of manufacturing knitted cotton articles with improved dimensional stability.

SUMMARY OF RESULTS
The washing shrinkage of knitted cotton fabrics is largely dependent on the drying method, line drying normally giving lower length shrinkage than tumbler. The shrinkage increases with the number of washes to a reasonably constant level after about 10 washes.

Greige fabrics show generally relatively high dimensional changes in washing, showing up as length shrinkage and either shrinkage or extension in width. These changes are influenced by the contractions or extensions taking place during dyeing/bleaching and finishing. Normally, extensions increase washing shrinkage, and contractions reduce it. In conventional finishing, i.e. processes not involving deliberate setting, low shrinkage fabrics may be obtained by letting the fabrics approach the shrunk dimensions, e.g. by allowing relaxation during tubular finishing, especially drying, and/or by using low tension and overfeed in stenter operations.

The relaxed dimensions are correlated with the reciprocal stitch length and probably also with the yarn count. Even conventionally finished fabrics show relaxed dimensions which are different from the corresponding greige. Precalculations of relaxed dimensions presuppose, thus, different equations for different finishing routes. The "Starfish" system developed by the International Institute for Cotton, on the basis of collected empirical data, contains such equations for interlock, rib and single jersey. Approximate calculations may be made in other cases on the basis of determinations of greige fabric shrinkage values, corrected for the influence of finishing by using approximate correction factors.

Spirality, which occur in single jersey and constructionally similar fabrics, may be counteracted by setting treatments. It may be practically eliminated by using S- and Z-twist yarns alternately, which in three thread fleece fabrics may be done without influencing the fabric's appearance.

SESSION D

UPGRADING OF LINEN TECHNOLOGY

Introduction by the Session Chairman
M. VAN LANCKER, Centexbel

Growth and preparation of flax fibre

Improvements and innovations in processing techniques
- New technologies for preparation and spinning and
the modification of products and processes

Improvements and innovations in processing techniques
- New approaches in the finishing of linen

Summary and conclusions by the Session Chairman

Contractors' sheets

UPGRADING OF LINEN TECHNOLOGY

M. VAN LANCKER
Centexbel

Flax is one of the rare vegetable textile raw materials produced in the European Community.

Here are a few figures to illustrate the role of flax in general and its importance to the textile industry in particular :

- flax represents less than 2 % of textile materials,
- in 1984 Community production of flax fibre (scutched flax and tow) was 79,881 T (137,000 T in 1962 and 82,100 T in 1982), and the production of classical spun flax more than   33,500 T (80,000 T in 1962 and 30,400 T in 1982).
- the balance of trade is strongly positive :
  . approximately 145 million ECU in 1984 for France,
  . approximately 70 million ECU in 1984 for Belgium.

The imbalance that has existed for a number of years between the production of flax fibre and the requirements of the classical spinners has forced dealers to turn to markets outside the EEC and the trade to seek new outlets.

For several decades the use of linen had been in decline and the various interests concerned, grouped in the I.L.H.C. (International Linen and Hemp Confederation), had worked out a programme of recovery and expansion that envisaged a vigorous drive in the areas of research, marketing and promotion.

The research programme, called "EUROLIN", was aimed at improving the methods of converting the flax straw into finished products so as to :

- economise on the raw material,
- create more satisfactory working conditions,
- exploit the intrinsic qualities of linen,
- cut production costs,
- produce article better suited to the consumer's needs.

On the basis of an analysis of the situation in the Community linen industry at the end of the '70s, its problems and shortcomings, an assessment of the measures already taken, and the results of previous studies, it was decided to urge the EEC, within the context of the 2nd Programme, to make proposals for financial assistance to textile research with a view to accelerating the completion of the EUROLIN project then in progress.

Many factors contributed to make this an especially well chosen moment for giving new impetus to this programme :

- the revival of consumer interest in natural fibres,
- the development of mixtures,
- the introduction of special finishing treatments,
- the new spinning techniques,

- the urgent need for the linen industry to re-invest,
- the existence of an extensive export market.

The research programme submitted to the EEC was drawn up by the Technical and Scientific Board of the I.L.H.C. working through a research and development committee, the Billaux Committee. This Committee includes the research centres interested in linen established in the countries of the EEC : ATPUL, ITL, ITF-Nord and ITF-Boulogne in France, Centexbel and OVLT in Belgium, TNO and IBVL in Holland, LIRA in Great Britain, Textil-Forschung Bielefeld in Germany,and the Stazione Sperimentale per la Cellulosa in Italy.

Each of these centres was invited to contribute its expertise in its particular area of specialisation but, for the purposes of the over-all programme, this made it necessary to provide for the coordination of the various individual contributions. Community participation accentuated the need to maintain a high level of cooperation. This was achieved through the Technical and Scientific Board of the I.L.H.C. and the Billaux Committee, a procedure which ensured that the work was oriented towards the real needs of the industry and that the technical and scientific results were rapidly disseminated and introduced into industrial practice.

An updating of the objectives made it possible to define the following technical goals. These have been translated into research programmes incorporated into the 2nd Community Programme of Research and Development in the fields of Textiles and Clothing under the general heading "Upgrading of linen technology".

TECHNICAL OBJECTIVES :

Raw material supplies and fibre extraction

To ensure a permanent supply of suitable raw material, the research effort must concentrate on improving cultivation techniques, fibre yield, the value of the byproducts and the presentation of the materials, as well as on labour-saving mechanisation.

Conversion of the fibre into yarn

In this stage it is necessary to adapt the fibre to permit the use of either high-productivity equipment or the new spinning techniques : OE spinning, friction spinning, Novacore, etc...

Finishing

Technical improvements and innovations in the productions and processes must be tried out and developed; it is also necessary to solve the general problems posed by linen finishing.

Product qualification

The qualification of linen in all its forms is necessary to establish the characteristics needed to determine the suitability of the linen for various purposes. Objective quality control must be achieved and supported by precise standards and specifications.

Fibre morphology

A better understanding of flax fibre morphology and the influence of production and working conditions is more than ever necessary in order to exploit all the potential resources of the fibre.

The "EUROLIN" programme, of which the "Upgrading of linen tech-nology" programme formed an integral part, is being pursued in the insti-tutes in conjunction with new promotional activities at the international level.

Thanks to these efforts in the fields of research and promotion, the linen industry of the EEC will be able to take its place in a modern technological and socio-industrial context, thus meeting the requirements of the EEC's industrial strategy.

The fruits of the new initiatives are already appearing at all levels and more especially in connection with the development of new out-lets and the production of high-quality products. The linen industry is preparing to meet the challenge and offer products sought by the customer on the world's markets.

GROWTH AND PREPARATION OF FLAX FIBRE

W.W. FOSTER
Lambeg Industrial Research Association

SUMMARY

This section of the presentations concerns recent work carried out at the European Research Centres under the Eurolin programme on flax and linen processing. It deals with the treatment of flax straw before, during, and after harvesting, and with the subsequent extraction and treatment of fibre. Developments described in this section have been aimed towards achieving the principal objectives of improving overall yield and quality of flax fibre, and reducing processing costs.

For convenience of presentation of results, research work on the processing of flax is generally described stage by stage, and in the order in which the process stages occur. It is emphasised, however, that the division of overall systems into separate process stages is rather artificial, since the seperate processes actually interact and relate to each other, and cannot strictly be considered in isolation. In this presentation, therefore, the results of work at any particular processing stage are described and then viewed in the light of the impact which they might have further along the processing chain. In particular, special note is made of process changes which could have an advantageous, or even an adverse, effect on the final product quality and cost.

In the same way that quality fabric results from good warp and weft, successful technology for its part results from an optimum blending of the interactions between the separate process stages. With this concept in mind, the work described below centres on the following four activities.

(a) Prevention of lodging
(b) Improvements in retting processes
(c) Reduction of contaminating materials in flax fibre
(d) Improved utilisation of all grades of fibre.

INTRODUCTION

Work carried out under the two-year Eurolin programme has been a team effort, as can be judged from the table below. By referring to this table, industrialists who wish to benefit from the work done at Technology Centres under the Programme can see which Centre to approach for more information about the technical work done, and about the commercial value of specific technologies. Ultimate success in all this work should come from the two-way flow of thinking - outwards from the Centres, and back again from industry - to ensure that work carried out is both relevant and becomes increasingly applied. This positive feed-back process must then continue, and one purpose of these presentations is to start the ball rolling.

ACTIVITIES	Research Centres				
	ITL	LIRA	Wagen-ingen	ATPUL	Centex-bel
1. In-field applications					
- anti-lodging chemical sprays	x				
- chemical spray desiccation/ dew-retting	x				
- chemical spray desiccation/ stand-retting		x			
- cutting/dew-retting for short staple fibre			x		
- straw-teasing techniques	x				
- contaminating materials				x	x
2. In-factory applications					
- scutching improvements				x	
- fibre treatments :					
. enzymes		x			
. chemicals		x		x	
- short fibre systems				x	
- lin-total				x	

IN FIELD APPLICATIONS

Five of the European textile research organisation have been active in work aimed at getting the best out of flax crops, and in extending the agricultural opportunities for industry throughout Europe. Thus, strategic factors as well as tactical factors have played a part in defining the work which was to be done. Sub-sections of this in-field work are outline below, which covers flax processing from quite early stages of growth, through the stages of retting, and finally to the stage of harvesting ready for scutching.

Anti-lodging chemical sprays

A considerable amount of knowledge has been built up over the years concerning the causes of lodging in flax crops and the factors which reduce the tendency to lodge. As a result of all the work, certain general rules are well-known. In spite of this knowledge, lodging occurs every year, resulting in wasted materials and wasted effort - simply a drain on Community resources. Crops are rarely a simple write-off, so that lodging in any crop has an impact at later stages of the process, usually increasing the costs and the variability of material, and reducing the general level of quality.

For some years it has been known that certain chemical sprays help to reduce lodging of flax. Figure 1 illustrates the effect of anti-lodging sprays. Work in this area under the Eurolin project has now made a positive contribution towards defining the best conditions for applying sprays, though it is fair to say that a great deal more remains to be learnt.

Figure 1

Three chemical products which are considered to be safe to use – Cerone, Atheverse and Terpal – have been tested. These are based on éthéphon. Wetting agents, which increase the effectiveness of the chemicals, have been tested, including Citowett, Sandovit, Etaldyne 95 and Rosemox, the last two producing a better yield.

These chemicals work by stunting and slowing the growth of plants and by thickening their stems; this results in some loss of fibre yield, and a general lowering of quality. However, these losses are not nearly as great as they are when crops actually lodge. Nevertheless, it is recommended that anti-lodging sprays should only be used as a last resort, so any action taken is very much a matter of judgement.

Treatment has been tested over six stages of height of flax from 30 cm onwards, but three stages have proved to provide adequate phases at which to apply control, namely 30/40 cm, 45/55 cm and 60/70 cm. Late treatment tends to lessen the shortening effect and is recommended. Furthermore, the higher the concentration of the liquid (for a given amount of chemical per hectare) the greater the retardation.

The greatest benefit is obtained from the treatment when equal parts of chamicals and wetting agent are used. All chemicals tested to date have been found to be effective, though the best conditions for using them are still to be determined. Bearing in mind the fact that 10 % of the total French flax crop was successfully treated in 1984, and that lodging remains a constant threat to crops, research in this important area will continue.

Chemical spray-desiccation as an aid to retting

The retting of flax as a means of freeing the fibre from the woody material prior to the physical separation by beating and combing is a centuries-old process which has remained essentially unchanged until relatively recent times. In the last two decades, chemicals have been tested to determine whether these could be used in the field to speed up and to control retting. Initial tests in the 1960s were based on chemical defoliants such as Diquat, but these proved to be unsatisfactory because the plant died only where the spray touched – at the top – leaving the rest of the plant alive. This merely reduced yields and brought no benefits so far as retting was concerned.

About five years ago work started on the use of systemic herbicides like glyphosate. In this process, crops are sprayed with a glyphosate preparation such as Roundup, and the plant absorbs and transfers the chemical within itself causing it to die slowly and to dry out in such a way that it continues to stand up in the field. As a result of the drying process, the plant becomes receptive to natural micro-organisms which are preent on the surface of the plant, these micro-organisms invade it, and the chemicals which they in turn produce (enzymes) proceed to attack the pectins and resins which bind the fibre to the woody material. At the end of this process, the plant is both dry and retted before it is pulled.

The first trials using this technique took place in Northern Ireland, where, for a variety of reasons, flax had not been grown for 30 years. Dew-retting had never been a practical possibility because the weather was too damp and cool, so the mechanisations which were possible on the continent were not at that time open for development in Northern Ireland or in the rest of the U.K. The traditional process for Northern Ireland - water-retting - had become too labour intensive in spite of attempts to mechanise, it was too energy-consuming, and it created problems with effluents.

As a result of close collaboration with other UK research organisations and with industry, progress was sufficiently encouraging to justify EEC research expenditure. Figure 2 shows the effect on flax of spraying with glyphosate.

Figure 2

The optimum rate of application was achieved with a 1 % solution of glyphosate in water, applied at the rate of 400 litres/hectare. It was found to be best to spray about two weeks after the mid-point of flowering, the actual date varying with the weather in the period some weeks before spraying, as well as around the time of spraying. In a relatively dry growing season when plants tends to go dormant, early spraying is necessary in order to achieve adequate translocation of the chemical throughout the plant. In a wet season on the other hand, spraying may be delayed as the plant is still active at a later date, and fibre is still being developed.

Using this process, retting of the drying plant on foot can occur. However, the rate of retting depends again on weather conditions. In very dry weather, retting proceeds very slowly, but when it is misty or wet, and the weather is warm, retting proceeds rapidly. Once retting has got well under way, the flax is usually in such a physical state as to dry rapidly when weather conditions turn dry, and the flax must then be pulled ready for baling and storage, otherwise over-retting might occur if the crop gets wet again.

The work has shown that the microbial population on desiccated flax retted on-foot was similar to that found on dew-retted flax, but there are some differences from year to year due to variations in weather conditions.

A further important finding is that the application of glyphosate retards lignification, which ceases between 1 and 3 weeks after spraying. This helps towards producing fibre which behaves well in later processing. Another positive aspect of glyposate-initiated retting of flax on-foot is that the fibre-to-fibre adhesion is reduced, thus enhancing its spinning properties. On the other hand, it would appear that care must be taken to avoid using more than 400 litres/hectare of glyphosate solution, otherwise there is a danger of defects occurring which could make it difficult to get the best out of fabric finishing treatments.

In parallel with research trials, commercial quantities of flax have been grown in Northern Ireland reaching 400 hectares in 1985, and in Scotland which has now reached 200 hectares. Scutching facilities are also planned to deal with future crops. However, there is still a great deal to be learnt about the on-foot retting process, and as is usual with agricultural products, it is strongly dependent on a wide variety of natural conditions, particularly the weather. But the work has opened the way to flax growing in the United Kingdom, and only time will tell whether a satisfactory retting process or processes can be devised.

Trials with chemical sprays on the continent have so far been at the non-commercial level. These trials were aimed at reducing the times of drying and of dew-retting of flax, and at improving the fineness of fibre. The process would be an aid to normal dew-retting, particularly in wet weather conditions. Work carried out so far has been in the dry seasons of 1983 and 1984, when it was found that the drying time and the dew-retting times were not reduced, though there is evidence of some improvement in fibre fineness.

Two chemicals have been used in these trials, Roundup and Harvade, Roundup being the better of the two. There is some evidence that Harvade appears to kill the plant within a few days, but that the plant then recovers.

Roundup on the other hand takes about three weeks to show its full desiccant effects, but once a plant has been affected, it does not recover. It should also be noted that there are now some legal restrictions on the use of Harvade in the U.K.

Removal of contaminant materials from flax straw

Dew-retted flax is prone to contamination with weeds, some being pulled with the flax, and others growing through the pulled straw as it lies on the ground. Stones are also caught-up in the straw during turning and subsequent lifting for baling. The weeds adversely affect scutching and also lead to contamination of the fibre by foreign materials which themselves interfere with spinning, weaving and finishing processes; all these factors increase the costs of processing and reduce the quality of final products. Stones can also damage scutching equipment and are a serious hazard.

Work has been carried out directly on the removal of the weeds and stones which contaminate dew-retted straw, and simple mechanical devices have been developed which lead to a significant improvement in this situation. The first device which is shown in Figure 3 is designed to lift up dew-retted flax as it lies in the field, so teasing out the living weeds, shaking out the stones, and also enabling air to flow through the straw and allowing drying to take place. About 200 machines are already in commercial use, a prime example of a practical, simple machine which has been designed to carry out a valuable job.

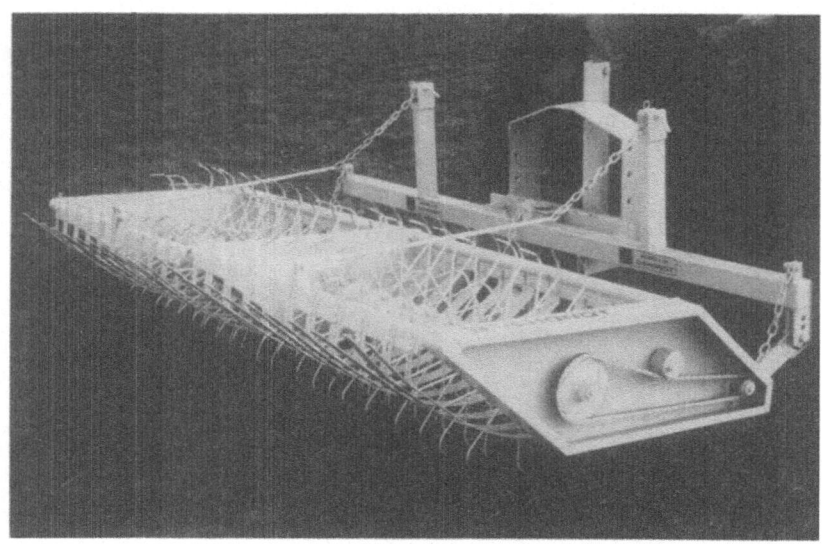

Figure 3

A further device shown in Figure 4 is an attachment which was designed to be fitted to the baler, enabling 85 % of the stones to be shaken out before the flax is baled. A normal bale can obtain 8-10 Kg of stones, so this reduction in the quantity of stones provides great benefits further along the process chain.

Figure 4

Other activities concerning the removal of contaminants have centred on polypropylene and on the identification and removal of the great variety of weed fragments which occur in flax tow. Work in these areas straddle a range of processes from the sowing of the flax seeds to the final finishing of fabric. For convenience, this work is discussed in this section.

So far as polypropylene is concerned, this material come from a wide variety of sources – not just from binder twine – and its removal from the agricultural scene must surely be a pipe-dream. Recognising this fact, methods of showing the presence of polypropylene in flax fibre have been developed, but it must be admitted that ways of removing it are yet to be devised. Meanwhile it will remain a serious and vexing problem.

So far as weed fragments are concerned, extensive work has been done in identifying the great variety of natural weed impurities in flax tow, and in investigating how these impurities might be eliminated at various stages of industrial production. A catalogue has been prepared, indicating some typical morphological and chemical characteristics for each type of impurity, and the methods of identification have been carefully described. The final document, which will include photographs and an explanatory text indicating some typical characteristics for each type of impurity, will become extremely valuable for use in industry and research.

Studies of the behaviour of natural impurities to mechanical and chemical treatments indicate that some contaminants cannot be completely eliminated by the usual mechanical treatments, although certain mechanical operations such as pre-carding, carding and combing can minimise the risks.

This conclusion is of industrial importance, since subsequent chemical treatments such as alkaline boiling, scouring and bleaching are not sufficient to de-colourise the major types of impurities. Further studies of the behaviour of contaminants during dyeing (direct, reactive and vat dyeing) over a range of three colours and four concentrations indicate that most contaminants cannot be dyed to the same intensity and hue as normally retted flax fibres.

It is concluded that the only real solution to the problem of weed contaminants in flax is to reduce the amount present by means of weed control in the field, coupled with the elimination of weeds during pick-up from the field (like the systems already described above), and intensive mechanical treatment of the straw and fibre when the presence of impurities might be expected. In other words, prevention is better than cure. These solutions are not usually available, of course, for flax tow purchased on the world market.

Cutting and dew-retting for short fibre

Work has been devoted to the development of techniques which would dispense with the use of specialised flax machinery such as pullers, and enable instead normal farming machinery to be employed up to and including the stage of baling. In essence, the flax straw would be cut (mowed), not pulled, and the resulting swathe would be matured and dried on the ground, then threshed by means of a conventional combine to remove the seed, the straw then being left to dew-ret in tangled mass in the field, and finally baled. In subsequent processing, the bales would be teased apart, short fibre separated, then refined and used in short-fibre outlets (see figure 5).

BLOCKDIAGRAM OF A UNIVERSAL HARVESTING AND EFFICIENT PROCESSING OF FLAX TO MIXED-YARN-FIBRE.

AGRICULTURAL LINE.

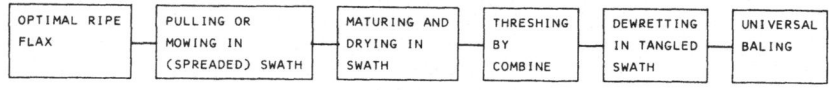

OPTIMAL RIPE FLAX	PULLING OR MOWING IN (SPREADED) SWATH	MATURING AND DRYING IN SWATH	THRESHING BY COMBINE	DEWRETTING IN TANGLED SWATH	UNIVERSAL BALING

FINAL PRODUCTS: -LINSEED
-DEWRETTED STRAWBALES

SHORT FIBRE-STRAWPROCESSING.

BALE OPENING AND FIBRE LOOSENING	(SHORT)FIBRE/SHIVE SEPARATION	FIBRE CLEANING	FIBRE REFINING/SHORTENING	SPINNERY PREPARATION

PRODUCTS: -STAPLE FIBRE (COTTON SYSTEM)
-SHIVES

Figure 5

Preliminary field trials and laboratory studies have been carried out in 1983 to verify that the basic concept is satisfactory. In 1984 it was shown that flax can be mown and that combine-threshing yields first-class linseed. It should also be mentioned that pulled flax threshes well. Dew-retting of the tangle straw dropped back onto the field was not entirely satisfactory, there being a good deal of over-retting due to wet weather.

A pilot line is being developed, but meantime, trials using improvised bale-opening and fibre cleaning systems produced short fibre for use in flax blends. Preparing and spinning tests on 20 Kg quantities of short flax fibre, designed to produce 20/80 flax/cotton yarn to Nm 20 are promising. First results indicate that the process could be technically and economically sound, producing materials which are competitive with those currently available.

IN-FACTORY APPLICATIONS

Three of the research organisations have been actively involved in studies aimed at improving the performance of scutching turbines, at increasing fibre yields and quality through physical, chemical and biological processes, and at developing better short-fibre systems. Some of the work done on short-fibre systems is described immediately above, and is not dealt with again in this section.

Scutching improvement

A considerable amount of valuable work has been done under the programme on improvements to scutching turbines, including modifications to the feed mechanisms to accept round or rectangular (high-density) bales, better breaker rollers (as shown in Figure 6) and beating arm sections (Figure 7).

Figure 6

Figure 7

A machine has been devised for eliminating seed-capsule "branches" from the straw pad entering the scutching machine so as to upgrade the quality of tow. This has been one of the most significant recent developments, but details of the equipment have not yet been made public. Scutchers seem anxious to use the system as soon as possible.

It has also been shown that an increase in the yield of long fibre by as much as 15 % can be achieved if a scutching turbine is carefully tuned so as to modulate the intensity of breaking by adjusting the number of fluted rollers, and by accurately regulating the speed of the turbines. Scutchers have already benefitted from these findings, and are applying the techniques successfully.

New fluted rollers made a special steel for longer life, and redesigned profiles, have also proved more efficient, and have been adopted commercially by a number of users.

A better gripping action by the scutcher conveyor belts has been designed, leading to increased yields of long fibre. A new form of flail scutcher has also been tested, indicating some good features, but highlighting the need for further development in order to get the best out of the ideas.

Fibre treatment

Detailed studies of the micro-organisms present on Roundup-desiccated flax-straw, and of the interactions between fungal colonies and specially prepared bacterial suspensions on flax, have opened up possible ways of treating flax straw and fibre under factory conditions. The work under the Eurolin programme did not reach an advanced stage of development but it did show that enzyme mixtures might be developed for speeding up retting and fibre separation. These processes would be likely to take hours in-factory rather than weeks in the open air, and to produce more uniform and more easily processed fibre of higher quality.

At present, commercially available enzymes are too expensive to make the processes economic, but further studies are under consideration with a view to bringing down costs, and to quantifying the commercial advantages of producing higher quality, more uniform fibre which can be more completely utilised than at present.

Similar work has also been carried out on the chemical treatment of fibre, with the same objectives in view.

Short-fibre and lin-total systems

Scutchers and spinners have expressed interest in the lin-total route, so this work has been reinstated, and one of the most important features is a new apparatus for removing seed capsule branches from flax tow.

Progress has been made at a number of research centres in the preparation and sub-division of flax fibres which are destined to be processed on cotton and wool systems. One programme which has been pursued concerned the use of low-grade flax tow for spinning blended yarns. These materials contain many natural impurities and coarse flax fibres, so to achieve better utilisation it is necessary for the tow to be cleaned and refined. Cleaning and fibre refining efficiency have been studied as a function of machinery parameters, numbers of successive operations, and divisibility of the fibres.

It was found that mechanical preparation of low grade flax tow could modify considerably the fibre length distribution. Flax tow whose fibres were difficult to divide could not be made significantly finer, but the mean fibre length was considerably reduced.

Fibres cut mechanically to short lengths contained high percentages of very short fibres which were lost in the process stages before spinning, resulting in low yields. Because of the wide distribution in fibre length from mechanically treated low-grade flax tow, direct use of these treated materials on conventional machinery in the spinning mill was not possible.

Two different process options for treated low-grade flax tow have been considered, one for material going to a cotton system, and the other for material going to a wool system.

Blends with different types of fibres (viscose, acrylics, polyester and polypropylene) at three levels - 25 %, 50 % and 75 % of flax - have been prepared. Trials are not yet complete, but it is likely that the use of low grade flax tow will be limited to blends with a maximum of 50 % of lowgrade flax fibres.

CONCLUDING REMARKS

In this presentation, the point has been made that successful technology results from an optimum blending of the interactions between separate process stages, in much the same way that quality fabric results from good warp and weft. Examples have been given of machinery and processess which industry has adopted, and others will almost certainly evolve from the work done.

The success of the project should not be judged in isolation. The project started on the firm foundations of decades of technological development, it was carried out in parallel with other related activities, and the thoughts it raised will gradually be absorbed into future work.

It is hoped that as the rest of the work which has been done is described, the value of continuing integrated involvement will to be truly recognised.

IMPROVEMENTS AND INNOVATIONS IN PROCESSING TECHNIQUES

**NEW TECHNOLOGIES FOR PREPARATION AND SPINNING
AND THE MODIFICATION OF PRODUCTS AND PROCESSES**

M. VAN LANCKER
CENTEXBEL

INTRODUCTION

   The principal activities evolved by four research centres; Bielefeld, Centexbel, I.T.F. and LIRA within the "upgrading of flax" programme are given in summary form in the table below. This work was undertaken with the aim of bringing about technical and economic improvement in the processes of converting the fibre into finished end-products in order to effect economies in raw material, utilise the intrinsic qualities of flax, reduce production costs, and make available products more suited to consumer needs.

   The various projects have been undertaken by the research centres at several complementary levels to cover the priorities expressed by the linen industry.

ACTIVITIES OF THE RESEARCH CENTRES

Activities	Research Centres			
	Biele-feld	Centex-bel	I.T.F.	LIRA
1. Utilisation of low-quality raw materials	x	x		
2. Problems created by new production systems :				
- optimisation of stretch-break converting		x		
- preparation and divisibi-lity of the fibre			x	x
- roving treatment, fibre dissociation			x	
- economic aspects of a new processing line			x	x
3. New spinning techniques				
- rotor spinning	x			
- friction spinning		x		
- Novacore			x	
- zero-twist spinning				x
4. Modification of products and processes			x	x

1. - UTILISATION OF LOW QUALITY MATERIALS

1.1. Foreword

Every year, a proportion of the textile flax crop is unusable for conventional flax spinning. This proportion consists largely of coarse fibres and low quality scutcher tow. Despite what means are available at the present time for controlling flax cultivation and ensuring high fibre quality, inevitably a proportion of the harvest is unsuitable for conventional flax spinning. The principal reason for this is the that climatic conditions over which one has no control have a critical influence on the quality of the flax.

In the spinning of tow on wool type machinery, to obtain clean raw material the slivers from the first or second gilling are processed on the comb. This operation produces a particular type of fibre waste called "noils". This material which has up to the present been destroyed or sold into the merchanting trade does, however, contain a lot of fine fibres of which use could be made in the textile cycle.

1.2. Materials of low quality

A conversion system developed for these materials consists of mechanical reduction (refinement) and cleaning on a Laroche machine (Figure 1), and drawing and spinning on the tow or short staple fibre system.

After processing a cut scutcher tow on a Laroche machine in a planned way with optimised machine settings, the percentage of spinnable fibres (Figure 2) has increased significantly and fibre fineness (Figure 3) has also improved along with the fibre length diagram (figure 4).

Figure 1

Laroche system for the treatment of flax fibres

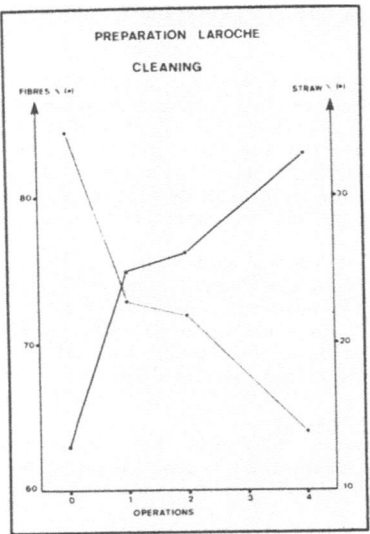

Figure 2
Laroche fibre preparation : cleaning of fibres

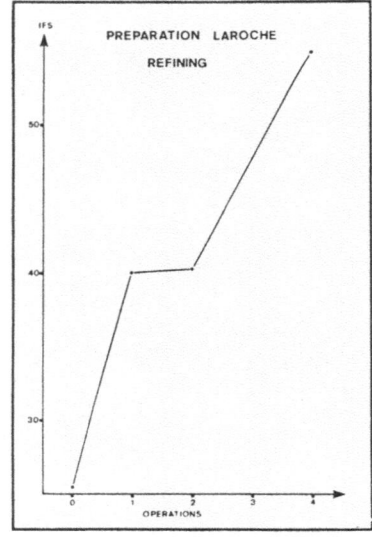

Figure 3
Laroche fibre preparation : Reduction (refining) of fibres

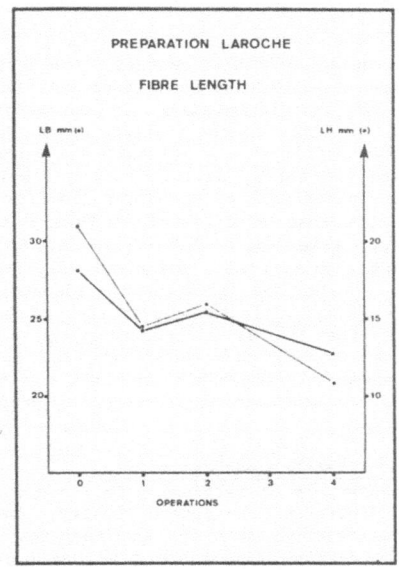

<u>Figure 4</u>
<u>Laroche fibre preparation : Fibre length</u>

At the present state of progess of research, it may be concluded that the Laroche system combined with a fine opener and cleaner/beater permits a significant degree of reduction to be achieved together with a considerable level of cleaning.

The second stage of research, drawing prior to spinning and the spinning of the processed materials has, however, revealed three basic problems. These are lack of balance in fibre length distribution, a significant heterogeneity in fibre fineness, and an impurity and dust content too great for viable drawing on cotton type short-staple machinery. The traditional "tow" system therefore appears to be the most appropriate.

The results of dry spinning trials have demonstrated that economically viable spinning cannot be achieved with 100 % flax and for materials of very low quality the percentage of flax shall be limited to 50 %. Fibres that may be envisaged for blending are viscose, polyester or acrylics.

1.3. Flax noils
A significant percentage of noils is obtained in the combing operation. This material contains predominantly short fine fibres. Utilisation of this material in the textiles processing cycle has been studied in the first place on a cotton type short-staple system with rotor open-end spinning (see 3.2.1.) and secondly on a wool-type staple system and friction spinning (see 3.2.2.)

2. - PROBLEMS ARISING FROM THE NEW PRODUCTION SYSTEMS

2.1. Foreword

Since the beginning of this century, the principle of drawing machinery - drawframes - has hardly changed at all. So-called traditional machinery possesses the twofold advantage of progressively and thoroughly dividing the fibre bundles in the faller zones and calendering the attenuated sliver into a thin web to initiate the subsequent splitting of the bundle. Here are two terms, divisibility and splitting, which will play a dominant role in the work of preparing flax for spinning.

It is unfortunate that traditional machinery also has its draw-backs : it is heavy, takes up a lot of space, is of relatively low productivity, is costly to maintain, and above all, is expensive to buy.

New solutions sought and studied by industry assisted by the research institutes have included conversion of the long fibre by stretch-breaking in order to allow processing on wool-type machinery, providing a satisfactory way of processing flax tow; designing by a textile machinery maker of compatible machines modified for processing the long fibre, and the improvement of spinning properties by chemical treatment of the roving.

2.2. Optimisation of stretch-break converting

As the flax industry is of marginal significance on a world scale by comparison with, for instance, the cotton idustry, most makers of textile machinery are not interested in flax, and therefore the technology of some machines for converting long-fibre flax has hardly changed at all during recent decades. In view of this situation, it would appear that the future of long-fibre flax and consequently of flax as a whole is at risk in the long term. In an attempt to find a viable solution to this, an alternative processing routine called the "high productivity system" has been developed as a result of research.

The stages of the process are the formation of a sliver at the scutcher delivery, equipment for which has been studied and developed by research, followed by stretch-break conversion, and by drawing on wool-type drawframes.

This system, conceived with the aim of finding a route halfway between the traditional long-filbre system and the tow system, has not yet aroused the interest that had been hoped for, because of a series of technical problems and because of product evolution.

The aim of the study undertaken within the framework of the EEC programme was to find viable solutions for the technical problems. On the basis of a theoretical analysis of stretch-break conversion and the results of preliminary trials, it has been possible to identify the parameters of importance relevant to the control of the system. These include the coefficients of friction (fibre-to-fibre, fibre-to-metal and fibre-to-rubber), the type of roller covering, the presentation and characteristics of feed slivers, the machine setting as a function of the properties of the raw material, and the total loading in stretch-break converting.

An experimental study performed on a "Duranitre R4" stretch-break converted has permitted the following investigations and conclusions to be made :

1. The influence of fibre additives was studied on the coefficients of friction and on the action at the stretch-break stage and more especially on the fibre length distribution. It was demonstrated that

the moisture regain of the slivers has a significant influence on the fibre-to-metal friction and that a slight reduction in the coefficient of friction results in an increase in the percentage of fibres escaping stretch-breaking. The maximum possible coefficient of friction is therefore desirable.

2. The influence of roller covering and in particular the hardness of rubber covering on the nip exerted on the fibres was investigated. The choice of type of covering is restricted especially by its useful life under industrial operating conditions.

3. The influence was studied of the various machine setting parameters (nip, pressure, size of draft, doubling, loading) on the characteristics of stretch-break converted flax. Some of these settings have a significant influence on the end result of the stretch-break conversion process.

This study has also enabled possible modifications to made to the stretch-break converter with the objective of controlling the process. Generally speaking, it appeared that the first generation of machines, despite a programme of action taken at the machine setting stage, had deficiencies especially : a badly balanced fibre length distribution of the stretch-break converted fibres, excessive nep formation, and inadequate dust removal.

In collaboration with a machine maker, a new stretch-break converter (Figure 5) has been studied and developed. A prototype of this machine is operating in industry. For economic reasons, the machine maker concerned has ceased all textile activity. The drawings of the machine have, however, been taken over by a large textile machinery manufacturer who plans further development. In a final stage of this study, spinning trials have been carried out to determine the spinning properties of the stretch-break converted flax and the quality of the end products. The progress achieved by the use of a second-generation stretch-break converter and optimised operation in stretch-breaking is clearly demonstrated in Table 1. In conclusion the principal technical obstacles have been overcome but the acceptance of this alternative system is an economic question : raw material costs, the market for the end-products, and the availability of the stretch-break converter which depends on a decision by the machine maker.

2.3. Preparation of flax and divisibility

As one of the participants in the project possessed both traditional machinery and the new high productivity machinery for long fibres, a study was carried out of the influence of different mechanical factors for each type of machinery. In addition, a comparison was made of the two types of machinery and of the characteristics of the end product. An attempt was also made to improve the operating conditions of the modern machinery.

2.3.1. Determination of divisibility

The factor which largely governs end-product quality is the propensity of the flax to be divided by mechanical action. This characteristic is given the term "divisibility" and is a basic property affecting the spinning. As a result of research, objective methods are now available for the measurement of this property. A visual estimate may

Figure 5
"Pierson and Digneffe" stretch-break converter

STRETH BREAKING LONG FIBRE	REFERENCE NUMBER					
	1*	1	26*	26	27*	27
Count (tex)	155.0	145.0	110.0	112.0	160.0	160.0
Tenacit (N)	19.5	20.1	21.2	18.9	30.5	28.0
CV (%)	19.0	22.4	15.9	16.8	17.2	18.5
RKM (CN tex)	12.6	13.8	19.3	16.9	19.0	16.9
U (%)	17.0	19.2	19.5	20.3	18.5	20.9
Defects (1000 m)						
. thin places	1,100	1,538	2,100	3,003	1,850	2,130
. thick places	55	94	100	115	210	147
. neps	40	88	65	95	75	321

* values for optimal stretch-breaking

Table 1
Yarn properties of yarns : stretch-break converted fibres

be obtained just by looking at cut fibre sections but this is inadequate and a numerical value is required. For determining the divisibility of a flax sample, the fineness of the fibre components is measured at two or more stages of production, using airflow apparatus by the ISO standard method. As the apparatus is calibrated, the result is expressed as index of standard fineness (ISF). The samller the ISF, the finer the fibre. To determine the fibre division caused by a process, the ISF fineness of fibres is measured before and after this process, and the rate of splitting is calculated by the formula :

$$\text{Rate of splitting} = \frac{\text{ISF before}}{\text{ISF after}} - 1$$

2.3.2. Experimental work
Different raw materials were processed on two types of long-fibre machinery. One the traditional "Mackie" machinery (Figure 6) with a double row of pins, includes calandering by wooden rollers on the drawframe and doubling at the front of the machine. The other high-productivity machinery of the GSL type (Figure 7), has a single row of pins and thick sliver drafting by rubber rollers with doubling at the back of the machine.

On each type of machine, study was made of the influence of the spinning, loading in the gill-boxes, loading on the drawframe and amount of draft.

Results and conclusions
The evolution of fineness (ISF) during preparation for spinning is presented for the various raw materials (Figure 8). It is found that the division of flax occurs principally in preparation and here the machine technology and the machine settings adopted are very important factors. The influence of spinning may differ markedly from one flax to another. Some types of flax are more responsive to pin action whilst others cannot be properly separated by mechanical action and chemical treatment has to be adopted.

Generally speaking, a good progression in pin density per unit area is beneficial. For the GSL type high-production machinery, it is not at present possible to obtain a high density per unit area despite a high lateral pin population.

Influence of the drawframe loading and the loading of the pins
The results show that with constant loading on the drawframe it is advantageous to have a minimum loading in the pins and a sufficient amount of draft. Yarn properties are consistently better when drawframe loading is at the lowest possible level : 0.4 g/m-cm on the the traditional machinery and 0.65 g/m-cm on the high productivity machinery.

As in the case of pin density, the influence of loading may differ markedly from one type of flax to another. The degree of retting and the origin of the raw material are the governing factors.

Comparison of the two types of machinery
In all cases with equivalent pin dessnity, the degree of reduction (refinement) and thus the rate of separation (Figure 9) are distinctly inferior on the GSL high productivity machinery.
The yarn properties (Table 2) are consistently better when the material has been prepared on traditional machinery.

254

Figure 6
Traditional spinning preparation machinery - Mackie

Figure 7
High-productivity spinning preparation machinery - NSC type GSL

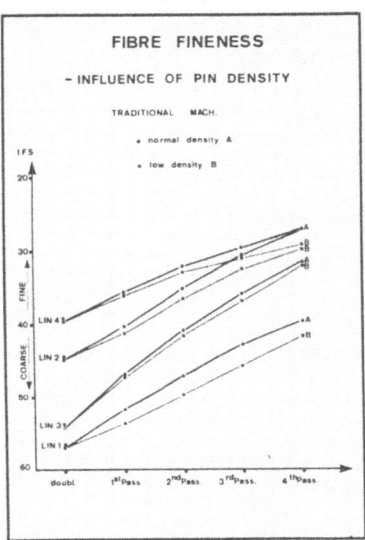

Figure 8
Evolution of fibre fineness in spinning preparation
traditional machinery

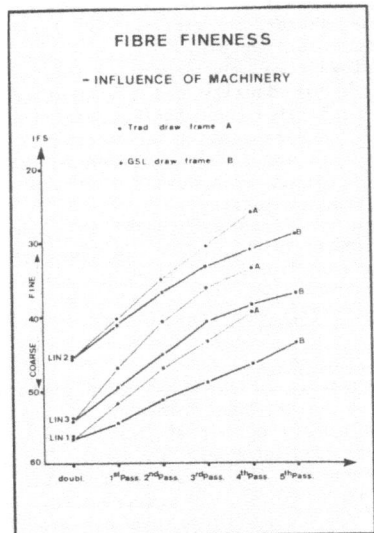

Figure 9
Evolution of fibre fineness in spinning preparation
machinery comparison

Table 2
Yarn properties (dry spinning) machinery comparison

(Nm) yarn	YARN PROPERTIES	TRADE MACHINE	GSL MACHINE
9 1/2	Ténacit (RKM)	30,9	27,4
	VC % forces	20,3	24,9
	CV % mass	27,3	31,2
	Defect / 1000 m	43,0	75,0
12	Ténacit (RKM)	28,7	25,8
	VC % forces	24,0	27,8
	CV % mass	29,6	32,9
	Defect / 1000 m	61,0	102,0

Conclusions on divisibility

For the trade in general and the spinner in particular, the thing of primary interest is to know what the numerous factors are on which the divisibility of flax depends.

At at the flax production stage these factors include the varietes, and the conditions of cultivation. Also involved are the bioclimatic conditions, and the extent of retting but these are less capable of control than the first two factors.

At the processing stage, the governing factors are the actual design of the machinery, the pin density and the loading. As a general conclusion it is not yet possible to use the GSL high productivity machinery as a complete set for the preparation of materials intended for the dry spinning of intermediate and fine counts and for the wet spinning of fine counts. Machinery of traditional design thus remains the most effective from the aspect of product quality.

In the development of new machinery, it is now possible to draw up the specification taking into account the conclusions of the studies and the observations made in industrial trials.

2.4. Conventional spinning processes

2.4.1. Treatment of the roving

Although wet spinning of flax has been performed on treated roving for at least fifteen years, the treatment of roving continues to be a subject of interest and is more than ever worthy of investigation for a variety of technical and economic reasons. The main purpose of this treatment is to loosen the individual fibres within the fibrous bundle. This dissociation is achieved by treatment of the roving before the spinning operation. In doing this, the cements welding the individual fibres together are partially removed. As the spinning characteristics are governed by the extent of this dissociation, it is vital to know the extent to which it has been achieved. For this purpose an original piece of apparatus, called the "Tenometre" has been developed (Figure 10). Its use permits the simulation of precisely what takes place on the wet spinning frame.

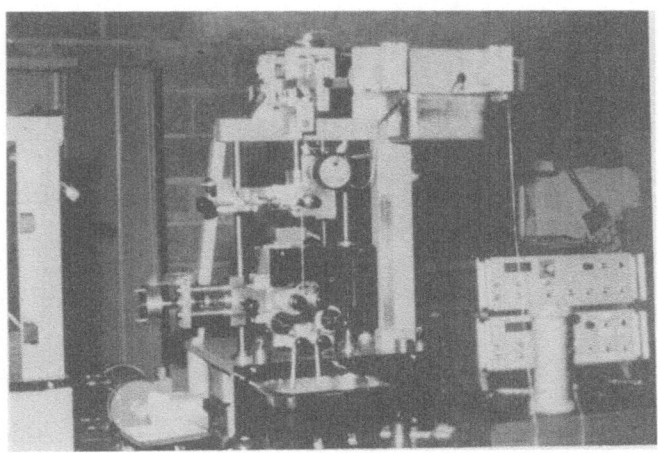

Figure 10

The resistance to drafting is measured at different ratchet settings, and it is then possible to trace the "dissociation curve" (Figure 11).

The interesting feature for the spinner is to know which factors govern fibre dissociation. They are numerous, but factors inherent in the raw material include the variety, the degree of retting, and the bioclimatic conditions. Factors relating to the chemical treatment are the process used and the conditions applied there in.

The influence of variety is important as is demonstrated by the example comparing the three varieties 1, 2 and 3. The experiments showed that the resistance to drafting of variety 1 is appreciably greater than that of varieties 3 and 2 (Figure 12).

A very large number of experiments following a factorial plan has enabled conclusions to be reached allowing the spinner to optimise roving treatment and to take account from the outset of the "raw material" factor. The basic conclusion is that it has been demonstrated by research and expressly acknowledged by the spinning industry that, for a given type of spinning frame, the attainment of good operating conditions in spinning is intimately related to a well defined form of the dissociation curve.

The tenometre is now available to incudstry and several models are in service in Europe.

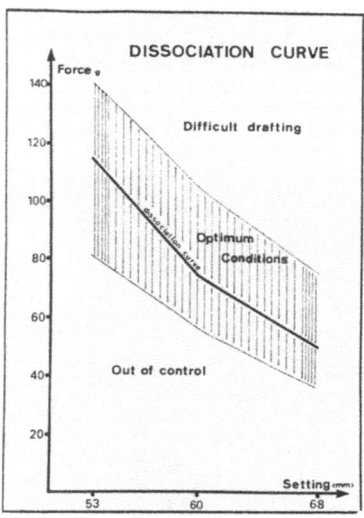

Figure 11
Dissociation curve

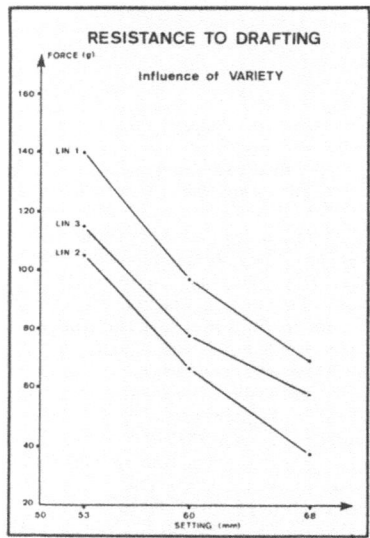

Figure 12
Resistance to drafting
Influence of variety

2.4.2. Wet spinning frame

One of the roles of research in this field has been the systematic comparison of the performance of the new machinery with that of the traditional machinery. It transpires that controlled drafting systems considerably improve yarn quality including tenacity, eveness and cleanliness as well as spinning performance as indicated by the number of end breaks (Figure 13).

In contrast with the case of drawing, the new spinning technology has the advantage over traditional machinery. In the near future it may be expected that controlled drafting systems will find general acceptance.

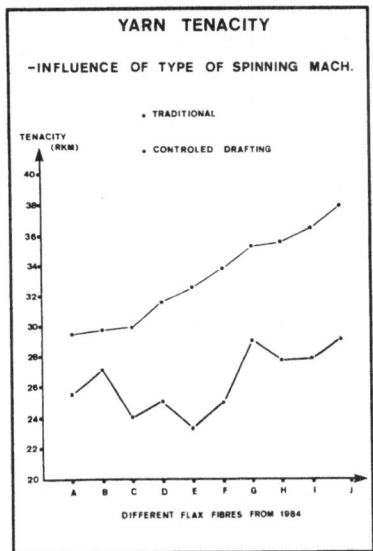

Figure 13
Yarn tenacity : influence of type of spinning frame

2.4.3. Economic aspects of a new production system

Studies on the divisibility of flax in preparation for spinning have shown the present limit of technological development in machinery. Neverthless, industry feels the need to renew its machinery stock. Looking at an entire replacement by new machinery, one of the research institutes has collaborated with a machinery maker in studying the capital costs involved and the production costs for several typical yarns. Although production would be very significantly improved by a new machinery line, the fact remains that amortisation would have to be spread over almost ten years, which causes the industry to resist such a major investment.

3. - NEW SPINNING SYSTEMS

3.1. Foreword

During the past ten years the European spinning industry has experienced a reduction in production, forcing merchants to turn to markets outside the EEC and spinners to seek new outlets. With its two preferred methods of spinning, dry and wet, both for long-filament and tow, flax has remained in the background of technological change.

Apart from one unique case involved in "Eurolin" research, trials carried out to promote the high productivity technologies have remained limited as far as traditional flax is concerned. The spinning systems for short-staple fibres show such high production potential that conversion to cotton techniques appears to be an economic necessity for flax. Once converted to the state of a reduced (finer) shortened fibre or a single fibre, the use of flax on cotton machinery may be envisaged.

3.2. Open-end spinning

Ever since the introduction of the first rotor open-end spinning machine, spectacular development has been the feature of this type of spinning. Other so-called open-end techniques have since been developed, including friction spinning.

Generally, open-end techniques permit much higher productivity to be attained than with ring spinning. Furthermore, some of these techniques enable special yarns to be produced by a direct route, such as core-spun yarns, effect yarns and composite yarns.

These modern techniques have been developed especially for short-staple spinning or for the spinning of wool-type fibres. The use of flax in these production lines, particularly on the cotton system, demands special attention in the selection of the flax to be reduced and in the preparation of the fibre product.

The research project include the preparation of an optimum fibre product, and the use of the converted fibre in pure or in blend form in new spinning processes.

3.2.1. Rotor open-end spinning

As far as reduction (refinement) is concerned, research has enabled a 33 % increase in machine productivity to be achieved by the adoption of suitable roller coverings and settings. Improvement in the quality of the reduced fibres is being pursued in the EEC research programme in three directions aimed at countering the deficiences of retting by mechanical, chemical and physico-chemical, biochemical means.

In one study the research results have led to an application for patent protection. Some classification criteria for the reduced fibres are now available (fineness, length, strength and cleanliness).

Experimental work : results and conclusions

Various processes for the preparation of the fibre product have been studied. The various routines differ according to the raw material and the sequence of operations.

1. Raw material : comber noils.
   Operations : opening, cleaning, reduction, blending (with other fibres), drawing and rotor open-end spinning).
2. Raw material : scutcher and comber tow.
   Operations : blending, carding, cutting, cleaning and reduction, blending (with other fibres), drawing and rotor open-end spinning.
3. Raw material : scutched long-fibre flax.
   Operations : cutting, cleaning and reduction, blending (with other fibres), drawing, and rotor open-end spinning.
4. Raw material : technical flax fibre.
   Operations : chemical treatment, drying, opening, blending,

261

drawing, and rotor open-end spinning.
5.   Raw material : separated fibre.
      Operations : opening, blending, drawing, and rotor open-end
      spinning.

      The yarns produced were of yarn counts Nm 12 (84 tex) to Nm 40 (25
tex),and were compose of :
      . 15 to 70 % flax noils / 85 to 30 % cotton.
      . 50 to 70 % flax noils / 50 to 30 % viscose
      . 40 to 60 % scutcher tow / 60 to 40 % cotton
      . 50 to 80 % comber tow / 50 to 20 % cotton
      . 60 to 80 % comber tow / 40 to 20 % viscose
      . 50 to 30 % reduced scutched flax / 50 to 70 % polyester

      On the basis of the results of the experimental work, performed in
part on the SKF Spinntester, the BD 200 and the Autocoro, it has been
possible to determine the spinnability criteria and the quality of the
end product.
      In some cases the relationship between the yarn quality factor and
the percentage flax in the blend is determined experimentally (Figure
14).

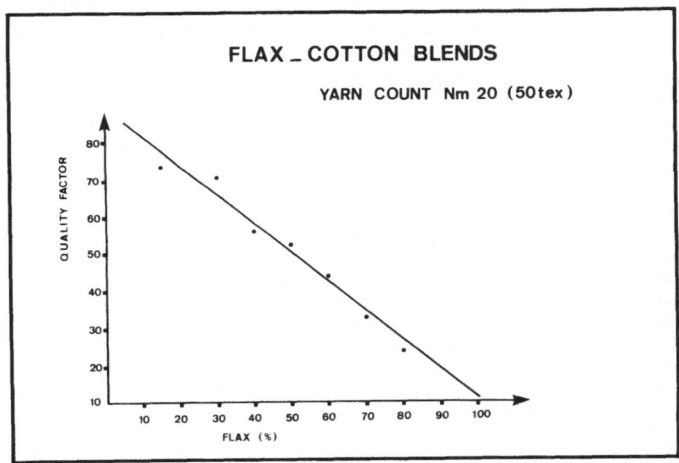

Figure 14
Open-end yarn quality factor :
influence of percentage of flax in flax + cotton blends

The open-end spinning trials have permitted a better view to be obtained of the possibilities of yarn manufacture on a cotton production line using reduced flax fibres.

Hovever, the following points must be considered :
- rotor open-end spinning demands that the flax must be a fine fibre, clean and with a suitable fibre length diagram, which means that very particular attention must be paid to the selection of the flax types to be subjected to reduction and to product development,
- quality control of the reduced flax with well defined standards and specifications is necessary for assuring high product quality,
- modern machinery for the drawing and spinning of cotton-type short-staple fibres can be used by applying suitable machine settings; Laroche, Clean-Star-System (Temafa) or Trutzschler machines for the early stages of preparation, Ingolstadt and S.A.C.M.T. for drawing, and Autocoro for spinning,
- for intermediate and heavy counts, a minimum of 20 % cotton, viscose or synthetic fibres is needed for correct performance in spinning and for a product of satisfactory quality; for relatively fine yarns (e.g. Nm 40) manufacture may only be envisaged using distinctly minor percentages of flax (e.g. 25 %),
- a comparison based on physical properties and/or value scales between these products of the open-end system and traditional flax, cotton or synthetic products is not feasible. These are novel products for which new end products and/or new markets should be considered.

3.2.2. Friction open-end spinning.
Within the EEC programme, the possibilities of the DREF friction spinning machine for the flax fibre were investigated. The principle of operation of this machine is a patented aerodynamic/mechanical system. To define the possibilities of this technique for the spinning of flax, research studied the following three points : the research involved study of the action of fibre seperation, experimental study of the formation of yarns of various types, and the technical evaluation of yarns and products manufactured from these yarns.

From a model of yarn formation an equation has been developed which permits the twist in the yarn to be studied as a function of different machine setting parameters and to identify those factors influencing actual yarn formation.

Because a feature of the flax fibre is its very high elastic modulus, the introduction of twist into the yarn by means of friction will be limited. Practical trials have confirmed this hypothesis. In fact, spinning of flax on the Dref II is not possible under economically acceptable conditions. The application of the theoretical results to the programme of practical and systematic trials has permitted identification of the various possibilities for use of the Dref II machine for the spinning of flax.

Possibilities exist for spinning blend yarns : flax/wool, flax/polyester, flax/polyamide, flax/acrylic and flax/viscose. Core and "effect" yarns may be obtained from different types of flax raw materials such as tow, noils and stretch-break converted fibres. A list of the various yarns gives an idea of where these yarns can be used within the furnishing and wallcovering sectors. Weaving trials have enabled the properties of these products to be assessed.

An overall assessment of the results, taking into account the technical properties and production costs, enables the conclusion to be reached that the Dref II technique can be used for spinning blends with a maximum flax content of 50 %, core yarns and effect yarns. For economic reasons, yarn count is limited to Nm 6, and to Nm 8 for core yarns. The areas of application lie mainly in the furnishing fabrics sector.

3.3. "Novacore" spinning

The first trials, with a yarn with a pure flax covering, have been carried out on a machine equipped with a standard wool-type drafting unit. These trials have enabled an indication to be gained of the possibility of producing fine yarns by a "dry" spinning technique. However, certain properties of the yarns produced were such that normal use in weaving and knitting was not possible. The major cause of this problem is the fact that flax fibres used had a badly balanced fibre length diagram. In view of the economic advantages of the process, spinning trials have been carried out using long filament flax. The drafting system of the RS 2000 machine, designed for drafting wool-type fibres, was unsuitable for the long filament flax. Investigations were carried out that led to the successful modification of the machine by increasing the ratchet whilst retaining the original controls for the drafting unit (Figure 15). The total length of the drafting units was converted from 220 to 465 mm.

Figure 15
Novacore drafting unit - flax type

For fine flax-wrapped yarns, the standard system of applying twist was unsuitable. The modified solution which was developed during this research was to produce a single core yarn around which the finest possible continuous filament wraps itself by self-twist (Figure 16). The intermittent roller producing the self-twist action has been modified accordingly.

The experimental trials consisted essentially of producing a yarn of resultant count 25 tex (Nm 40); this yarn being made up of a core yarn, textured polyester core of 50dtex and flax cover of 176 dtex with a continuous filament polyester yarn wrapping thread of 24 dtex. The composition of this yarn is 70 % flax and 30 % polyester and production speed is 150 m/min.

In conclusion to this study it may be said that the Novacore technique opens up new horizons. In fact, thanks to the support given by continuous filaments it becomes possible to produce relatively fine yarns by dry spinning. A 25 tex yarn is produced in the laboratory under fully acceptable operating conditions.

However, the machine modified during this research needs to be developed still further by the machinery manufacturer in order to assess its commercial possibilities.

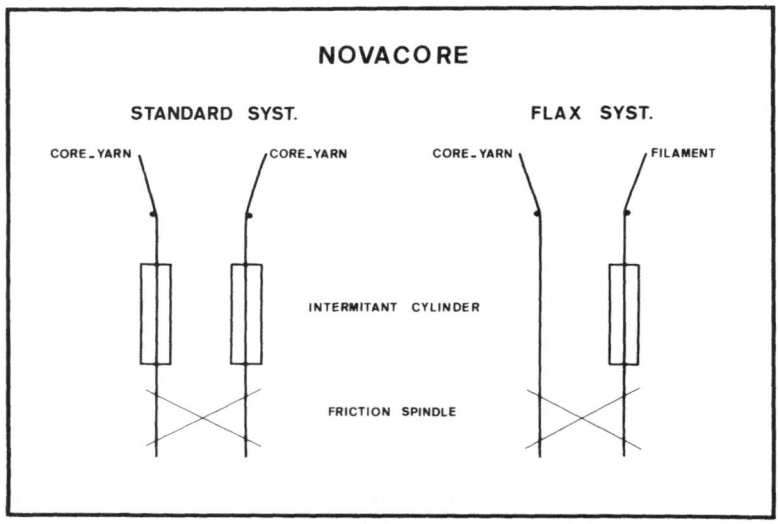

Figure 16
Principle of Novacore twist application
Wool type and flax type

3.4. Zero-twist spinning
The possibility has been examined of applying the zero-twist spinning system developed in the late 1960's by TNO and taken up by the company Signaalapparaten Fabriek in Eindhoven. A machine designed for the spinning of cotton was used. Several yarns were produced from flax dried in the standing crop and treated in roving form. The results are satisfactory. The advantage of this spinning system lies in the production rate (120m/min) and in the better covering power of the yarn in woven structures. It should, however, be pointed out that the yarn is of a unique structure and that spinning is only possible using chemically treated roving. A more detailed technical and economic study is required before drawing definite conclusions.

4. - MODIFICATION OF PRODUCTS AND PROCESSES
The study of new spinning processes and the development of novel yarns has resulted in research being undertaken into the modification of products and processes and the manufacture of novel fabrics.
Woven and knitted goods have been produced from Dref II, Novacore and short-staple flax yarns.

4.1. Using Dref II yarn
Weaving trials were carried out using the different types of Dref II yarns. In general, the performance of the Dref yarns during the weaving trials was satisfactory. Nevertheless, some precautions in thread guiding need to be taken, especially in the case of effect yarns, high-bulk yarns and high-twist yarns. The fields of use for these yarns lie essentially in the furnishing fabrics sector.

4.2. Using Novacore yarn
Weaving and knitting trials were carried out using the Novacore reference yarn Nm 40, 70 % flax / 30 % polyester. Weaving characteristics were considered to be acceptable. Fabric construction has a marked influence on the principal characteristics. In the case of the knitting trials, mention must be made of the various problems raised by the numerous irregularities in yarn mass and the intrinsic yarn structure. To improve the knitting performance it appears necessary to reduce the tendency of the yarn to unravel and to improve its unwinding characteristics.

4.3. Using short-staple flax yarns
Several woven and knitted fabrics were manufactured. As far as woven fabrics are concerned, the open-end yarns are capable of entering the immense shirting, lingerie and sheeting markets. In knitwear, the yarns may be used in garments such as T-shirts, and summer and sports vests. The weaving trials ran satisfactorily and the products manufactured are of high quality. The knitting trials were also satisfactory but the use of these yarns in knitting has shown the necessity for appropriate yarn clearing. The properties and user trials have shown that these products possess the principal qualities looked for in their respective fields of use.
The research institutes are now co-operating with interested industrial firms in developing more specific products, including those incorporating ternary blends. Some of the products manufactured using these novel yarns are eminently suitable for the high-class sector of EEC and world markets and have already been launched commercially.

IMPROVEMENTS AND INNOVATIONS IN PROCESSING TECHNIQUES
NEW APPROACHES IN THE FINISHING OF LINEN

M. SOTTON
Institut Textile de France

INTRODUCTION

The principal acticities evolved by four research centres within the framework of Eurolin programme are given in summary form in the table below. These various projects have been conducted on several complementary levels of research in order to provide answers to general questions arising in the finishing of linen through a better knowledge of the raw material and to bring about technical improvements or innovations in products and processes in the different flax sectors of member states.

Activities	Research centres			
	Centex bel	I.T.F.	LIRA	Staz. speri.
1. Improved knowledge of the intimate structure of the fibre, state of techno- logy		x		
2. Importance of alkaline pretreatments in fabric finishing	x		x	x
3. Innovations in finishing : - of apparel fabrics - of furnishing fabrics : -- chairs, wallcoverings, household linen -- control of flamma- bility of upholstered chairs	x		x	x
4. Prospects : - short-liquor treatments (foam finishing) - mechanical finishing		x	x	

IMPROVED KNOWLEDGE OF THE INTIMATE STRUCTURE OF THE FIBRE, STATE OF TECHNOLOGICAL PROGRESS

This stage of flax research, of a somewhat basic nature, has been undertaken with the aim of reviewing certain opinions held for a long time concerning this fibre which is acknowledged to possess outstanding properties. This improved knowledge of flax is intended as a source for long-term and short-term progress and is exploitable in different ways. Information will be provided for those authorities requiring constantly revised technical and scientific back-up to maintain the position of linen in competition with other materials, for understanding and solving problems encountered in processing the fibre and optimising the processing equipment, and for suggesting new products and processes and for opening up new markets.

Research effort has been directed essentially on the individual flax fibre since its properties, especially mechanical, which are the basis of the fibre's reputation, are largely the product of the specific characteristics of the ultimate fibres. The research programme, conducted with advanced apparatus, has enabled a deficiency of basic knowledge of this raw material to be remedied.

In a first stage, numerous mechanical tensile tests were carried out on the ultimate fibres taken from the tow by an original method of micro-dissection. The wide variation found in the specific mechanical characteristics of the single fibres may be explained by the influences of morphological phenomena interfering with the fibrous structure in a random manner from fibre to fibre.

In a second stage, structural investigations using scanning electron microscopy have permitted the location of these phenomena which take the form of visible imperfections on the surface of the fibres. These vary in appearance (folds, cracks, dislocation, etc.).They pre-exist in the plant stalk and their dimensions increase as processing proceeds as illustrated in the micrograph of a bundle scutched fibres shown in Fig. 1.

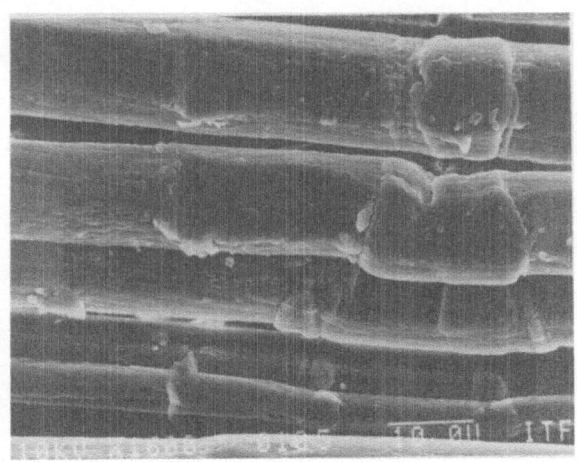

Figure 1
Scanning electron micrograph showing several flax fibres
from a bundle, in which there are structural dislocations
all located at the same level in the fibres (x 1,600)

The most important findings relating to these morphological phenomena are as follows :
- fracture of a single fibre always occurs at the site of an imperfection visible on the surface;

- the structural dislocations may be amplified or multiplied by subjecting the fibre to axial compression; this is shown in the flax fibre seen in Figure 2 which exhibits characteristics compression rings following rapid recovery after tensile fracture. Only flax fibres exhibit these rings so distinctly, the presence of which is accompanied by lateral dislocations and longitudinal fibrillations (fibrils from 2 to 5 microns thick), Figure 3.

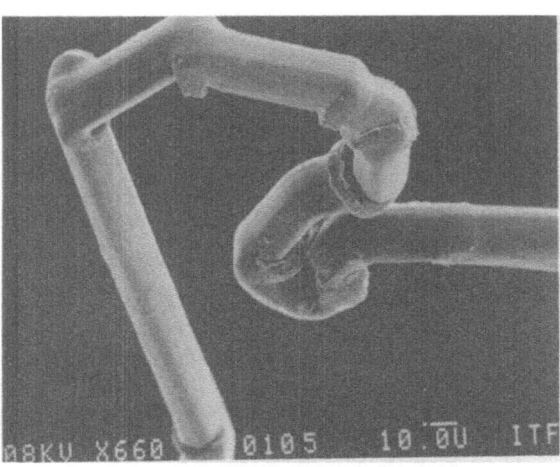

Figure 2
Electronic micrograph illustrating the creation of numerous dislocations
following axial compression loading of a single fibre (x 660)

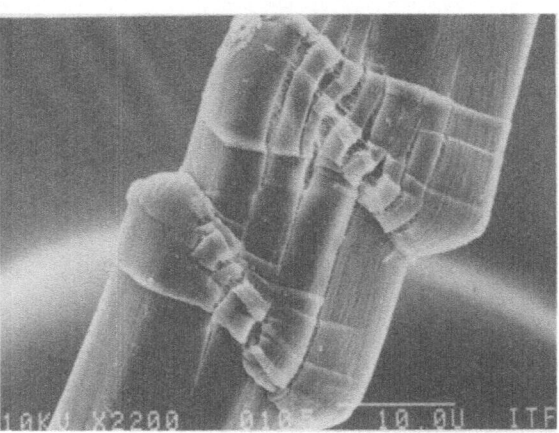

Figure 3
Detail view at a dislocation site on a single fibre with
longitudinal fibrillation of the structure following
axial compression (x 2,200)

- it is possible to create such phenomena intentionally by applying lateral stress to the fibre (bending, shear), Figure 4;

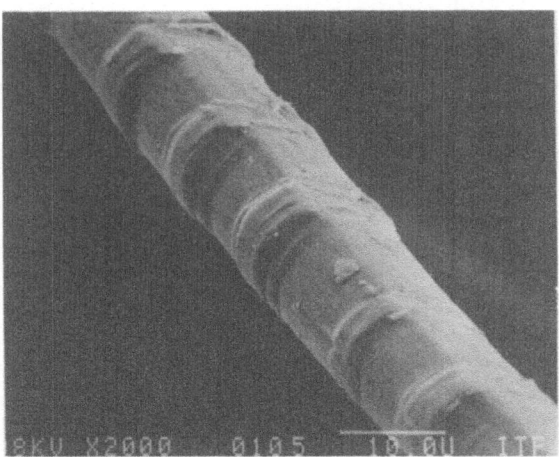

Figure 4
Scanning electron micrograph of a single fibre in which dislocations have been created intentionally by flexing (x 2,000)

- the most marked dislocations occur in the deep inner strata of the fibre as shown by the micrograph of Figure 5 relating to an ultrathin section taken longitudinally at the site of a dislocation.

Figure 5
Transmission electron micrograph of a longitudinal section of a single fibre at a dislocation site showing the profound disruption of fibrillar orientation at this point (x 6,500)

270

The perfectly oriented fibrillar structure on either side of the fold becomes completely disturbed in the fold with dislocation lines running across the fibre;

- if these lateral folds are extensive, they may be demonstrated by preferential staining and be statistically analysed by optical micros- copy incorporating an image analyser. Such a method has enabled one fold to be counted every 80 microns alongs the length of fibre that has been subjected to industrial mechanical dressing treatment, whereas no fold appears along the length of a native fibre after microdissection. However, examination by polarised light reveals all the faults, even those involving potential dislocations. Thus a native fibre possesses about one orientation fault every 100 microns against one every 40 microns in a mechanically dressed fibre;

- the native fibres possess mechanical properties coming very close to those of some aramid fibres and thus may be truly described as "cellulosic Kevlar".

Finally, this study of the "flax" fibre has enabled information to be collected on the distribution of non-cellulosic encrustation matter. The electron micrograph of Figure 6, a lateral section of a single fibre, shows the lamellar strata of cellulose separated by hemicellulose encrus- tation. The amount of encrustation increases at the centre of the fibre and correspondingly the crystalline order decreases from the periphery to the centre.

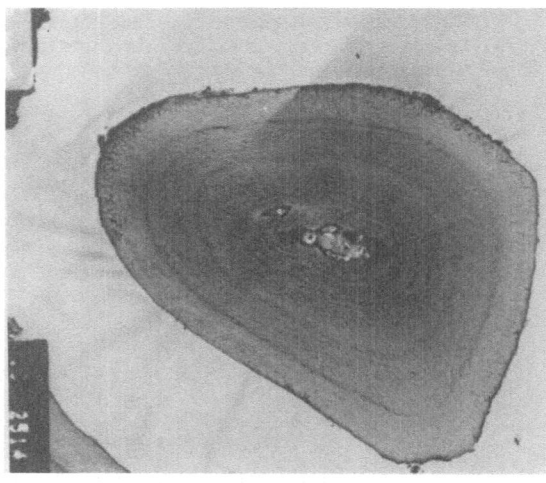

Figure 6
Electron micrograph of a transverse section of a fibre showing
the concentric strata of cellulose separated by encrusting
hemicelluloses occurring with greater intensity at the centre
than at the periphery (x 4,000)

Beyond this research on the single fibre, some work has been carried out on processed fibre bundles. It is at this level that lack of knowledge remains greatest and it is here that the problems are the most complex, since the material is a composite consisting of fibres bonded together by a complex pectin-lignin cement. Mathematical modelling of these fibre bundles demands not only knowledge of the individual fibre, but also of the physico-chemical nature of the interface about which we still only have vague ideas, either concerning the topochemical location of the constituents (lignin, hemicellulose, mineral salts...) or their relative reactivity. However, it is when the morphology and mechanics of this structure are understood that technologists will be able to control retting, divisibility and bundle regularity.

The principal results obtained about the structure of the bundle relate to the fact that in dry drawing, dressing is initiated by the breaking of a fibre from the bundle at a dislocation site, but never by splitting of the bundle (an important phenomenon in dry spinning). Also a better understanding is obtained of the action of desiccating agents (glyphosate) which cause specific modification of the pectin-lignin interfaces.

The knowledges of the flax fibre can be used immediately for modification of conversion processes, an improved understanding of the performance of linen products, and in the promotion of new applications.

One of the most significant results of this research which needs to be understood and made known throughout the flax processing cycle is that the flax fibre is a particularly anisotropic and crystalline fibrillar structure of low deformability which is only able to withstand significant dry strain by the creation or amplification of a dislocation. These dislocations play the role of articulations. The deformation propensity of the fibre by this mechanism is already evident during fibrogenesis in the plant (compression of fibres into bundles, wind effects, etc.) and is accentuated at various stages of preparation (scutching) and conversion including spinning and weaving. These articulation points are also sensitive zones and the most accessible to chemical agents such as dyestuffs finishing agents. It is thus understandable that crosslinking agents will act particularly at these sites and cause tendering (loss of abrasion resistance). Table I summarises the Centexbel results and illustrates this loss of abrasion resistance following resin treatment.

Table I
Loss of weight in abrasion

RESIN/FABRIC	100 % FLAX	BLEND
Untreated	3.5 %	2.5 %
DMDHEU	13.1 %	12.5 %

The micrograph in Figure 7 shows the face of a fracture at the site of dislocation in a flax fibre to which 7 % resin has been applied and has been broken in abrasion. For comparison, Figure 8 shows the face of an untreated fibre broken by abrasion.

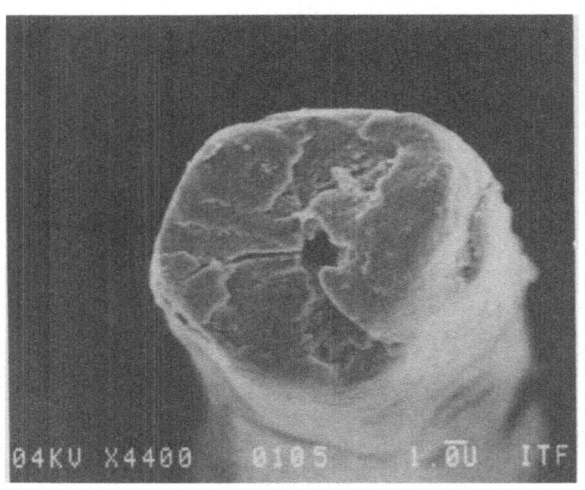

Figure 7
Appearance of characteristic fracture of a single fibre
by abrasion of a finished fabric (7 % resin);
brittle fracture of the fibres occurring at the
site of dislocations where the resin has acted. (x 4,400)

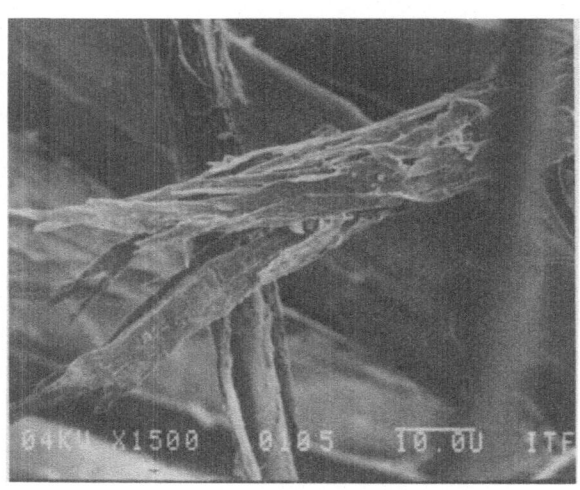

Figure 8
Appearance of characteristic fracture of a fibre
by abrasion of a fabric with no finish applied;
fibrillated aspect (x 1,500)

Further research is in progress to exploit in new finishing processes dislocations produced by specific mechanical treatments aimed at
making linen products more supple. The exceptional mechanical properties
of native flax may lead to applications in the reinforcement of composites by replacing man-made fibres in products such as cement and
bitumen.

It is possible today to explain more easily the stiffness of linen
goods (influence of fibre modulus), their susceptibility to wrinkling and
the ability to accept smart pleats (influence of dislocations) along with
their comfort qualities involving their cool nature and their ability to
absorb perspiration, both of which are related to the structure of the
fibre bundle. Little air is trapped between the fibres and water is
retained by capillary action between the fibres.

IMPORTANCE OF ALKALINE PRETREATMENTS IN FABRIC FINISHING

The specific morphological properties of flax explain the importance
of preparation treatments, usually alkaline, which have the effect of
homogenising the fibre structure by attenuation of flexure zones. The
fibres become more supple and stable and are made more accessible to
reagents by producing both a physical transformation of the molecular
structure by mercerisation and a chemical modification by the total or
partial removal of encrustation matter.

Alkaline pretreatments have been investigated in the programme to
improve the available knowledge. It is essential to approach flax in a
different way from cotton. The work has resulted in the establishment of
precise recommendations for industrial applications which differ between
countries depending on the products involved such as outerwear, underwear
and furnishing fabrics.

In the work directed towards furnishing fabrics, study has been made
of the various alkaline preparation treatment parameters including
boiling off. This has led to a recommendation of the optimum method. The
original feature has been the determination of possible interactions
occuring between two successive alkaline treatments : boiling-off (soda
or carbonate) and bleaching prior to resin application.

In this context the following conclusions have been reached :

- caustic soda treatment (6 - 10 g/litre) at 95°C can cause fairly significant weight losses depending on the duration of treatment, with risk
of fibre damage;

- the weight losses at the bleaching stage (chlorite + hydrogen peroxide)
are lower, the greater the weight losses occurring in boiling-off;

- oxygen consumption (hydrogen peroxide) is highly dependent upon the
nature of previous alkaline treatment (soda or carbonate) and on
temperature;

- the interaction between boiling-off and associated bleaching is
revealed by a very variable degree of fibre dissociation. The best
results have been recorded with two successive boiling-off treatments
performed quickly with 10 % and 5 % sodium carbonate.

In Italy where linen goods are experiencing considerable growth in
the clothing sector, the Stazione Sperimentale of Milan has also been
studying the effects of alkaline pretreatments. For this type of goods,
alkaline preparation represents the most delicate phase of the finishing

routine, during which all the parameters should be well under control if the handle, appearance and wear characteristics of the end product are to be preserved. The conclusion is that it would be difficult to replace caustic soda treatment in industry with other alkaline treatments or by purely mechanical processes because they do not impart to flax fibres the required increased accessibility to chemical agents.

In all the cases of linen textile products processed in co-operation with Italian processors, the conditions adopted for alkaline pretreatment reduce the initial differences between raw materials, yarns and fabrics. Laboratory research has provided confirmation of the results for furnishing fabrics, namely that alkaline treatment performed at high temperature (95°C) and low soda concentration (4 - 20 g/litre) produces a high weight loss and degradation of the molecular structure of the fibre without, any significant changes in mechanical strength. On the other hand, treatments at ambient temperature (25°C) and high soda concentrations (140 g/litre) produce a significant loss of mechanical strength in the industrial fibre. X-ray diffraction studies suggest that the crystalline transformation of flax cellulose occurs partially starting at soda concentrations of 10 - 12 % (at 25°C), a transformation which is accompanied by molecular disorientation and increased accessibility of the structure to water and dye molecules.

INNOVATIONS IN FINISHING

Apparel textiles

To maintain the upward trend in the market for linen apparel fabrics it is important to ensure high product quality. To attain this, quality control must be carried out at all stages of processing from spinning to fabric finishing. In the apparel sector, quality may be assessed from product appearance, the intrinsic properties of the product which may be evaluated by mechanical and physico-mechanical tests, and the comfort and fit properties of the garment.

A multi-parameter approach of this type to linen quality has been conducted in Italy with the active participation of the processing industry and of retailers.

Four pure linen fabrics for men's outerwear, women's outerwear and underwear were produced using different yarns, bleached or boiled off in the sliver. Weaving was performed on different machines and the fabrics (approx. 3,000 linear metres) were sent to ten finishing plants for preparation, dyeing (3 specified shades) and application of finish in accordance with the standard formulae and routines used in each establishment.

Analysis of this ambitious experiment has produced a rich data base enabling the setting of a specification for pure linen apparel fabrics from the informative labelling aspect. This specification, summarised in Table II, defines the minimum criteria of acceptability and the test methods for fabrics of 75 to 300 g/m^2 (outerwear) and 75 to 150 g/m^2 (underwear). This experiment has also produced definition, as mentioned in the previous paragraph, of the fundamental role of alkaline pretreatment and the need for combining shrink-resistant treatments and alkaline pretreatment for fabrics intended for underwear.

Furnishing fabrics

The aim of the research carried out at Centexbel on these materials concerned the optimisation of preparation and finishing processes including resin treatments to improve their resistance to abrasion and to soiling, two fundamental characteristics within the quality spectrum of furnishing fabrics. A means of measurement has been developed to enable specifications to be defined and attained.

TABLE II

Summary specification for pure linen fabrics for apparel use

```
1. Appearance :
      no bars, stripes, stains, holes

2. % Resins :
      maximum 2 % (underwear)
              3 % (outerwear)

3. Free formaldehyde :
      1,000 ppm (outerwear)
        300 ppm (underwear)

4. Dimensional stability :
      saturated steam = ± 2 % (outerwear)

5. D.P. :
              1,500

6. Colour fastness :
      -   Xenotest :
          . vat dye              5/6
          . reactive dye         4/5

      -   washing fastness at 40C   4

      -   fastness to dry rubbing   4

      -   fastness to wet rubbing   4
          . reactive dye         2/3

      -   fastness to solvents
          . vat dye              4
          . reactive dye         4
      -   perspiration fastness  4
```

The result in Table I clearly illustrate the losses in abrasion resistance which usually accompany resin treatments. Centexbel has suggested an original formula of a combination of conventional resins with elastomeric substances of the dymethylol-polysiloxane type which permits a partial solution to this problem by reducing the loss in abrasion resistance whilst giving acceptable wrinkle recoveries.

With the aim of improving the abrasion resistance of 100 % linen furnishing fabrics, various substances were applied by padcure at different concentrations. Some of these substances enabled the abrasion resistance to be improved appreciably. This is so for dispersions of polyacrylate (6 to 7 % concentration), polyvinyl (9 %), pva (2 %), silicone emulsions (9 %) and polyurethane dispersions (8 %). Nevertheless, the "handle" of the fabrics finished in this way is modified significantly.

With regard to the improvement of soiling resistance of linen fabrics, various substances (silicone, vinyl, acrylic, fluorine compounds) were studies at several concentrations. These additives were applied by full bath, by vapour phase on one face, or by lick roller on one face.

The fluorine compounds provided the most interesting results with the time of absorption of liquid soil being more than 300 sec. Application by lick roller was the best although the rate of deposition of active substance per unit area must be adjusted fairly precisely according to the surface and hydrophobic nature of the substrate.

Control of flammability of upholstered chairs

A very important project was undertaken at Lira with the aim of producing fabrics in linen or linen blended with other fibres for upholstered chairs, which are able to meet the requirements of legislation recently introduced in the United Kingdom on resistance to ignition (BS 5852 Parts 1 and 2).

This legislation states that chairs for domestic use must resist ignition by a weak source such as a burning cigarette. Chairs for contract use in public assembly rooms and multi-storey office blocks must satisfy the double requirement of resistance to ignition by a flame source plus a high abrasion resistance.

Within this context, the experiment conducted at Lira is especially instructive because firstly, it is desirable that the techniques used in product development can be quickly transferred into industry, and secondly, there is a strong probability that such legislation will be extended to other EEC countries.

This project has clearly shown that fabrics of linen or linen blended with other cellulosic fibres and with a weight per unit area less than 240 g/m^2 withstand the cigarette test, whereas heavier fabrics subjected to the same test allow progressive combustion within the underlying upholstery foam. An important fact has also been demonstrated on these cellulosic products, concerning traditional flame-resist treatments such as Proban NX which, although they increase the resistance to ignition in the flame test, may considerably reduce the resistance in the cigarette test.

Blends of flax with man-made fibres pass the cigarette test without requiring additional treatment. The satisfactory blend proportions have been determined and new opportunities are thus opened up to industry for introducing new products in the market.

Resistance to the flame test for products in linen or linen/man-made fibres may be acquired by the use of flameproofing agents at low rates of application, by using fibres of a flame-resist nature or by back-coating with a flameproofing agent. This work has also demonstrated the potential of linen/polyamide fabrics which possess excellent wear properties, especially abrasion resistance and which may be suggested for chairs required to meet contract specifications provided that a back-coating of a flameproofing agent has been applied under carefully controlled conditions. Several 80/20 linen/polyamide fabrics were thus produced by a back-coating of Mydrin TUX 84 257 applied by the doctor-blade technique with a coating rate of 250 - 300 g/m^2 and adjusted according to the surface characteristics of the fabric. The doctor-blade coating technique was adopted in preference to the foam method because of the need to reduce costs in view of the relatively high cost of the polymer system used. These trials are being extended into industry on a large scale to evaluate the performance of linen/polyamide fabrics (80/20) in upholstered chair applications.

POTENTIAL FOR SHORT-LIQUOR TREATMENTS

The aim of this project was to evaluate the possibilities of finishing treatments using the foam technique for the application of finishing agents and dyes to linen and to compare the properties of the products with those obtained by more traditional finishing methods.

Laboratory plant was designed for foam preparation and application; the rates of foam applied were 20 - 25 %. The performance of "wrinkle-resist" foam-finished fabrics is similar to that of fabrics finished by conventional means; but drying costs are considerably reduced. Softening agents have also been applied by this technique on wetted-out fabrics; homogeneity of application is especially good. However, no advantage appears to be gained by the foam technique in dye application, except perhaps for pastel shades with direct dyes.

The low penetration of flameproofing finishes applied by this method produces less good results than those obtained by padding. Sizing of unfinished flax yarns by the foam technique does not at present appear to show any promise.

CONCLUSIONS

This ambitious research programme carried out jointly by the principal European laboratories involved in linen finishing has proved to be especially productive. Besides the positive collaboration existing between the research workers in the different countries during these past two years, the multi-disciplinary and complementary nature of this research is today translated into an abundance of results which are already a source of progress in the linen industry and which are opening up surprising horizons for the remarkable flax fibre.

SUMMARY AND CONCLUSIONS BY THE CHAIRMAN OF THE SESSION

M. VAN LANCKER
CENTEXBEL

The papers prompted several questions which highlighted a number of aspects of interest to both the flax-based and the general textile industries, i.e. :
- alternative retting processes;
- the use of new manufacturing techniques;
- product classification;
- the morphological structure of flax.

As regards the question of how long it will be before the alternative retting processes (chemical and enzymatic) can be considered really efficient, it should be noted that, as they stand, the new processes are not adequate to cope with the problem fully under unfavourable climatic conditions. However, the chemical retting processes mean that flax can be cultivated in areas where the conditions are generally unfavourable, and results to date indicate that in the near future it will be possible to achieve adequate retting by chemical means and thus obtain high-quality textile fibre in bad weather and to reap benefits from such processes under normal conditions.

The use of new manufacturing processes, especially stretch-breaking of long staple and the spinning of short fibres, were particular points of interests. The number of breaking zones in the stretch-breaking of long staple was largely dependent on the maximum acceptable length and the raw material in question. Tests suggested that there was an advantage to be gained from working with a third zone.

The processes used for short fibres are generally similar to those traditionally used for cotton. However, if flax is used in such processes, particular attention must be given to the choice of the materials to be dressed and the manufacture of the fibre product. Under certain conditions carded wool machinery can be used for working short fibres. However, few industrial tests have been carried out.

Objective assessment would appear indispensible for the non-flax-based textile sectors, which mainly use flax in conjunction with other fibres.

For several years now research has been underway with a view to developing analysis and working methods aimed at classifying materials and products. At present there are three tests, two involving simulation, which can be carried out although they are not regarded as completely reliable. One permits an assessment of the hackling yield, the second enables the mechanical divisibility of the hackled material to be determined, while a third permits measurement of the fineness of the fibre by means of an air-flow technique. Nevertheless, these techniques cannot match organoleptic analyses by experts. Products (i.e. filaments, fabrics, etc.) are normally classified on the basis of standard tests. However, in certain cases tests which are specific or adapted to flax must be used (e.g. measuring the resilience of the fibres or the degree of polymerization).

Particular interest was shown in improved knowledge of the

morphological structure of flax fibre. Two important aspects emerged :
firstly, the difference between the reaction of flax fibre and cotton
fibre to alkaline treatments and, secondly, the mechanical properties of
the basic virgin flax fibre, which are similar to those of Kevlar.

When flax fibre is treated with caustic soda (mercerization or
causticization) the highly crystalline structure of the fibre and a
number of other factors present barriers. The caustic soda penetrates
very unevenly into and through the flax fibre compared with cotton, where
the action of the caustic soda is homogenous. Gradually the entire flax
fibre is converted into mercerized cellulose. At this stage the effects
are the same as in the case of cotton fibres, i.e. penetrability to
chemical products and stability are increased. An important aspect of
caustic soda treatment of flax fibres is that the joints and dislocations
in the structure of the fibre are retained.   Different results are
obtained if liquid ammonia is used as mercerizing agent. In this case the
structure of the fibre is transformed more slowly and less drastically
than if caustic soda is used.

In view of the mechanical properties peculiar to the virgin fibre,
it could possibly be used in composite materials in an organic or mineral
matrix. The difficulty of proposing flax as a substitute for reinforcing
fibres such as Kevlar in composite applications is due to the fact that
the flax fibre will not have similar intrinsic mechanical properties
unless existing extraction techniques are modified. Also, there is no
flax-based fibre which could offer the same thermal resistance as
Kevlar. This will restrict the possible use of flax to composite
materials in specific non-thermal applications. Moreover, flax occurs
naturally in the form of fibres measuring a maximum of 6cm in length,
while other reinforcing fibres such as Kevlar are supplied to users in
the form of continuous filaments.

As regards the question of whether flax may have a place in future
markets in composite materials, the answer is that even now it would be a
good thing if research, in collaboration with industry were to seriously
examine this possibility and this will become essential in the very near
future.

280

SESSION   D

UPGRADING OF LINEN TECHNOLOGY.

*   *   *

Contract 007-TEX-D - Dr.Ing. A. FUNDER

Contract 008-TEX-F - C. SULTANA

Contract 009-TEX-F - J. PRUNIER

Contract 010-TEX-F - J.P. BRUGGEMAN

Contract 011-TEX-F - Dr. M. SOTTON

Contract 017-TEX-I - Dr. B. FOCHER

Contract 025-TEX-B - M. VAN LANCKER

Contract 029-TEX-UK - W.W. FOSTER

Contract 030-TEX-UK - W.W. FOSTER

Contract 031-TEX-UK - W.W. FOSTER

Contract 039-TEX-NL - H.J. LEUTSCHER

CONTRACTOR :TEXTILFORSCHUNG
           Bielefeld
ADDRESS : Am Bahnhof 6,
         Postfach 3008
         D - 4800 BIELEFELD 1

PROJECT LEADER : Dr.Ing.A.FUNDER

CONTRACT N° :007-TEX-D

STARTING DATE : 1/10/1982

FINISHING DATE : 30/9/1984

TELEPHONE : 521/ 61.187

TELEX

TITLE OF PROJECT

PROCESSING OF FLAX ON OPEN-END ROTOR SPINNING SYSTEMS

AIM OF PROJECT

Novel flax blend yarns through the use of modern spinning technologies.

SUMMARY OF RESULTS

In attempts to make the processing of flax or linen more efficient and economical, one question investigated is the extent to which open-end rotor spinning techniques developed for cotton and man-made fibres can be applied to flax fibres. It is known that special characteristics of the flax fibre relating to structure and fibre length distribution are obstacles to direct spinning by this technology. However, numerous experimental variants have shown that novel yarns can be spun very well with appropriate machine settings by carefully planned preparation processes, adaptation of fibre lengths to the requirements of the spinning rotor, cleansing the fibre from undesirable contaminants and the production of suitable slivers. This method of production is also particularly suitable for blends with other short fibres because of the preparation machines used. The proportion of blending was predominantly with 15 to 80 % cotton or viscose fibres. The yarns, which are of distinctive appearance, in counts of Nm 12 (84 tex) to Nm 34 (30 tex), differ from conventional flax or blend yarns by having different fibre lengths and configurations.

It was also possible to use as high-quality raw material, after appropriate cleaning, flax noils from tow spinning, which was previously a waste product not usable in the textile field. Because of its use of raw inexpensive material, its few processing stages and its efficiency, the newly developed production technique has very good economic prospects.

CONTRACTOR : INSTITUT TECHNIQUE
AGRICOLE DU LIN
ADDRESS : 5 Rue Cardinal Mercier
FR - 75008 PARIS

PROJECT LEADER : C. SULTANA

CONTRACT N° : 008-TEX F

STARTING DATE : 1/10/1982

FINISHING DATE : 31/12/1984

TELEPHONE : 1/ 42.80.40.56

TELEX

TITLE OF PROJECT

UTILISATION OF FLAX : IMPROVEMENT OF CULTIVATION AND
HARVESTING TECHNIQUES IN RELATION TO FIBRE YIELD AND
SUITABILITY FOR EXTRACTION AND PROCESSING.

AIM OF PROJECT
- Counteracting the laying of the crop by chemical treatment in the
field.
  -- Study of the effectiveness of treating agents and the conditions
  of their application.
- Chemical desiccation of flax before cutting.
  -- Study of the influence of desiccation on retting time and fibre
  quality.
- Improvement of presentation of raw material in scutching.
  -- Study of a swath lifter to aerate the flax and allow it to dry
  more quickly before harvesting.
  -- Study of a harvesting device to remove stones and weeds.

SUMMARY OF RESULTS
The study relating to counteracting the laying of the crop has
allowed the effectiveness to be demonstrated of several agents, three of
which are now authorised for use on flax thanks to the references thus
established. The conditions for their application have been defined but
remain to be precisely set to ensure success of the treatment under
optimum economic circumstances.

Chemical desiccation of the standing crop before cutting of the flax
has essentially demonstrated the potential for improving fibre fineness.
The climatic conditions of the years of the experiments did not permit
the profit to be shown of the time of retting on the ground.

The study of crop lifting has resulted in the development of equipment, the line production of which has started and several hundred
machines will be in service by 1985.

The study of a harvesting device removing stones and weeds has been
directed towards rolling presses which are the cause of the greatest loss
in flax harvesting. The results have shown removal of 90 % of stones by
the process. It has not been possible to carry out quantitative control
of weeds but observations indicate a positive result. The process is
currently being exploited industrially, no doubt in simplified form.

CONTRACTOR : ITF-ATPUL   CONTRACT N° : 009-TEX F

ADDRESS : ATPUL   STARTING DATE : 1/10/1982
          Lagny le Sec
          FR - 60330 LE PLESSIS BELLEVILLE   FINISHING DATE : 30/10/1984

PROJECT LEADER : J. PRUNIER   TELEPHONE : 4/ 460.50.37

                                                  TELEX

TITLE OF PROJECT

TECHNICAL AND ECONOMIC IMPROVEMENT OF SCUTCHING FOR
SPECIFIC SPINNING AND NEW USES.

AIM OF PROJECT

Development of new equipment for scutching, conditioning and use of
scutched flax in preparation for different types of spinning.

SUMMARY OF RESULTS

With regard to scutching equipment, the performance of most of the
experiments on industrial sites has enabled almost immediate transfer of
the results attained within the programme.

Several scutching plants have improved the settings of their
machines following two very instructive trials. Gains in yield amounting
to more than 10 % have been reached.

Seven large companies have acquired the new crushing rollers that
have been developed. Amongst them, the crushing devices have been
completely modified and simplified on 6 scutching lines.

Development of a new form of transporter belt reducing fibre losses
by slippage has been undertaken and one scutching line has adopted it.

A new flail crushing-scutching device has been developed.
Complemented by the action of a completely redesigned scutching turbine,
it permits the production of scutched flax with better separation and
improved yield. Installation of this new process in a scutching plant is
the subject of a current project.

The apparatus for the formation of sliver of scutched flax has seen
its use extended to the evening of slivers of combed flax. Several
machines have been ordered for this purpose.

The "Total Flax" scutching line has been brought into service
following development of a new device elimating the capsule supports. The
spinning trade has shown much interest in the type of material produced
and instruction is being given on installation projects for new lines.

With regard to the preparation of flax intended for short staple
spinning, research has enabled a 33 % increase in machine productivity to
be attained by the selection of roller coverings and settings.

Improvement in the quality of prepared fibres is being pursued by
research in three directions to counteract the inadequacies of retting :
chemical or semi-chemical methods appear to be the most promising.

Specification criteria for prepared fibres are now available.

CONTRACTOR : I.T.F.

ADDRESS : 35 rue des Abondances
B.P. 79
FR - 92150 BOULOGNE BILANCOURT

PROJECT LEADER : J.P. BRUGGEMAN

CONTRACT N° : 010-TEX F

STARTING DATE : 1/10/1982

FINISHING DATE : 31/10/1984

TELEPHONE : 1/ 48.25.18.90

TELEX : 250.940

TITLE OF PROJECT

UTILISATION OF FLAX

AIM OF PROJECT

Technical and economic improvement of stages of processing of fibres into end-products.

SUMMARY OF RESULTS

Preparation and fibre separation

Setting of the pins plays a vital role, both on the GSL "high-production" machine and on the conventional machine, but must be well matched to each type of flax if optimum results are to be attained.

On the GSL machine, fibre separation is little influenced by the load. On the other hand, on the conventional machine a reduction in load at the drafting unit improves fibre separation, especially with a flax of high separation potential.

With equal pinning density, fibre separation is always more restricted on GSL "High-production" machines than on conventional machines, and in order to attain the same level of preparation quality at least one additional passage is required.

Obtained improvement in performance of the GSL machine may be by a calendering system at the delivery and better distribution of pinning.

Conventional spinning techniques

As wet spinning is governed by fibre separation brought about by chemical treatment, it is important to be aware of the extent.

The "TENOMETRE" apparatus conceived and developed by ITF-Nord enables this property to be investigated perfectly. The trade in Europe has acknowledged its usefulness and is at present acquiring this test equipment. This has enabled a better knowledge to be gained of the phenomena occurring during drafting, and has resulted in the development by M.A.B. of a new 3-roller drafting system which has also been put to industrial use.

The trend in dry spinning is towards increasingly finer counts. The admixture of a low percentage of man-made fibres has proved to give a significant improvement in spinning properties.

New spinning processes

The Novacore technique, thanks to support by continuous filaments, permits the production of fine yarns by dry spinning.

On the modified machine, problems that remain to be resolved are dust formation and premature wear of drafting aprons. These problems are within the competence of the machine maker who is studying whether or not to undertake the production of an industrial prototype.

With regard to short flax fibres, the special routine developed by research, and especially the spinning by ITF-Nord, has taken the process to the stage of industrial application.

Economic aspect of a new production line

The study of fibre separation has shown the present limits of technological development of equipment. Nevertheless, the industry feels the need to renew its machine stock with a view to making an entirely new introduction. MACKIE and ITF-Nord have studied the corresponding capital costs. Although production would be significantly improved, the fact remains that amortisation of almost ten years would have to be envisaged, which makes the industry hesitant.

Modification of products and processes

The study and development of novel yarns have provided a range of fabrics by both knitting and weaving, in products as varied as sportswear, apparel, furnishing and leisure. The development of manufacturing routines for spinning short fibres has led to the study of end-products in blends of linen-cotton, linen-polyester and linen-acrylic which are now being marketed commercially in most member states. Further specific products, including those made from ternary blends, are now being developed.

Specification

The production of fine wet-spun yarns requires chemical treatment to permit better utilisation of the raw material. This separation treatment removes the pectin-lignin cements which hinder the inter-bundle slippage of fibres during spinning.

New methods have been developed for the quantitative analysis of flax-viscose and flax-polyester blends.

CONTRACTOR : ITF-Paris

ADDRESS : 35 rue des Abondances
B.P. 79
FR - 92150 BOULOGNE
Billancourt Cedex
PROJECT LEADER Dr. M. SOTTON

CONTRACT N° : 011-TEX-F

STARTING DATE : 1/10/1982

FINISHING DATE : 30/9/1984

TELEPHONE : 1/ 48.25.18.90

TELEX : 250.940

TITLE OF PROJECT

UTILISATION OF FLAX

AIM OF PROJECT

Relation between the fibrous structure of flax, the conditions of use, finishing treatments and the specification of products.

SUMMARY OF RESULTS

This basic research carried out on flax has permitted reconsideration to be made in a highly novel and elegant manner of long standing concepts on a number of significant characteristics of this natural fibre. This better understanding of flax must certainly in the short or long term be a source of progress and lead to exploitation at the levels of promotion, conventional processing and innovation.

Morphological studies performed with powerful investigative tools on bundles and single fibres have enabled numerous detailed aspects very specific to flax to be demonstrated and explained, which had previously remained unclear and unexplained. On the basis of the morphology of the fibre bundle and the properties of the single fibre it is now possible to establish in a fairly scientific manner comfort models of linen products for comparison with cotton (coolness, moisture retention and absorption), and to understand the stiffness of linen fabrics and their creasing propensity. With its exceptional mechanical properties and its structure, the natural flax fibre is proved to be a true "cellulosic Kevlar".

One of the most signficant pieces of information from this research which must be broadcast and understood throughout the whole linen trade is as follows :

The individual flax fibre is able to withstand a high degree of deformation only by the formation and/or amplification of structural transverse dislocations. This behaviour is evident right from firbrogenesis in the plant, and is accentuated in retting, in scutching, in dressing and up to weaving. The sites of dislocation in the structure are, in fabrics, the most accessible to chemical finishing agents and to dyestuffs. They also form the sites of preferential breakage and their presence in large numbers or their extreme fragility (reticulation of these sites by chemical crosslinking agent for example) may permit the low abrasion resistance of these fabrics to be explained. It appears possible to create these structural dislocations at will, a feature which seems favourable to the development of new mechanical finishing processes for linen fabrics.

CONTRACTOR :STAZIONE SPERIMENTALE
PER LA CELLULOSA, CARTA E FIBRE
ADDRESS :Piazza L. da Vinci 26
     I - 20133 Milano

PROJECT LEADER : Dr. B. FOCHER

CONTRACT N° : 017-TEX-I

STARTING DATE : 1/10/1982

FINISHING DATE : 31/3/1985

TELEPHONE : 2/29.29.57 / 60

TELEX : 350.572

TITLE OF PROJECT

UTILISATION OF THE VARIOUS TYPES OF FLAX RELATION BETWEEN
STRUCTURE OF THE FIBRE AND STRUCTURE OF THE PRODUCT

AIM OF PROJECT

Evaluation of the influence of the structure of the raw material on the quality of the end-product and identification of the optimum conditions for alkaline treatment.

SUMMARY OF RESULTS

The quality of linen apparel fabrics is governed by several factors such as the raw material used, the geometry of the yarn, the weave of the fabric and the finishing treatment. Four pure linen fabrics have been studied, for men's wear, women's wear and shirting, produced with yarns of various origins and finished in ten finishing establishments.

The quality of the products has been assessed by the physico-mechanical parameters (tensile strength, abrasion resistance, tear strength and dimensional stability), physico-chemical parameters (degree of polymerisation), subjective tests (assessment of "handle", presence of spinning faults, weave and finish) and investigation by optical and electron microscopy.

For all the products considered, the conditions adopted in alkaline finishing treatment had a determining influence by cancelling out the initial differences in raw material, yarn and fabric.

Alkaline treatments carried out at high temperatures and low concentrations cause high weight losses and large differences at molecular level without, however, affecting the strength of the products.

Treatments at room temperature and high concentrations of caustic soda on the other hand cause a considerable loss in strength of the fibre structure.

Finally, somewhat mild alkaline treatments, especially when carried out for a short time, do not allow consistent products to be produced with adequate dimensional stability.

CONTRACTOR : CENTEXBEL

ADDRESS : St.Pietersnieuwstraat
nr. 43
B - 9000 GENT

PROJECT LEADER : M. VAN LANCKER

CONTRACT N° : 025-TEX-B

STARTING DATE : 1/10/1982

FINISHING DATE : 31/10/1984

TELEPHONE : 230.93.30

TELEX

TITLE OF PROJECT

UPGRADING OF LINEN TECHNOLOGY

AIM OF PROJECT
  Technical and economic improvement of the processing techniques.

SUMMARY OF RESULTS
  Fundamental study has been made of certain problems related to the presence of vegetable impurities in flax fibres.
  After the setting-up of a classfication scheme, the determination of the morphological and chemical characteristics, the drafting of a list, and the study of the use of herbicides the research work resulted in characterizing the behaviour of the vegetable impurities during mechanical and chemical treatments.
  The knowledge of the overall behaviour of the impurities during mechanical and chemical treatments permits the indication of the various industrial means for resolving the problems related to the presence of vegetable impurities in flax fibres.

Upgrading of low quality materials during spinning.
  After having examined the behaviour of certain raw materials (scutched tow) treated by various cleaning and refining operations, it was found that these materials cannot be used, without modification in the conventional process.
  Accordingly the study was pursued along two lines :
- cutting, refining and cleaning, blending and carding, and processing on the cotton system,
- carding, breaking, blending and processing on the woollen system.
  The technical and economic balance sheets provide information about the possibilities for using low quality material in industry.

Upgrading of the breaking process
Following the study of the dressing agents and their influence on the fibre to fibre and fibre to metal coefficients of friction, the various breaker setting parameters were examined.
This study formed the basis for a new type of machine.
Spinning trials were carried out in order to study the influence of the various parameters on the spinning process and the quality of the yarns.

Flax spinning on the DREF-system
A theoretical study resulted in the elaboration of a model for the yarn formation. The degree of refining and shortening of the fibres was determined in relation to the setting parameters.

The results and the theoretical model allowed to specify the conditions for spinning flax fibre on the DREF system.

The production of core-yarns and "fancy" yarns was examined and their use was studied in particular for as ornamentations, home furnishings and wall coverings.

Upgrading of finishing treatments.

New products and finishing techniques have provided fabrics with defined or improved quality.

The efficiency of certain treatments for improving wear and soil resistance of furnishings fabrics has been demonstrated.

The results also allow to specify the parameters, related to the fabric structure, influencing the preparation and finishing treatments.

290

CONTRACTOR : L.I.R.A.

ADDRESS : Lisburn, Co Antrim
Northern Ireland
BT27 4RJ

PROJECT LEADER : W.W. FOSTER

CONTRACT N° : 029-TEX-UK

STARTING DATE : 1/10/1982

FINISHING DATE : 31/3/1985

TELEPHONE : 8462/ 2255

TELEX : 74.74.25

TITLE OF PROJECT

UPGRADING OF LINEN :
IMPROVEMENTS IN QUALITY OF THE END PRODUCT

AIM OF PROJECT

Objectives

The specific aim was to retain and improve the quality of linen
end-products particularly in the up-market region through a deeper
understanding of the basic properties of the linen fibre and textile
structures and through need-oriented product development. In particular
the work considered improvements in wallcoverings, and curtains and
drapes.

SUMMARY OF RESULTS

1. Wallcoverings

Wallcoverings treated with non-substantive flame-retardant
treatments can be wiped with a limited amount of water sufficient to give
some cleansing action without significantly affecting the flame retardant
properties. Since dry-cleaning solvents contain no significant amount of
water, it is probable that such solvents could also be used for
cleansing.

2. Drapes

It has been found that linen fabrics commercially treated with
permanent flame-retardant finishes (Proban and Pyrovatex) show a wide
range of properties, both before and after cleansing (washing and
dry-cleaning). This indicates that there is room for improvement in the
application and control of such finishes and this points to the need for
further research and investigation in this field.

BENEFITS TO INDUSTRY

1. Wallcoverings

The main benefits here are related to marketing. Fabrics for this
end-use when suitably treated could be claimed to be wipeable with a
limited amount of water and also with dry-cleaning solvent.

2. Drapes

The reasons for the variable results found require further research
and investigation to produce more uniform products. This work is required
before commercial exploitation can be undertaken.

291

CONTRACTOR : L.I.R.A.	CONTRACT N° : 030-TEX-UK

CONTRACTOR : L.I.R.A.

ADDRESS : Lisburn, Co Antrim
Northern Ireland
BT27 4RJ

PROJECT LEADER : W.W. FOSTER

CONTRACT N° : 030-TEX-UK

STARTING DATE : 1/10/1982

FINISHING DATE : 31/3/1985

TELEPHONE : 8462/ 2255

TELEX : 74.74.25

TITLE OF PROJECT

UPGRADING OF LINEN
IMPROVED METHODS OF PROCESSING FIBRE, YARN AND FABRIC

AIM OF PROJECT

Objectives
The broad aims of the work were to look at new or modified processes for the production of linen goods at competitive prices, and by extending the range of possible products. In particular the work considered the use of new machinery for traditional flax and flax treated by herbicides, the use of foam techniques in sizing and finishing, the durability of bed linens, and the development of flame retardant upholstery fabrics.

SUMMARY OF RESULTS

1. Processing
The new high-speed continuous hackling system has been found to be suitable for low-medium counts up to 40 lea (40 Tex). For finer counts, the intermittent action of the traditional hackling system is preferred. The traditional machine is therefore being upgraded mechanically to improve output without impairing quality.

High-speed monohead line preparing machinery was found to give inadequate control of fine fibres, and rubber drawing rollers gave inadequate fibre subdivision. Thus, progress has been made in the multi-head screw-gill system, using wooden pressing rollers to achieve fibres subdivision, and using triple-threaded screws to increase output.

The new generation of wet-ring spinning frames uses higher drafts, and higher spinning speed, and gives higher yarn strength than the older frames. Auto-doffing is also possible. These features all lead to improved output and improved yarn quality.

Automatic winding of wet-spun linen yarns has been achieved successfully by two winding machine manufacturers.

Fibre produced by the desiccation retting system has been processed in the normal way at all stages of manufacture, using the equipment described above.

2. Foam sizing
Due to problems associated with (1) viscosity of the concentrated size solution required (2) the lack of wettability of grey flax yarns and (3) stability of the foamed size, it was not found possible to size grey flax yarns satisfactorily by a foaming technique. Foam sizing of treated rove yarns was found to be rather more successful.

3. Foam finishing

The performance of linen fabrics treated by foamed crease resist resins is similar to that of fabrics treated by the padding.

Softening agents foamed and applied to wet fabric result in very even application.

The flame retardancy of linen fabric treated by foamed soluble finishes was not as good as that of fabric treated by normal padding due to lack of penetration. Application of dyes by foam seems to offer no advantage over normal procedures.

4. Bed linen

No successful method could be found to improve the abrasion resistance of bed linen.

5. Non-Flam upholstery

Resistance to smouldering ignition without the use of an additional fire barrier has been achieved by the introduction of synthetic fibres either as warp yarns or as intimate blends with flax. Flame ignition resistance of flax based upholstery fabrics suitable for domestic and contract markets has been achieved in three ways; use of low concentrations of flame retardants; use of inherently flame retardant fibres in blends with flax and by means of a flexible flame retardant back coating.

Benefits to industry

High-speed high draft spinning, and automatic winding of fine wet spun linen yarns have been adopted widely within the Northern Ireland Industry during the last two years, with consequent increase in both productivity and yarn quality. This in turn has facilitated the increased usage of high-speed rapier looms.

Linen finishers can take advantage of the savings in energy and chemicals and of higher productivity by the use of foam finishing techniques.

Flame retardant linen and union fabrics can be produced to meet present and proposed flammability regulations.

CONTRACTOR : L.I.R.A.   CONTRACT N° : 031-TEX-UK

ADDRESS : Lisburn, Co Antrim   STARTING DATE : 1/10/1982
          Northern Ireland
          BT27 4RJ   FINISHING DATE : 31/3/1985

PROJECT LEADER : W.W. FOSTER   TELEPHONE : 8462/ 2255

                                         TELEX : 74.74.25

TITLE OF PROJECT

UPGRADING OF LINEN : IMPROVEMENT OF PROCESSING PROPERTIES OF
FIBRES BY THE USE OF SYSTEMIC HERBICIDES AND ENZYMES

AIM OF PROJECT

Objectives
    To develop new methods for freeing flax fibres from woody material
in the plant without recourse to conventional of retting. The new method
employed commercially available systemic herbicides. Improvements in the
separation of fibre from the plant were sought using enzyme methods to
upgrade the quality of the fibre.

SUMMARY OF RESULTS

Desiccation retting of flax
    The optimum dose rate for glyphosate was established to be 4 litres/
hectare, applied in 400 litres of solution. The optimum application date
is weather dependent. Early spraying gave better translocation than late
spraying in a dry season, but early spraying reduced fibre yield. Early
pulling is necessary to avoid adverse weather conditions. This tends to
lead to under-retting, which may require rove-boiling or rove-bleaching
to yield optimum yarn strength.Yarns have been spun from desicccated
flax, and have been woven into commercially acceptable fabrics. Further
work is necessary to improve uniformity of spraying, uniformity of
retting, and overall fibre yield.
    The application of glyphosate was found to retard lignification,
which ceased between 1 and 3 weeks after spraying.
    The microbial population on desiccated flax, retted on foot, was
similar to that found on dew-retted fibre. Some differences resulted from
year to year due to variations in weather conditions.

Enzyme treatment
    Enzyme extracted from organisms grown on flax by-products, and
commercially available enzyme, have been found to promote fibre subdivision in both straw and scutched fibre, green or partially-retted.

Benefits to industry
    In consequence of the work on desiccation and subsequent retting on
foot, extensive commercial trials are in hand. Two companies, one in
Northern Ireland and one in Scotland, have been established. During 1985,
approximately 1000 acres (400 hectares) will be grown and scutched in
each country. The successful outcome of these trials could lead to
substantial increase in flax production in the United Kingdom.

CONTRACTOR : I.B.V.L.

ADDRESS : P.O. Box 18
NL - Wageningen

PROJECT LEADER : H.J. LEUTSCHER

CONTRACT N° : 039-TEX-NL

STARTING DATE : 1/2/1984

FINISHING DATE : 1/8/1985
1/2/1986

TELEPHONE : 8370/ 19.043

TELEX : 45.371

TITLE OF PROJECT

RESEARCH PROGRAMME FOR A RATIONAL HARVESTING AND
PROCESSING METHOD FOR FLAX

AIM OF PROJECT

The processing of tangled straw, obtained from a new combine-harvesting method, by means of a cheap alternative process, into a competitive short fibre for the mixed-yarn-industry.

SUMMARY OF RESULTS

In the last two years preliminary field- and laboratory-work has been carried out, to verify ideas about the universal combine harvesting of flax by combining and about the processing of straw, direct for the short fibre market.

Over the 1984 season culture and harvesting tests have been carried out in variations of mowing and pulling in spreaded swath. Better maturing was obtained and the combine-threshing turned out to give good production of first class linseed. The tangled straw was dew-retted in the swath. With this experience it can be concluded that this way of harvesting flax with the aid of universal agricultural machines does not meet severe problems and gives openings for flax in general crop-rotation.

In this way the linseed and dew-retted tangled straw bales are the final products of arable farming (for Dutch conditions).

The main target of the research is to fill in the missing link of a modern and cheaper processing of the straw. The tangled straw is one of the main products in this conception, instead of a waste product; and it has to be processed in total into the textile fibre condition.

A pilot production line is under development. In an improvised opening and cleaning process a useful short (staple) fibre has been obtained for mixed yarn spinning. Preparing and spinning tests on "tracer" samples (20 kg) have given positive indications about feasibility and quality. Some results :
- efficiency of dew-retting of straw : 6 tons/ha;
- opening and cleaning : efficiency 57 %;
                        after purifying 36.6 %;
                        after carding and combing : 26.6%.

First spinning tests into Nm 20 yarns of 20 % flax and 80 % cotton have given good values for tensile strength and draft. Preliminary evaluation of processing and quality indicates that the flax fibre obtained by the new process will command better prices in the market place.

CLOSING SESSION

Highlights of the European textile research symposium
    D. FINLAY-MAXWELL, Chairman of the Scientific Re-
    search Committee of Comitextil

Likely technological developments in the textile industry and their impact on the future competitiveness
of the industrialized countries
    R. VERRET, Werner International

General Conclusions of the Symposium
    D. FINLAY-MAXWELL

Closing address
    J.H. LEACH

HIGHLIGHTS OF THE EUROPEAN TEXTILE RESEARCH SYMPOSIUM

D. FINLAY-MAXWELL
Chairman of the Scientific Research Committee of Comitextil

I should now like to draw together some conclusions from the Symposium and particularly to suggest where the future lies.

The opening address yesterday was given by our President, Harry Leach. We are fortunate in having a president, who like his predecessor is highly competent technically and acutely aware of the part that research and innovation will play in determining the survival and future prosperity of our industry. He urged us to be "dynamically active" and reminded us that "participation is essential for success".

I agree entirely and I have been encouraged by the degree of participation during the seminar. While accepting the need for dynamic activity, I hope Mr President that you in turn will accept that we expect highly qualified dynamic leadership in the field of science and research from Comitextil.

Mr P. Rutsaert gave us an excellent economist's review in which he included a scenario of world trade movement, with a comendable knowledge of our industry. A simplistic interpretation of this scenario is that if world consumption in a comodity is relatively static, advantage will move first to low labour cost countries and then back to highly capatalised technologically advanced countries. Principle I am sure we can accept this philosophy.

Mr P. Rutsaert finished by saying that he was "afraid that there was not much certainty" and the message was that "the future will be all right". I am sure, he will not take offence if I take this statement slightly out of context and say that "the future will not necessarily be all right" and there is "at least one certainty" - namely, that unless we significantly increase our innovation and Research & Development effort, it will not be the European Community which will be the first to recapture the market; it will be one or more of our highly capitalised competitors.

Success will require significant effort; in my opinion it will not be achieved without Community assistance. I will seek to justify this later.

Mr Ph. Bourdeau, with characteristic modesty, ommitted to mention the significant personal contribution he made in rescuing this programme from the bureaucratic chasm that it slipped into eight or nine years ago. When I joined Comitextil six years ago, we had unsuccessfully been trying to fund this programme for about three years. We even had the ludicrous situation of the entire programme being rejected because one or more member country objected to an individual element of the programme. It was at this stage that we decided that we should not hesitate to sacrifice part of the programme, if necessary, in the interests of the whole. Fortunately, as mentioned by Mr Ph. Bourdeau, we were able to divert the offending elements into the energy or environment programmes.

I am not in any way seeking to criticise anyone for this situation. At that time we were pawns in a political game, we were pioneering and the market pressures on our industrial leaders at Comitextil were such

that research was considered as "any other business", if at all. We have
come a long way since those dark and difficult days and Mr Bourdeau has
been a stalwart and invaluable supporter of our industry, and in particular, of this programme. I know he will be a sound adviser for us in the
future but remember he finished his discourse with the words "the future
depends on you".

In Dr. H. Tent, we are indeed fortunate in having a man utterly
dedicated to the philosophy of research in industry and who will work
with exceptional tenacity of purpose to achieve these ends. I shall
choose just one item from his very interesting discourse and that is the
question of demonstration projects. Few things can be more convincing
than a working demonstration model and it is possible that we could give
more attention to this possibility in the future.

Arising from Mr. Verret's presentation, I conser the problem of
proprietary software is very disappointing and one would have thought it
possible to write a reasonably universal and flexible Community-orientat-
ed programme, allowing the option of a range of local parameters to be
inserted and having a reasonably standard output such as RS 232 or some
more modern equivalent.

It is my pleasure to thank those of you who have contributed so much
to the success of this conference. We record ou thanks to our four guest
speakers for their interesting, informed and thought-provoking contribution. We also thank our four topic chairmen, rapporteurs and those
involved in the preparation of such an interesting and stimulating set of
papers - without which we would have had no conference. We are indebted
to those of you who have participated in discussions. Interested,
constructive criticism is a vital element in the encouragement and
motivation that our research work must receive if it is to develop on
the paths desired by industry. We express are gratitude to the organi-
sers, Dr. Wurm and his colleagues, Camille Blum and his colleagues, for
the enormous amount of detailed preparatory work so vital if a conference
is to be a success. The the exceptional quality of the Commission
hospitality has been greatly appreciated. We acknowledge too, the efforts
of President, Harry Leach, who has kindly agreed to close the conference
when I finish, not only for his extremely active interest and support of
the conference but also for his leadership as the driving force in
formulating the future organisation necessary if research is to survive
and to prosper in the future. Finally, may I thank you in advance for
what you intend to do when you return to your organisations.

I believe this Symposium has been a considerable success, but not a
complete success. To complete this success may I ask you to do three
things :

1. To do your utmost to disseminate the knowledge that you have gained
    at this conference with the intention of actively utilising the
    results.

2. To carry the message that collaborative research within the Community
    has proved a success. Further, that this formula for research is one
    of the most important ways if not the only way, of bringing together
    the multi-disciplinary expertise so vital for future research.

3. Will you consider what personal commitment you can make to assist in
    furthering these objectives.

I believe that dynamic innovation, research and development with the implied application of advanced technology is the corner-stone for the future prosperity and profitability of our industry. It will not happen by chance – it will require even greater dedication on our part as well as that of Comitextil if we are continually to out-pace our worldwide competition.

Ladies and Gentlemen, the future is in your hands; this conference has provided a springboard into the future.

I have confidence, that if you take action now, you will succeed that thereby complete the success that this conference has richly deserved.

LIKELY TECHNOLOGICAL DEVELOPMENTS IN THE TEXTILE INDUSTRY AND
THEIR POTENTIAL IMPACT ON THE FUTURE COMPETITIVENESS OF THE
INDUSTRIALIZED COUNTRIES

R. VERRET,
Werner International

SUMMARY

The "Textile Industry" encompasses a very complex and diverse
group of industries with very different parameters to ensure their
viability. The sub-sector that is mostly affected from imports from
low wage countries is apparel of both woven and knitted goods. It
represents close to 50 % of the EEC 12 textile consumption. The
viability of the primary textile sectors of the European community
depends on their ability to compete with imports of yarns and
fabrics from low wage countries, but also to a greater extent on the
ability of the apparel sector of the Euopean community to compete
with imports of apparel from low wage countries. This is illustrated
with costing samples of a cotton dress shirt made in a high wage EEC
country and in South East Asia. It shows that to be competitive
manufacturing costs of producing a shirting fabric must be reduced
by 10 % and the manufacturing costs of making-up a shirt by 30 %. It
also shows that in order to achieve these savings, technological
developments are required by decreasing order of priority in making-
up, in weaving, in dyeing and finishing and in spinning. An analysis
of the textile patents granted in the last 18 months in the top
seven textile patent producing countries, shows that due to a lack
of research in the past in Europe, the new technologies for making-
up garments that will be introduced in the next three to five years
will come from Japan or the U.S.A. since they had 85 % of the
patents granted in the last 18 months for that sector. A more
detailed examination of the number and type of patents granted,
permits the identification of theprocesses where technological
improvements will be made. We can safely predict that the yarn and
fabric sectors in the European Community will be basically competitive worldwide in four to five years from now, provided that the
industry will be able to generate the funds needed to acquire the
new technology and provided the textile world operates on the basis
of fair trade.
While substantial progress will be made, the remaining weaknesses of
the making-up sector may put a large section of the primary textile
industry in difficulty, unless a form of MFA agreement is implement-
ed in the community until the early 90's when the apparel sector
should become as competitive as the yarn and fabric sector.

The subject of my presentation today is "the likely technological
developments in the textile industry and their potential impact on the
future competitiveness of the industrialized countries".
We all know that in the past decade the textile industry in the
industrialized countries has suffered greatly, their share of their
domestic consumption has been reduced substantially and the blame has

been generally allocated to imports from low wage countries. People outside the industry always talk about the textile industry as if it was one
industry with a problem that could be solved with one solution. We all
know that it is far from being that simple, and that the so-called
"textile industry" encompasses a very complex and diverse group of
industries with very different parameters to ensure their viability.
Table 1 shows a simplified textile process flow chart, which identifies
the various sectors of the industry. The sub-sectors named between
brackets, are at least up to now not affected by competition from low
wage countries. They include : textured yarns, non-wovens tufting,
carpets and industrial textiles.
The sub-sector named in normal type, of home furnishings is partially
affected and the most affected, (names in heavy type), is apparel, of
both woven and knitted goods.

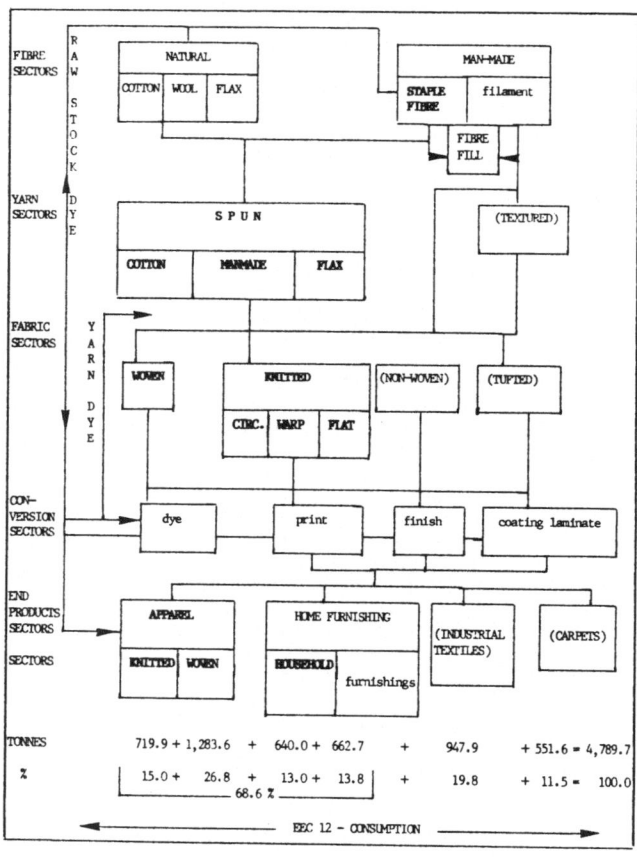

Table 1
Textile process flow chart

In the 12-EEC countries in 1983, of the total textile consumption of 4,800 thousand tons, 45 % went into the sectors that are not affected : carpets, industrial textiles and furnishings, while 55 % went into the threatened sectors : knitted and woven apparel and household textiles.

The viability of the primary textile sectors of the European Community depend on their ability to compete with imports of yarn and fabrics from low wage countries, but also and to a greater extent on the ability of the apparel sectors of the European Community to compete with imports of apparel from low wage countries.

In order to illustrate this, we have selected as an example a product that faces fierce competition : a cotton dress shirt. The following sample costing (see table 2) is representative of the cost breakdown to produce a cotton dress shirt in a high wage European country and a low wage South East Asia country.

As you can see the difference is considerable, the Far East total cost being 71 % of the European cost, and if we look only at the manufacturing cost, the difference is even more considerable, with the Far East cost being only 61 % of the European cost.

Our example shows that for a high wage EEC country, the largest cost item is the labour cost which represents the cumulative labour cost in spinning, weaving, dyeing and finishing, and making-up. It adds up to 41 % of the total cost. The total cost difference of 4.59 DM/Shirt is equal to the difference in the labor cost content alone, the other advantages in raw material, manufacturing overhead, depreciation, selling and general administration being offset by disadvantages in energy and financial costs.

	EEC-High Wage DM/shirt-chem.	Far East DM/shirt-chem.	Difference DM/shirt-chem.
1. Raw material	2.88	2.60	- 0.28
2. Labour	6.59	2.05	- 4.54
3. Energy	0.74	1.17	+ 0.43
4. Manufacturing Overhead	1.77	1.39	- 0.38
5. Depreciation	0.95	0.64	- 0.31
6. Total Manufacturing cost	12.93	7.85	- 5.08
MANUFACTURING COST RATIO	100.0	61.00	39.00
7. Financial Costs	0.96	1.66	+ 0.70
8. Selling - General Administration	2.23	2.02	- 0.21
9. Total cost	16.12	11.53	- 4.59
TOTAL COST RATIO	100.00	71.00	29.00

* From raw cotton to made-up shirts.
No yarn or cloth profit included.

Table 2
Sample costing : Cotton Dress/shirt

You should note that these sample costings do not include any profit on the transfer prices of yarn or cloth, nor any shipping and handling charges, and are on the basis of fair trade with no hidden subsidies, which is not always the case.

We believe that if the total cost ratio (Far East/Europe) could be raised from 71 to 85 the vulnerability of the European industry would disappear, the difference being not sufficient to cover the additional expenses involved and the risks inherent in longer lead times, when importing from the Far East. To raise the ratio to 85 in our sample costing, would mean reducing the total cost by 2.56 DM/shirt.

So the question is, will likely technological developments permit this overall cost reduction ? If we assume that the general overhead costs cannot be greatly modified, it means that all this cost reduction will have to be made within the manufacturing costs.

Can the foreseeable technological developments permit the overall cost reduction needed ? Before answering this, let us see where these technological improvements are required, on the basis of our sample costings.

Table 3 compares the manufacturing costs at each stage of production. Of the total cost difference of 5.08 DM/shirt, 1.49 DM or 30 % is accounted for in the cost of the finished fabric and 3.59 DM or 70 % in the cost of making-up the shirt. It means that for our example, technological developments are required to reduce the manufacturing costs of producing a shirting fabric by 10 % and the manufacturing costs of making-up a shirt by 30 %.

	EEC – High Wage		Far East		Difference	
	DM/shirt-	Ratio	DM/shirt	Ratio	DM/shirt	Ratio
1. Spinning	2.94	100	2.59	88	– 0.35	12
2. Weaving – Added value	2.22	100	1.48	67	– 0.74	33
3. Dyeing/Finishing Added value	1.28	100	0.88	69	– 0.40	31
1 + 2 + 3	6.44	100	4.95	77	– 1.49	23
4. Making-Up – added value	6.49	100	2.90	45	– 3.59	55
TOTAL MANUFACTURING COST	12.93	100	7.85	61	– 5.08	39

TABLE 3
Sample costing : Cotton dress/shirt – manufacturing costs

In the relative cost analysis of the fabric, the spinning sector is more competitive than the weaving and finishing sectors and it explains why there is limited cotton yarn imports into the EEC from Hong Kong, Taiwan or South Korea.

We have seen before that the total cost difference in our sample, was equal to the cost difference in labour cost alone.

In Table 4 we see that of the total 4.54 DM/shirt cost difference in labour, 1.30 DM or 29 % of the difference is accounted for in the cost of the finished fabric and 3.24 DM or 71 % in the labour cost portion of making-up the shirt.

Nearly one half of the labour cost difference in the fabric is due to weaving, the rest of the difference being equally divided between spinning and dyeing and finishing. From all of this, again for our specific example, a cotton dress shirt made in Europe with fabric made in Europe would require a labour cost reduction of 45 % in the making-up sector, of 30 % in the weaving sub-sector and of 25 % in the spinning and dyeing and finishing sub-sectors.

	EEC – High Wage		Far East		Difference	
	DM/shirt	Ratio	DM/shirt	Ratio	DM/shirt	Ratio
1. Spinning	0.59	100	0.24	41	– 0.35	59
2. Weaving	1.08	100	0.48	44	– 0.60	56
3. Dyeing/Finishing	0.56	100	0.21	38	– 0.35	62
1 + 2 + 3	2.33	100	0.93	42	– 1.30	58
4. Making-Up	4.36	100	1.12	26	– 3.24	74
TOTAL LABOR COST	6.59	100	2.05	31	– 4.54	69

TABLE 4
Sample costing : cotton dress/shirt – labour cost analysis

We now have a pretty clear idea as to where the technological developments are needed for the EEC textile industry to become more competitive with the Far East in the manufacture of articles that represent the largest share of the European textile consumption.

These can be summarized as follows in order of priority : (Table 5)

1. In making-up with the emphasis in the sewing room, and in automated material handling, with the objective of reducing the labour cost content of making-up by 45 %.

2. In the weaving sub-sector, with emphasis on the fabric formation process and in second instance on the warp preparation process, with the objective of reducing the labour cost content of weaving by 30 %.

3. In dyeing and finishing, with emphasis on automation and flexibility with the objective of reducing the labour content by 25 % or an equivalent reduction through a combination of labour cost reduction with dyes and chemicals cost reduction.

4. In the spinning sector, with emphasis on the spinning process for finer counts.

| TECHNOLOGICAL IMPROVEMENTS |
| NEEDED |
| |
| PRIORITIES : |
| |
| 1. MAKING-UP |
| a) Sewing |
| b) Material Handling |
| |
| 2. WEAVING |
| a) Weaving |
| b) Warp preparation |
| |
| 3. DYEING / FINISHING |
| |
| 4. SPINNING |
| a) Finer counts |

TABLE 5

At this point I guess I could reach for my crystal ball and speculate about the future technological improvements that may come about and whether these improvements will be sufficient to reach the objectives that we have defined.

Instead, I have tried to use a different approach that I believe is more rational. It is usually recognized by machinery manufacturers that from the time a patent is granted for a new machine or process it usually takes three to four years before a commercially viable product can be sold on the market, so we decided to make an analysis of the patents granted recently, reasoning that this should give us a pretty fair idea of what will most likely become operational in the next four to five years.

In the past this would have been a huge task, but today with computers and the availability of data banks it is relatively easy if still fairly complex. We made this analysis in the top seven textile patent-producing countries, namely West Germany, Great Britain, France, Italy, Switzerland, Japan and the United States. Furthermore we only took into account the patents that originated in these countries so that we could determine with better probability where the research was actually performed.

There are eleven broad classifications for textile patents, and in order to relate to our specific example, we did not include in our analysis patents granted in knitting and non-wovens, and regrouped the remaining nine classifications into the four categories of spinning, weaving, finishing and making-up.

Table 6 summarizes the results of this analysis. During the period from the 1st of January 1984 to the 30th of July 1985, a total of 2,673 patents were granted and originated in the top seven textile patent producing countries of the world.

PROCESS OPERATION	SOURCE COUNTRY	COUNTRIES								
		D	GB	F	I	CH	S.TOT.	JAPAN	USA	TOTAL
1. Spinning		127	15	31	6	23	202	229	87	518
2. Weaving		61	25	67	3	57	213	235	124	572
3. a) Dyeing / Printing		67	10	13	0	39	129	404	125	658
b) Finishing		30	7	8	0	1	46	63	52	161
4. Making-Up		78	9	11	2	15	115	465	184	764
TOTAL		363	66	130	11	135	705	1396	572	2673
RATIO		14.%	2.%	5.%	1.%	5.%	27.%	52.%	21.%	100.%

* PATENTS ISSUED IN THE TOP SEVEN TEXTILE PATENT PRODUCING COUNTRIES

TABLE 6
Patents analysis * : 1.1.84 to 30.7.85

As you can see of that total 1,396 patents or 52 % originated in Japan, 572 patents or 21 % originated in the U.S.A. and 705 patents or 27 % originated in the five European countries. Even though quantity does not necessarily mean quality it does give us a clear indication as to where the research efforts are concentrated. It seems reasonable to conclude from this data, that Europe lags behind in research in the making—up area and that the technology that will become operational in the next few years in that area will most likely come from Japan or the United States. In the European countries, research is concentrated on the spinning and weaving sectors, where they already hold a technological lead which they will maintain but for how long ?

It is also interesting to relate the number of patents granted in each textile sector, to the added value generated per sector in our sample costing. (Table 7)

The research in Japan seems to be conducted in a way that is more directly related to the needs of the sectors. It does not mean that there is less research in spinning and weaving, but rather that there is much more in the making—up processes.

	ADDED VALUE IN MADE UP SHIRT		PATENTS GRANTED / SOURCE COUNTRIES			
	EEC—HIGH WAGE		5 EUROPEAN	USA	JAPAN	TOTAL
	DM	%	%	%	%	%
1. Spinning	1.19	12	35	19	23	26
2. Weaving	2.10	21	37	28	24	28
3. Finishing	1.04	10	8	12	6	8*
4. Making—Up	5.72	57	20	41	47	38
TOTAL	10.05 **	100	100	100	100	100

* EXCLUDING DYEING AND CHEMICAL FINISHING

** EXCLUDING INTEREST, SALES, GENERAL ADMINISTRATION COSTS

TABLE 7

Following up on the assumption, that it is on the basis of the textile patents granted in the last 18 months, that the technology available in the next three to four years will depend, we have looked in more detail on the actual patents granted to see where the emphasis of the research in each textile sub-sector is being directed. On the basis of this further analysis, I believe we can safely predict the following :

In pre-spinning :
Two areas will have some definite improvement : the carding process both from a quality and automation point of view and the tow converting process. Apart from these two areas, no major improvement is expected.

In spinning :
The accent is on three main areas :

- the spinning process itself with improved technology for open-end spinning, air jet spinning and friction spinning;

- new and improved automated piecing-up, and cleaning devices;

- improved stop motion devices, responsive to change in tension, failure of supply or yarn breaks.

In weaving preparation :
Some further improvement can be expected in two for one twisting and in the warping process.

In weaving :
There is a very big emphasis on patents for air jet looms and to a lesser degree on other types of shuttleless looms. The second largest area of interest is in the automation of the supply of weft yarns to the loom. There should also be some significant improvement in jacquard and dobby attachments.

In dyeing and printing :
The main emphasis is on better utilization of dyes and nearly all types of dyes are involved; in diminishing order of importance they include : reactive dyes, azoic dyes, dispersed dyes, dyes for dyeing leather and acetate dyes.

In chemical finishing :
There will be further improvement in this area, with two-thirds of the patents in the last 18 months coming from Japan.

In mechanical finishing :
The overwhelming emphasis is in the area of napping and raising machines. But some improvements can also be expected in tentering machines, and in energy saving apparatus in the heating or cooling of fabrics.

In making-up :
The majority of recently issued patents seem to concentrate on two main areas : robotized or programmable sewing machines for specific seams configuration and the automation of material handling and work feeding to and from the sewing machine.

We have seen earlier that on the basis of our specific example, in order for a plain shirting fabric manufactured in a high wage European country to be competitive with the Far East, the manufacturing cost would have to be reduced by 10 %.

On the basis of the state-of-the-art technology existing today, and what improvements we can safely expect to happen within the next three to five years, there is a very strong probability that the objective will be achieved and even surpassed. But we have also seen that for the making-up sector to be competitive with the Fast East, the manufacturing costs (excluding raw material costs) would have to be reduced by 30 % which would also translate into an overall gain of labour productivity of 80 %. On the basis of the state of the art technology existing today, and the improvements we can safely expect to happen within the next three to five years, this objective will not be reached. It is more likely that the technology currently in the pipeline will bring an overall labour productivity increase of 40 to 45 %. This is a substantial improvement but not sufficient to reach the stated objective, which will eventually be reached in eight to ten years from now.

Technical developments have progressed to the point where today the concept of the automated textile mill is now possible and most likely will occur in some form by the end of the 80's and early 90's. By definition the automated mill will consist of a series of modular manufacturing systems, each consisting of electronically and mechanically connected components with appropriate communications interlinking to enable mostly labour free and cost-effective operations within the modules.

Furthermore the modules will be interlinked to enable sequential processing of the required conversion steps. To the automated textile mill, four basic elements are essential. These involve the following factors :

- machinery and process technology,
- process interlinking,
- management information and control system,
- human resources training.

All four elements have to play a vital role if the benefits of full automation are to be realized. There is still a very serious problem for textile mills in reaching their ultimate objective of automated operations, and this is the problem of computer compatible software. As new modular manufacturing systems are combined and assembled by mills from a wide range of machinery sources, it is certain that the on-board computers and micro-processors installed by the respective machine builders will be incompatible, that is, they will not communicate or integrate. This presents several potential problems :

- The communications systems to interlink machinery from several suppliers may have to be designed by each mill user.

- The cost of new machines include micro-processors which may have to be replaced or reprogrammed if at all possible. This would involve discarding a capability already paid for with additional costs to systemize and standardize all computer components which is likely to be quite expensive.

- Given that software systems tend to be proprietary, it is not expected that machine builders will be able to adapt a potential user's system into its equipment when assembling the machinery.

Consequently, despite the widespread availability of sophisticated textile machinery, the computer question remains unresolved and is, in fact, a major stumbling block at this time to achieve automation.

As it now stands, each new purchaser of equipment will be saddled with the challenge to intra-and interlink systems and generate separate programs to handle the equipment in question. Without a proficient overall design and integrated computer system, the full potential of the new machinery will not be achieved. Also most textile companies do not have personnel on board which have combined knowledge and expertise in textile and machinery technology, in the computer sciences and in management systems.

However, this hurdle can and will most assuredly be overcome in time. In any case, a fully automated mill is not necessary, and even not desirable at this point in relation to the additional cost required to achieve 100 % automation. A much more valuable objective to reach would be to achieve flexibility without adding to the manufacturing costs.

In summary : we can safely predict that the yarn and fabric sector in the European Community will be basically worldwide competitive in four to five years from now, provided that the industry will be able to generate the funds needed to acquire the new technology and also provided the textile world operates on the basis of fair trade, which is not necessarily the case today.

But even with the primary textile industry being competitive, the remaining weaknesses of the making-up sector may put a large section of the primary textile industry in difficulty.

GENERAL CONCLUSIONS OF THE SYMPOSIUM

D. FINLAY MAXWELL
Chairman of the Scientific Research Committee of Comitextil

As far as the scientific achievements are concerned, you have had, at this Symposium, a preview presentation, four detailed sessions, four highlights presentations and one highlight of the highlights. The message must now be very clear. We have achieved significant successes in each of the themes. We have proved, beyond question, that with the assistance of the Commission, we can successfully undertake collaborative research between complementary research units, and across national borders.

Why communal research? Why not attempt to carry out this work on a national basis? The ever increasing rate of change of technology means that more and more frequently research in textiles must be multi-disciplinary.

Let me give you a hypothetical example, for example the dyeing or surface treatment of linen or for that matter, any fabric. One needs to study the effect of a wide variety of chemicals, probably a range of heat treatments, including the use of radio-frequency to cause a thermal gradient inwards or alternatively microwave or infra-red radiation to help to move foam applied additives outwards. There is always the possibilities of using alpha or beta radiation or even lasers. We can also use KAWABATA analysis techniques for optimising these treatments.

Can any one of our organisation be expected to have expertise in all these fields? If not, how is this type of research to be carried out? One of the major strengths of community research lies in the ability to bring together an ever-increasing range of disciplines - each specialist unit having a high level of competence.

Such programmes have an additional value in that they are more frequently subjected to constructive appraisal and self-criticism. I have heard from one or two delegates that here and there part of work presented has already been carried out elsewhere. This will always happen, particularly where national programmes proceed in relative independence. Here again, community programmes provide a very important role in eliminating or substantially reducing costly duplication.

In broad terms, therefore, I see the role of Community programmes for the larger, broadly based projects. National programmes are perhaps better orientated towards the development of particular expertise within this complex. The two routes of development can be made complementary.

We have discussed the successes that have been highlighted at this Symposium. What do we intend to do with the results?

First, we must undertake to ensure the maximum possible dissemination and distribution of these results. As a first stage, within 6 months, the Commission willl produce and distribute a complete set of the papers to all delegates. It will then be up to us to re-distribute the knowledge as widely as possible throughout our organisations and Federations.

With determination, we must endeavour to ensure the implementation of much of the completed work. However, there remains the incomplete work which we may well consider worthy of the effort necessary to bring it to a conclusion. To monitor and guide this I should like to see some form of a "programme of implementation" encouraged by Comitextil.

I am fully aware of the ever increasing workload that will be required to be undertaken by our technical staff but I have to say that this has been evident for some time. Indeed from a personal viewpoint I considered that my attendance at meetings, once or twice a month, as an amateur chairman, based in Huddersfield was totally inadequate for your future requirements. It was for this reason that I recently stepped down to leave the field clear for a successor to implement the changes that seem to be necessary.

These changes will require a re-appraisal of a significant part of our time and resources. I believe this can be achieved and that an appropriate organisation can be established to ensure that programmes are developed on a continuing basis. Programmes appropriate to the community requirements are needed - programmes of the highest technical quality that our combined available resources can produce and carry out.

*   *   *

312

CLOSING ADDRESS

J.H. LEACH
President of Comitextil

My role is quickly to draw to a conclusion this Symposium on "Competitivity through innovation".

The chairman of this session, Dr Finlay Maxwell has already commented on the summaries of the technical material of the presentation and I now wish to comment on how I believe that Comitextil and the Commission together must continue the good work already done in order to maintain the dynamic interest of the textile and clothing industries of Europe over the coming years.

I must comment on what Mr Verret and Mr Tent have said.

Mr Verret has given specific information on how the gaps between Europe and the sophisticated Far Eastern World can and must be bridged. I take this opportunity to point out that these prospects constitute one of the reasons that make it necessary to maintain a framework for regulating trade for a certain period of time to come. The Multifibre Agreement may represent to some a form of restriction of trade but it is only one of many that exist throughout the world and I am sure that much more time is necessary to simplify the complex framework that at present exists.

A form of trade stability will lead to a better future for all, rather than the chaos that would result from the termination of the Multifibre agreement. There is enough chaos in this world already. The trend is in fact towards highly automated production equipment with robots and computerized sensors providing automatic manufacturing and quality monitoring.

Mr Verret described the problems of interlinking the various types of software and the problems that we face in introducing new technologies, into our companies and into our industry.

These developments will result in optimum production conditions thanks to zero or minimum defects, on-line monitoring of both quality and colouring production, non manual handling of materials and automatic information retrieval. That is why technological progress can and must make Europe again a viable location for a textile industry able to serve Europe and the rest of the world. Not only is the continent a source of modern technology but we possess a level of industrial sophistication to meet the demand for efficient maintenance and operation which is a prerequisite for the assurance of high quality.

A commitment to quality has to be the basis of our future strength. Garment manufacturers need "easy to cut" quality fabrics to be able to minimize their own production costs.

We have seen sector by sector that there is room available in spinning, weaving, dyeing and finishing, and garment making for these improvement to be made. But at the same time buying and selling are frequently much more than a matter of price. Market proximity is still of great importance to retailers. They are not on their own able to satisfying increasingly volatile customer requirements and pressure from consumer associations.

Mr Tent , I am sure, understands that the textile industry is essentially composed of small and medium sized enterprises and that we have had difficulties in complying with the general terms for applications under the BRITE programme.

In addition the problems of the textile industry do not easily fit in the topics that have been presented. The projects selected are likely to have some impacts without really covering the entire range of problems that we are facing.

Major projects are underway in Japan and the USA which will help to form the future shape of the world industry. In both countries the projects are based on the active participation of leading retailers, garment makers, finishers, fabric makers, spinners, fibre producers, machinery and equipment suppliers, and Government.

Comitextil is willing, indeed is anxious, and must contribute actively to such an approach but it will need the support that I believe it is possible to obtain from the Commission, to ensure the work goes ahead in a time scale that is meaningful, because time will run out. Essentially we have a period of time now while we face the fact that the pressure for many national governments is towards free trade and that the concept of the Multifibre agreement, or whatever its successors will be called, will not continue forever.

I know that Mr L. Gros, who has taken over the role of Dr Finlay-Maxwell within Comitextil, will actively pursue the opportunity that is now in front of us.

There are two other duties for me to perform

The first is to thank those who have made the Symposium in Luxembourg such a success.

First of all thanks to the Commission whose support of the research programme and financial and moral support of the Symposium itself has done so much to make it all possible.

To Dr Finlay-Maxwell who within Comitextil has maintained his activities as Chairman of the Research Committee for almost six years.

To the Committee who have organized the Symposium :

From Comitextil particularly, Mr C. Blum, Miss J. Candries and for many of the domestic arrangements Mrs P. De Wilde.

From the Commission Mr J. WURM, and Mr D. NICOLAY, who has fed us and arranged buses for all the right times and for all the other members from the Commission who had been with us during these days. Also to the speakers, researchers, and the contributors at the sessions.

Also to our special lectures, including Mr Boden on behalf of the Luxembourg Government, and Mr Bourdeau, Mr Rutsaert, Mr Verret, Mr Tent, and the session Chairmen Dr Jeffriies,Mr Stryckman, Dr Bona and Mr Van Lancker. We also thank the translators who made it all understandable for us and "tous les autres".

Finally to all of you for coming, participating and helping to make this Symposium a success and for your future work in disseminating the results and pursuing the opportunities.

And now "bon retour" and good wishes for the future.

I declare the Symposium closed.

ZUSAMMENFASSUNGEN IN DEUTSCHER SPRACHE

Bekleidungsphysiologie und Konstruktion von Kleidung

Neue Spinnereitechnik in der Wollindustrie

Qualität von unkonfektionierten und konfektionierten
Maschenwaren

Qualitätsverbesserung von Leinenerzeugnissen

Allgemeine Schlussfolgerungen des Symposiums

BEKLEIDUNGSPHYSIOLOGIE UND KONSTRUKTION VON KLEIDUNG

R. JEFFRIES
Shirley Institute

Diese Sitzung beschreibt die Forschungsarbeit auf zwei Gebieten : Physiologie und Komfort von Kleidungsstücken, Stoffbeschreibung und Kleidungsaufbau in Bezug auf Kleidungs-Herstellungsmethoden und Kleidungsstil und Ästhetik. Diese beiden Gebiete waren Teil des Zweiten Textil-Forschungsprogrammes der EWG wegen ihrer Bedeutung für die Textilindustrie und die Hersteller und Einzelhändler von Kleidungsstücken. Es besteht kein Zweifel darüber, daß die Verbesserung im Komfort von Stoffen und Kleidung gegenwärtig für Hersteller und Einzelhändler ein wichtiges und in Entwicklung befindliches Thema darstellt, und deshalb ist eine bedeutende Gruppe von Projekten bezüglich dieses Themas heute äußerst willkommen. In ähnlicher Weise besteht ein wachsendes Interesse an der Entwicklung von Kleidung mit gutem Stil, guter Qualität und gutem Sitz, sowie an der Automatisierung der Kleidungsherstellung; in beiden Bereichen ist die Bezeichnung, Einordnung und Gestaltung der Stoffe und daraus gefertigten Kleidungsstücke von entscheidender Bedeutung.

Es wird die Arbeit an fünf Projekten beschrieben. In jedem der Fälle ging man mit dem Ziel vor, Ergebnisse und Schlußfolgerungen von direktem und sofortigem Wert für die Textil- und Bekleidungs-Industrien der EWG zu produzieren, um diesen Industrien bei der Konzeption besserer Produkte und bei der Entwicklung verbesserter Herstellungsverfahren zu helfen.

Der erste Bericht beschreibt die Forschung bezüglich des Komplexen Themas von Kleidungskomfort, bezogen auf Beschaffenheit und Eigenschaften der betroffenen Gewebe, Kleidungsstücke und Kleidungszusammenstellungen. Zwei Arten von Komfort werden untersucht : "thermophysiologischer" Komfort (verbunden mit der Aufrechterhaltung des Wärme- und Feuchtigkeitsgleichgewichtes, das erforderlich ist, um den Körper auf einer angenehmen Funktionstemperatur zu halten), und "fühlbarer" Komfort (bezogen auf die tastbaren Eigenschaften direkt auf der Haut zu tragender Kleidungsstücke und die Art, wie sie mit der Haut zusammenwirken).

Drei Projekte befassen sich mit dem thermophysiologischen Komfort. In einem dieser Projekte erhielt man ausführliche Informationen über den thermophysiologischen Komfort von Kleidungsstücken und die Entwicklung von Bewertungsmethoden und Leitfäden. Es wurden Verhaltensmodelle der menschlichen Haut entwickelt, und die Ergebnisse der Wärme- und Feuchtigkeitsdampf-Übertragungsmessungen wurden in voraussagbare Formeln eingesetzt, um die Errechnung eines "Komfort-Wahlbeschlusses" zu ermöglichen.

Dieser "Komfort-Beschluß" steht in wechselseitiger Beziehung zu der Erfahrung von Trägern und ersetzt so zu einem großen Teil die Notwendigkeit für kostspielige Träger-Versuche. Ein anderes Projekt befaßte sicht mit Verbesserungen im Thermopysiologischen Komfort von Regenbekleidung. Es wurden Untersuchungen an den verschiedenen Arten von wasserdichten und wasserdampfdurchlässigen Stoffen durchgeführt, und besondere Aufmerksamkeit wurde der Kombination von Durchlässigkeit und Ventilation gewidmet. Eine Technik zur Untersuchung der Ventilationswirkungen wurde entwickelt. Es werden Anleitungen zur Konzeption von Regenschutzkleidung gegeben. Ein drittes Projekt beschäftigt sich mit physiologischen Messungen der Verteilung von Körperwärme und Schweiß in Verbindung mit Kleidungsstücken aus Wolle, Baumwolle, Wollmischungen und Baumwollmischungen.

Das vierte Projekt untersucht das Thema des "tastbaren" Komforts von auf der Haut zu tragenden Kleidungsstücken. Bei dieser Arbeit war es zuerst erforderlich, sorgfältige und umfangreiche Träger-Versuche durchzuführen, um festzustellen, welches die Empfindungen und physikalischen Faktoren sind, die den fühlbaren Komfort dieser auf der Haut zu tragenden Kleidungsstücke ausmachen. Auf der Grundlage der Ergebnisse aus diesen Versuchen wurde eine Reihe einfacher und subjektiver Testverfahren entwickelt, um solche Faktoren wie örtliche Reizung, Prickeln, Kitzeln, örtlichen Druck, Kältegefühl, Rauhheit, Nässehaftung, Faserschuppung und elektrostatische Wirkungen zu messen.

Der zweite Bericht beschreibt die Arbeit an der Festlegung und Klassifizierung der physischen Eigenschaften von Geweben in Bezug auf Kleidungsdesign, um bei der Optimierung des Designs zu helfen, was zu verbesserten Herstellungsverfahren und zur Fertigung von komfortablen und ästhetisch zufriedenstellenden Kleidungsstücken führt. Die Fähigkeit von Geweben, den Beanspruchungen und in verschiedene Richtungen gehenden Faltungen zu widerstehen, die bei der Kleidungsfertigung anfallen, wurde untersucht. Diese Art von Untersuchung erfordert den Einsatz hochmoderner Techniken, die sich vielleicht nicht zur routinemäßigen Benutzung von seiten der Kleidungshersteller eignen.

Der Kleidungshersteller haben sich mit der Überprüfung der Untersuchungsergebnisse befaßt. In einem zweiten Teil dieses Projektes wurden Arbeiten hinsichtlich von Problemen beim Zusammenwirken von Gewebe und Maschine in der Industrie durchgeführt; die Überwachung von Geweben hinsichtlich der automatischen Einstellung der Verarbeitungsmaschinen wurde berücksichtigt. Zwei spezifische Themen wurden experimentell studiert : Stoffvorschub beim Nähen (um Gewebe- und Maschinenmerkmale zu identifizieren die den Durchgang des Stoffes durch die Maschine negativ beeinflussen) und Unterdruck im Fertigungsverfahren (um die Kompromisse zu untersuchen, die erforderlich sind zwischen Produktionsgeschwindigkeit, Druckbedingungen und physikalischen Dimensionen).

MESSUNG UND BEURTEILUNG DES TRAGEKOMFORTS VON TEXTILIEN UND KLEIDUNG

K.H. UMBACH
Hohenstein

1. - THERMOPHYSIOLOGISCHER TRAGEKOMFORT

Um Wettbewerbsfähig zu sein, muß Kleidung heute neben guten mechanisch/technologischen Eigenschaften und guter Pflegbarkeit auch einen guten thermophysiologischen Tragekomfort besitzen. Gute Trageeigenschaften können jedoch als Konstruktionselement bei der Textil- und Kleidungsentwicklung nur dann eingeplant werden, wenn es möglich ist, den thermophysiologischen Tragekomfort quantitativ zu messen und zu beurteilen. Ziel des Projekts war es, dafür ein Analysensystem zu schaffen, das von der Industrie bei der Produktentwicklung eingesetzt werden kann.

Für die Beurteilung der thermophysiologischen Trageeigenschaften von Textilien wird ein Thermoregulationsmodell der menschlichen Haut (Hautmodell) eingesetzt. Es konnten mit diesem Modell Meßverfahren entwickelt werden, mit denen sich die Wärmeisolation sowie das Feuchtetransport- und Feuchtespeichervermögen von Textilien sowohl unter "normalen" Tragebedingungen als auch bei stärkerem und starkem Schwitzen des Trägers quantitativ bestimmen und in artikelspezifischen Kenngrößen ausdrücken lassen. In dem Vorhaben wurden Vorhersageformeln aufgestellt, in die diese Kenngrößen eingesetzt werden und die eine Bewertungsnote für den zu erwartenden thermophysiologischen Tragekomfort eines Textils liefern. Die Formeln wurden für verschiedene Artikelgruppen wie z.B. Unterwäsche, Hemden, Hosen, etc. getrennt entwickelt. Ausgedehnte Trageversuche mit Testpersonen haben gezeigt, daß die mit diesem Vorhersagemodell für ein Textil berechneten Tragekomfortnoten in guter Übereinstimmung mit dem vom Menschen beim Tragen subjektiv registrierten Komfortempfinden stehen.

Da diese Beurteilung der Trageeigenschaften von Textilien auf rationallen Labormeßverfahren basiert, bietet sich als direktes Ergebnis des Forschungsprojekts der Textilindustrie nunmehr die Möglichkeit, Tragekomfort bereits in der ersten Phase einer Neuentwicklung gezielt als Konstruktionselement einzuplanen. Dies insbesondere deshalb, weil die in dem Vorhaben durchgeführten Grundsatzuntersuchungen zu Konstruktionsleitlinien geführt haben, mit denen bestimmte Artikel (z.B. Sportwäsche, Wetterschutztextilien, Kälteschutztextilien) mit optimalen thermophysiologischen Eigenschaften ausgestattet werden können.

Auf Stufe 2 des Analysensystems werden die konfektionierten Kleidungsstücke mit einer beweglichen Gliederpuppe untersucht, die ein Thermoregulationsmodell des Menschen darstellt.

Es liefert spezifische thermophysiologische Kenndaten (Wärme- und Wasserdampfdurchgangswiderstand) des gesamten Kleidungssystems einschl. der im Mikroklima enthaltenen sowie an der äußeren Kleidungsoberfläche anhaftenden Luftschichten.

Mehrere hundert kontrollierte Trageversuche mit Testpersonen in der Klimakammer, die als Stufe 3 in dem Analysensystem Abb. 1 enthalten sind und in denen sowohl objektive Körperfunktionsgrößen als auch subjektive Empfindungen der Testpersonen erfaßt wurden, führten zu einem Vorhersagemodell für konfektionierte Kleidung.

Mit diesem Modell, in das die mit der Gliederpuppe gemessenen Kleidungskenndaten eingesetzt werden, ist die universelle und mit Praxiserfahrungen übereinstimmende Beschreibung des Tragekomforts von Kleidung unter jeder beliebigen Variation der Randbedingungen von Klima und

Tätigkeit des Trägers möglich.

So kann nunmehr nicht nur aufgrund weniger Labormessungen der Temperatur- bzw. Verwendungsbereich einer Kleidung angegeben werden, in dem der Träger bei einer bestimmten Aktivität weder friert noch unzumutbar stark schwitzt, sondern es lassen sich auch direkt die zeitlichen Verläufe verschiedener Körperfunktionsgrößen des Menschen sowie der Feuchte im Mikroklima beim Tragen von Kleidung unter bestimmten äußeren Klima- und Tätigkeitsbedingungen quantitativ berechnen.

Die in dem Projekt durchgeführten Untersuchungen haben gezeigt, daß diese Körperfunktions- und Mikroklimagrößen unmittelbar das thermophysiologische Komfortempfinden des Menschen bestimmen. Wie bei den Textilien allein konnten damit biophysikalische Formeln entwickelt werden, die direkt eine Tragekomfortnote für das komplette Kleidungssystem liefern. Auch für die Bekleidungsindustrie sind damit bei der Produktentwicklung zukünftig ausgedehnte und kosten- wie zeitaufwendige Trageversuche weitgehend verzichtbar geworden.

Eine Beurteilung der physiologischen Gebrauchstüchtigkeit von Kleidung ist vielmehr als Ergebnis des Forschungsvorhabens heute aufgrund weniger Labormessungen möglich.

2. - HAUTSENORISCHER TRAGEKOMFORT

Ziel des Vorhabens war die Ermittlung spezifischer Konstruktionsparameter, die den hautsensorischen Tragekomfort von Textilien bestimmen, hervorgerufen durch deren Berührkontakt mit der Haut. Es wurden einfache Labormeßmethoden entwickelt, mit denen sich die verschiedenen Aspekte des hautsensorischen Komforts erfassen lassen. Eine Literaturrecherche, eine Umfrage, bei der ca. 1000 Menschen nach ihren Kleidungsgewohnheiten befragt wurden, sowie Trageversuche mit Testpersonen, die 22 zu T-Shirts verarbeitete Textilmuster umfaßten, ergaben, daß die folgenden Einflußgrößen den sensorischen Tragekomfort bestimmen :

2.1. Lokale Paßform (local fit)

Der auf die Haut von zu eng sitzender Kleidung ausgeübte Druck kann alle anderen sensorischen Komfortempfindungen überlagern. Es wurde ermittelt, welche Drücke an verschiedenen Körperpartien gerade noch toleriert werden. Ein Meßgerät wurde entwickelt, mit dem der Anpreßdruck von Kleidungsstücken am Körper quantitativ erfaßt werden kann.

2.2. Hautirritationen hervorgerufen durch Etiketten
(local irritation by garment labels)

Scharfrandige Ränder und Ecken, wie sie insbesondere bei thermoversiegelten Etiketten vorkommen, können zu erheblichen Diskomfortempfindungen führen.

2.3. Pieken (prickle)

Wollhaltige Textilien werden "piekend" empfunden, wenn 3.5 % der Fasern einen Durchmesser von über 30u m bzw. wenn 0.6 % der Fasern einen Durchmesser von über 40u m besitzen.

2.4. Kitzeln (tickle)

Insbesondere bei schweißfeuchter Haut werden Textilien mit "haariger" Oberfläche als unangenehm kitzelnd empfunden. Aus der fotografischen Abbildung der von der Textiloberfläche abstehenden Faserenden läßt sich beurteilen, ob ein Textil zum Kitzeln neigt.

2.5. Flusenbildung (fibre shedding)

Flusen, die sich beim Tragen aus dem Textil herauslösen und auf der Haut ankleben, führen zu negativen Trageempfindungen. Mit einer Apparatur, bei der Textilien in definierter Weise ausgeschüttelt werden, läßt sich deren Neigung zur Flusenbildung durch Vergleich mit fotografischen Standards in einem "Flusenwert" quantitativ ausdrücken.

2.6. Kleben feuchter Textilien auf der Haut (wet cling)

Eine Meßapparatur wurde entwickelt, bei der die Kraft, die aufzuwenden ist, um einen Textilstreifen über einen Plexiglaszylinder zu ziehen, das unangenehm empfundene Kleben von Textilien auf der Haut wiedergibt. Die durchgeführten Trageversuche haben eine gute Korrelation zwischen diesen Klebekraftwerten und dem subjektiven Empfinden der Testpersonen ergeben.

2.7. Initial-Kältegefühl (initial cold feel)

Obwohl das beim Anziehen von Kleidungsstücken evtl. empfundene Kältegefühl nur wenige Sekunden andauert, kann es dennoch sehr wesentlich und nachhaltig den sensorischen Komforteindruck des Trägers prägen. Mit einem speziell entwickelten Meßgerät läßt sich dieses von Textilien hervorgerufene Initial-Kältegefühl quantitativ erfassen.

2.8. Kratzen (scratchiness)

Die Messung der Haftreibung zwischen einem Textil und einem Probenkörper ermöglicht die Beurteilung, ob ein Textil auf der Haut als kratzend empfunden wird.

2.9. Elektrostatische Aufladung (static electrical effects)

Funkenentladungen beim Ausziehen von Kleidungsstücken oder bei Körperbewegungen sowie das Ankleben von Textilien an der Haut durch elektrostatische Aufladung können zu schlechtem hautsensorischen Tragekomfort führen. Ob sich ein Textil hier kritisch verhält, kann durch die Messung der beim definierten Ausziehen von Kleidungsstücken auftretenden Ladungsmengen beurteilt werden.

Insgesamt wurden in dem Forschungsprojekt nicht nur Verfahren entwickelt, mit denen sich der sensorische Tragekomfort von Textilien messen läßt, sondern aus den gewonnenen Grundlagenkenntnissen konnten darüberhinaus Leitlinien aufgestellt werden, die es der Industrie besser als bisher ermöglichen, hautnah getragene Kleidungsstücke gezielt mit guten Hautsensorischen Trageeigenschaften herzustellen.

3. - VERBESSERUNG DES TRAGEKOMFORTS VON SCHUTZKLEIDUNG

Ein großer Teil von Schutzkleidung soll ihren Träger vor witterungsbedingter Nässeeinwirkung schützen. Aus physiologischer Sicht stellt sich bei dieser Art Kleidung das Problem, Wasserdichtheit mit gleichzeitiger ausreichender Wasserdampfdurchlässigkeit zu verbinden. Ist letztere nicht gegeben, resultiert ein schlechter Tragekomfort, der die Akzeptanz dieser Schutzkleidung herabsetzt.

Um eine physiologische Übersichtsanalyse der heute verfügbaren Wetterschutztextilien durchführen zu können, wurden in dem Forschungsprojekt zunächst spezifische Meßverfahren zur Bestimmung der Wasserdichtheit sowie der Luft- und Wasserdampfdurchlässigkeit von Textilien unter Praxisbedingungen entwickelt.

Die mit diesen Verfahren durchgeführten Reihenuntersuchungen haben gezeigt, daß insbesondere Laminate mit integrierten Membranen, spezielle mikroporöse Beschichtungen sowie Mikrofasergewebe die an Wetterschutztextilien zu stellenden technologischen und physiologischen Anforderungen gleichzeitig erfüllen können. Es zeichneten sich Möglichkeiten ab, durch eine Kombination dieser verschiedenen Technologien Textilien mit gegenüber den derzeitig verfügbaren Konstruktionen verbesserten Eigenschaften zu entwickeln. Verbindet man mit diesen Textilien noch eine geeignete Schnittkonstruktion der Kleidungsstücke, die bei Körperbewegungen des Trägers bzw. bei äußerer Luftbewegung eine Konvektion und Ventilation im Mikroklima ermöglicht, so ist das Ziel einer optimalen Wetterschutzkleidung erreicht. In die Praxis umsetzbare Möglichkeiten dafür wurden in dem Forschungsvorhaben aufgrund durchgeführter Trageversuche mit Testpersonen aufgezeigt, für die ein spezielles Verfahren zur quantitativen Messung der Ventilationsrate in der Kleidung entwickelt wurde. Diese Versuche führten darüberhinaus zu einem mathematischen Modell, das die physikalischen Wärme- und Feuchtetransportvorgänge in der Kleidung beschreibt und mit dem sich die Auswirkung der Wasserdampfdurchlässigkeit von Textilien in Verbindung mit schnittbedingter Kleidungsventilation auf die Tragekomforteigenschaften von Schutzkleidung unter bestimmten Klimabedingungen angeben läßt.

DESIGN- UND FERTIGUNGSOPTIMIERUNG VON KLEIDUNG

J. DESCHAMPS
CETIH

Der hier untersuchte Teil des Programms befaßt sich mit Überlegungen über die physikalischen Stoffeigenschaften beim Kleidungsdesign, um folgendes zu erreichen :

- Optimales Design des Kleidungsstücks mit möglichst wenig Veränderungen.
- Leichtere Aufmachung.
- Herstellung von Kleidungsstücken, die eine bequeme Form haben und ästhetisch zufriedenstellend sind.

1. - KLASSIFIZIERUNG VON STOFFEN AUF DER BASIS IHRER PHYSIKALISCHEN EIGENSCHAFTEN.

Die allgemeingebräuchlisten Methoden der Stoffklassifizierung beziehen sich auch heute noch auf Kategorien des Endgebrauchs. Sie sind mehr oder weniger subjektiv und beziehen sich direkt auf visuelle oder greifbare Informationen, die man bei der Inspektion und beim Griff des Stoffes erhält, obwohl diese gelegentlich von einfach meßbaren Merkmalen, wie z.B. dem Tuchgewicht, beeinflußt werden.

Die Modernisierung der Bekleidungsindustrie, die Automation sowie die kurz- oder langfristige Anwendung der Informatik für computerisierten Entwurf und Produktionsablauf erfordern eine objektivere Klassifizierung von Stoffen, die auf meßbaren Kriterien beruht.

Um in der Mehrheit der Bekleidungsfabriken Anwendung zu finden, müßte das objektive Klassifizierungssystem nur eine begrenzte Zahl von meßbaren Kriterien in Betracht ziehen; die optimale Klassifizierungsmethode wäre, Stoffklassen nach einer oder mehreren ausgesuchten Eigenschaften zu bestimmen.

Eine Untersuchung umfasste die in Frage kommenden Stoffeigenschaften
die jetzt mehr oder weniger intuitiv beim Design und der Herstellung von
Kleidungsstücken berücksichtigt werden aber die systematischer Untersuchung wert sind.

Die Eigenschaften, welche die Schwere der Aufmachung von Endprodukten am meisten beeinflussen, und die Produktion hochqualitativer Kleidungsstücke mit Komfort und Ästhetik, welche die Erwartungen der Designer
erfüllen, wurden untersucht.

Die höchstmöglichen Abweichungen bei allen Kleidungskategorien
wurden für alle Eigenschaften bestimmt, die einen Einfluß auf momentan
gebräuchliche Stoffe haben könnten.

Die Veränderlichkeit jeder dieser Eigenschaften innerhalb desselben
Tuchstücks und zwischen zwei verschiedenen Stücken wurde untersucht. Der
Zweck dieser Untersuchung ist es, die Veränderungen zu bestimmen, die
bedeutend sind.

Es hat sich als notwendig herausgestellt, Beschädigungen in der
Struktur des Stoffes, die bei der Aufmachung des Kleidungsstücks vorkommen, zu identifizieren.

2. - IDENTIFIZIERUNG DER STOFF-EINSCHRÄNKUNGEN UND DER BEIM AUFMACHEN ERLITTENEN BESCHÄDIGUNGEN.

Mit Hilfe fotografischer Studien von Veränderungen in der Stoff-
struktur bei den wichtigen Nahtvorgängen (Seitennaht, Schulternaht, Joch,
Ärmelnaht) konnt man jene Stoffeigenschaften identifizieren, die einen
Haupteinfluß auf die "Aufmacheigenschaften" von Stoffen haben.

Scherleichtigkeit und Falten in mehrere Richtungen beim Stoff
scheinen die Eigenschaften mit einem großen Einfluß zu sein, und die
Entwicklung einfacher Tests ist deshalb notwendig, um Herstellern eine
Ermittlung dieser Eigenschaften zu ermöglichen.

3. - VERSUCH ZUR FINDUNG VON METHODEN DIE EINE INDUSTRIELLE MESSUNG VON SCHERLEICHTIGKEIT UND FALTUNG IN MEHREREN RICHTUNGEN ERLAUBEN.

Die Scher- und Simultanfaltungseigenschaften von Stoffen scheinen
die Hauptaspekte bei der Aufmachung zu sein, und Messungen dieser Parameter erscheinen als äußerst wünschenswert.

Experimentelle Methoden wurden entwickelt, um diese Art von Messung
durchzuführen.

Es erschien passend, den Einfluß dieser Eigenschaften beim Design
und bei der Herstellung in der Fabrik zu untersuchen. Eine Untersuchung
"vor Ort" wurde deshalb durchgeführt.

4. - "VOR ORT" - BEWERTUNG UND PRÜFUNG DER BEDEUTUNG VON STOFFEIGENSCHAF-TEN IN IHRER KLASSIFIZIERUNG.

Die praktische Herstellung von Kleidungsstücken (Röcke, Kleider,
Mäntel, Jacken) nach gegebenen Spezifikationen und gegebenen Stoffen
wurde von Bekleidungsfabriken ausgeführt; sie folgten dabei ihren gewöhnlichen Design- und Aufmachungsverfahren, um die Berechtigung der entstan-
denen Hypothesen nachzuprüfen.

Experten der Bekleidungsindustrie und Verbrauchervertreter bewerte-
ten die erhaltenen Resultate (Produktionsvolumen - bequemer Sitz und
Ästhetik des Kleidungsstücks).

5. - ABSCHLUSS

Diese Studie stellt lediglich eine vorläufige Annäherung auf das
Problem dar, welchen Einfluß die physikalischen Eigenschaften des Stoffes
auf das Design des Kleidungsstücks haben.

Ein positives Design der gesuchten Form wurde bis jetzt weder vom ästhetischen noch vom bequem-sitzenden Aspekt erreicht. Eine bessere Kontrolle der Gleichheit von Stoffeigenschaften würde die Erfüllung dieser Aufgabe sicherlich beschleunigen.
Eigenschaften des Stoffes auf das Design des Kleidungsstücks haben.

Ein positives Design der gesuchten Form wurde bis jetzt weder vom ästhetischen noch vom bequem-sitzenden Aspekt erreicht. Eine bessere Kontrolle der Gleichheit von Stoffeigenschaften würde die Erfüllung dieser Aufgabe sicherlich beschleunigen.

324

ZUSAMMENFASSUNG UND SCHLUSSFOLGERUNGEN DES VORSITZENDEN DER SITZUNG

R. JEFFRIES
Shirley Institute

Die Sitzung über Bekleidungsphysiologie und Konstruktion von Klei-
dung war in zwei Teile unterteilt:
- Messung und Beurteilung des Tragekomforts von Textilen und Kleidung,
- Design- und Fertigungsoptimierung von Kleidung.
Als erste Schußfolgerung dieser Sitzung kann man sagen, daß diese
beiden themen gut gewählt waren, da sie für die heutigen und zukünftigen
Anforderungen der Textil- und Kleidungsindustrie der Gemeinschaft sehr
wichtig sind, wenn sie auf Neuerungen bedacht und konkurrenzfähig bleiben
will.
Der Komfort von Textilien und Geweben betrifft in allen Aspekten die
Funktionalität, die von größter Bedeutung für Hersteller und Benutzer von
Stoffen und Kleidungstücken sind, und in Zukunft aus verschiedenen
Gründen noch zunehmen wird. Zuerst, da die Durchführung von Schutzan-
sprüchen immer strenger werden,nehmen auch die Komfortansprüche zu;
zweitens gibt es zur Zeit eine stets zunehmende Forderung nach komfortabler, sportlicher Kleidung, besonders dann, wenn man übermäßig stark
schwitzt; drittens nimmt die Forderung nach komfortabler, einfacher all-
täglicher Kleidung stets zu.
In seinem Bericht über die Arbeiten am Komfort, beschreibt
Dr. Umbach die Forschung im Hohenstein Institut, Shirley Institute und
dem TNO Institute für Empfinden der thermophysiologischen Komforts,
sensoriellen Komforts und dem Komfort von halb durchlässiger ("atmungs-
aktiver") Regenkleidung. Das Projekt hat eine gesunde Wissensgrundlage
geschaffen, die diese Industrie beim Design von komfortabler Kleidung fü
gewisse Umstände unterstützen wird, wobei die Länge verkürzt wird und
kostspielige Trageversuche verhindert werden.
Man kam jedoch zu dem Schluß, daß trotz ausgezeichneter Fortschritte
auf diesem Gebiet noch viel zu machen ist. Zuerst sind mehr wissenschaftliche Hintergrundkenntnisse über alle Aspekte des Kleidungskomforts
erforderlich. Ein Teil dieser Arbeit muß natürlich darauf ausgerichtet
sein, die physiologischen, psychologischen und quasi medizinischen Aspekte des Problems zu erläutern. Obwohl man z.B. viel über die Schmerz-
schwelle der Haut weiß, ist dies nicht der Fall mit der Hauptempfindlich
keit gegen den sehr leichten Kontakt mit dem getragenen Stoff. Man muß
noch viel mehr lernen über das gesamte psychologische Konzept des
Komforts. Ist z.B. Komfort, die alleinige Abwesenheit von Unbehangen ?
Gäbe es da nicht einen positiveren Aspekt vom Kleidungskomfort. Könnte
man so z.B. Textilien und Kleidungsstücke entwickeln, die positive
Gefühle des Wohlseins hervorrufen ? Diese Möglichkeit ist vielleicht an
den Haaren herbeigezogen, aber vielleicht auch nicht ! Zweitens müssen
neue Stoffe entwickelt werden, die sogar bessere Komforteigenschaften
aufweisen, vielleicht Stoffe, die die positiven Empfindungen des
Wohlseins hervorrufen anstelle der o.a. Abwesenheit von Unbehagen.
Drittens müssen neue Gewebedesigns entwickelt werden, damit die Eigenschaften und Möglichkeiten dieser neuen komfortablen Stoffe voll und ganz
ausgenutzt werden können. Vielleicht müssen auch neue Methoden zur

Kleiderbelüftung entwicklet werden, um die gesamten Komforteigenschaften des Gewebes zu optimieren.

Dr. Umbachs Beitrag gab Anlaß zu einer guten Diskussion, die den Wert der Arbeit über Komfort für die Textil- und Kleidungsindustrie noch unterstrich. Die Forschungsarbeiten an diesem Projekt unterstützen die Industrie zuverlässig bei den Vorhersagen zum Komfortverhalten der Kleidung. Dr. Manni beschrieb die von Lanerossi unterstützte Arbeit über Komfort, mit der Hilfe von physiologischen Messungen von Wollstoffen im Vergleich zu Wollmischungen und von Baumwolle im Vergleich zu Flachs. Die technischen und wissenschaftlichen Eigenschaften von Regenschutzkleidung, die wasserdampf- aber nicht Regenwasserdurchlässig ist, riefen reges Interesse hervor. Als Antwort auf eine weitere Frage sagte Dr. Umbach, daß weitere Arbeiten am Hohenstein Institut die Wirkungen von Stoffertigungen auf den Komfort untersucht haben, jedoch wurde hierüber anläßlich des Symposiums nicht berichtet, da es nicht Teil des Projektes war (in Bezug auf die Qualität hängt sehr viel von der hydrophilen oder hydrophoben Eigenschaft der Ausführung ab). Man zeigte reges Interesse bei der Diskussion über die Natur der verwendeten Techniken zum Messen der komfortbestimmenden Faktoren; Sie bilden den Schlüssel zu dem gesamten Thema Komfort.

Im zweiten Teil der Sitzung beschrieb Herr Deschamps die Arbeit der CETIH und die in Nebenvertrag an das British Clothing Centre vergebene Arbeit zur Optimierung von Design und Herstellung von Geweben. Man kam zu zwei wichtigen Schlußfolgerungen. Erstens, daß dieses Thema von großer Bedeutung für die Zukunft der Textil- und Bekleidungsindustrie ist. Die Beziehung zwischen den Eigenschaften von Textilien und dem Design und der Herstellung von Kleidungsstücken muß viel besser verstanden werden, wenn die künftige Konkurrenzfähigkeit und das Wohlsein der Industrie gewährleistet sein sollen. Diese verbesserte Kenntnis der Grenzflächen zwischen Textilien und Kleidung ist von grundlegender Bedeutung, wenn man die bestehenden Design - bzw. Herstellungstechniken ersetzen oder zumindest ergänzen will. Diese beruhen größtenteils auf persönlicher Erfahrung und Fachwissen, die man nach jahrelanger Arbeit gesammelt hat und, obwohl sie sich als zufriedenstellend bis heute erwiesen haben, werden sie für die Zukunft auf keinen Fall genügen. Neue informationen über die Textil/ Gewebe-Grenzflächen werden bestimmt dazu führen, daß die mit den bestehenden Methoden erreichbare Qualität und Produktivität gesteigert werden können, aber, und das ist noch wichtiger für die Zukunft, daß sie ebenfalls einen bedeutenden Beitrag zur automatisierten Kleidungsherstellung und verstärkten Benutzung robotischer Systeme auf diesem Gebiet leisten werden. Die von Herrn Deschamps beschriebene Arbeit betraf einerseits die Untersuchung der physischen Eigenschaften von Textilien in bezug auf die Optimierung des Designs von Gewebearten, die funktional, komfortabel und doch stilgerecht und ästhetisch sind, und andererseits die Bereitschaft, die Herstellungsverfahren von Kleidungsstücken zunehmend zu automatisieren und von Robotern steuern zu lassen (d.h. die Wechselwirkungen zwischen der Stoff- und Gewebegerätschaft besser zu verstehen).

Die zweite Schlußfolgerung von Herrn Deschamps' Bericht, die unterstrichen werden muß, ist, daß die Arbeit über die Stoff/Gewebe-Grenzflächen erst ganz am Anfang ist. Untersuchungen über diese Projekt leisteten einen ausgezeichneten Beitrag, jedoch bleibt noch viel Arbeit über dieses komplizierte und schwierige Thema zu tun, um eine solide Wissensbasis, die für die Zukunft sehr wichtig ist, zu bilden. Die uneingeschränkte Zusammenarbeit von Wissenschaft, Technik und Industrie wird von grundlegender Bedeutung sein, wenn auf diesem Gebiet Erfolge verzeichnet werden sollen. Die vorzügliche Mitarbeit, die die Industrie

Herrn Deschamps geboten hat, ist in der Tat ermutigend; diese Zusammenarbeit muß weitergeführt und ausgedehnt werden.

Die beiden Themen der Sitzung über Gewebephysiologie und Konstruktion sind von grundlegender Bedeutung für die Textil- und Kleidungsindustrie der Gemeinschaft, wenn sie neuerungsbedacht und konkurrenzfähig bleiben wollen.

NEUE SPINNEREITECHNIK IN DER WOLLINDUSTRIE

M. BONA
Città Degli Studi

Seit vielen Jahren wurde in der Baumwollspinnerei eine bedeutende technologische Neuerung eingeführt, dank der Entwicklung unkonventioneller Techniken, durch die das klassische System der Düsenspinnmaschine teilweise ersetzt wurde; die wichtigsten dieser Techniken gehören zur Kategorie der System mit freiem Ende, zuerst vom Rotor-Typ und seit kurzem auch – zumindeste steht dies in Aussicht – mit Reibung oder Luftstrahl.

Es ist normal, daß von diesen wichtigen Entwicklungen in erster Linie der Sektor der Baumwolle und gewisser synthetischer Fasern betroffen wurde und zwar sowohl aus objektiven technischen Gründen, die die Behandlung dieser im Vergleich zu Wolle kürzeren und widerstandsfähigeren Fasern erleichtern, als auch aufgrund der Tatsache, daß die Forschungs- und Entwicklungsinvestitionen der Textilmaschinen-Konstrukteure sich in erster Linie auf einen potentiellen Austausch-Markt richteten, der mindestens zehnmal größer ist als der Wollmarkt.

Um auch die Übernahme neuer Spinn-technologien für Wolle zu begünstigen, mit dem ziel, die steigenden Herstellungskosten zu reduzieren oder einzuschränken, war es also erforderlich, eine Anstrengung auf dem Gebiet der Forschung zu unternehmen, indem man von dem tatsächlichen Bedarf der Industrie ausging und von dieser unterstützt wurde, diese Bemühungen mußten darauf abzielen, das Verhalten der Faser bei Einsatz der neuen, unkonventionellen Verfahren zu untersuchen und die mechanischen und textilen Bedingungen für ihre eventuelle Anwendung optimal festzulegen. Diese Forschung, mit der sich 8 Institute aus 5 Mitgliedsländern der EWG befaßten, hatte bereits auf Anregung einiger dieser Länder begonnen, wurde jedoch entscheidend unterstützt und vor allem koordiniert durch die Lancierung des Zweiten Textilforschungs-Programmes der Gemeinschaft und dessen Finanzierungen.

Die erste Phase des Programms bestand näturlich in der Durchführung von Versuchen auf den verschiedenen Stuhl-Modellen mit freiem Ende vom Rotortyp, die bereits auf dem Markt verfügbar sind, ohne jedoch andere unkonventionelle Lösungen unberücksichtigt zu lassen, wie z.B. das Friktionsspinnverfahren oder das Hohlspinverfahren.

In dieser Phase lieferte das Forschungsprogramm der Gemeinschaft zahlreiche hochinteressante Informationen, sowohl für die Kammgarn- als auch für Strichgarnspinnerei, deren praktische Bedeutung durch die Tatsache unterstrichen wird, daß es – dank der Zusammenarbeit mehrerer Institute – möglich war, Versuche auf zahlreichen Maschinentypen zu fahren.

Man wurde sich jedoch allmählich bewußt, daß der technische Charakter des Problems bis vor das Spinnen als solches Versetzt werden müsse, d.h. man müsse sich bereits mit der ersten Vorbereitung der Fasern und dem Streichen befassen.

Was die Vorbereitung betrifft, so hat das Programm gestattet, einige interessante Modifikationen der bei Rohfasern angewandten konventionellen

Verfahren vorzuschlagen, wie z.B. ein mechanisches Reinigen, das von der klassischen karbonisation begleitet werden kann und diese vereinfacht; besondere Aufmerksamkeit wurde natürlich dem Einsatz von Rohmaterialien aus der Rückgewinnung geschenkt, dem eine große wirtschaftliche Bedeutung auf dem Sektor der europäischen Streichgarnspinnerei zukommt und der hinsichtlich der anschließenden Arbeitsgänge die schwierigsten Probleme stellt.

Im Hinblick auf die künftige Anwendung der neuen Spinn-technologien, ob man nun von Rohmaterialien oder, vor allem, von rückgewonnenen Materialien ausgeht, war eines der wichtigsten Ergebnisse des Programmes zweifellos die Konzeption, die Entwicklung des Prototyps und die praktische Erprobung von Karden einer neuen Gestaltung, die die Vorzüge der traditionellen Woll- und Baumwollsysteme in sich vereinigen, mit dem Ziel, ein Band herzustellen, das zugleich homogen und sehr offen und sauber ist, um damit die Spinnmaschine mit freiem Ende zu speisen. Es ist wünschenswert, daß sowohl die Konstrukteure von Spinnmaschinen, als auch diejenigen von konventionellen Karden nützliche Kontakte mit den Forschungsinstituten aufnehmen, um aufgrund der neuen Konzeptionen industrielle Maschinen zu entwickeln, die andererseits auch zu einer interessanten Weiterentwicklung auf dem Gebiet der konventionellen Spinnerei führen könnten.

Ein letzter Aspekt der Forschungsprogrammes betraf eine Untersuchung des physiko-chemischen Verhaltens der Wolle während der Fabrikation, eine unerläßliche Bedingung, um die neuen Technologien auf Optimale Art anwenden zu können und dabei negative Begleiterscheinigungen zu vermeiden, die – wie z.B. die übermäßige Bildung von Staub und Ablagerungen in den Rotoren – eine wirtschaftliche Nutzung der Verfahren verhindern würden.

Von den Merkmalen der Wolle, die ihr Spinnen unter Anwendung der neuen Technologien schwieriger machen als zum Beispiel Baumwolle, wäre ihre größere Empfindlichkeit zu nennen, sowie die Notwendigkeit, Zusätze zu verwenden, die sorgfältig ausgewählt und dosiert werden müssen, wenn man auf einer Maschine mit freiem Ende spinnen möchte.

Man muß infolgedessen als wichtiges Resultat des Forschungsprogrammes ganz besonders die Entwicklung von Labormethoden betrachten, die in der Praxis zur Kontrolle der Herstellungsverfahrens dienen können, sowie zur Auswahl derjenigen Wollqualitäten, die sich der Verwendung der neuen Technologien am besten anpassen lassen.

Die drei nachfolgenden technischen Berichte liefern die Ergebnisse des Forschungsprogrammes. Sie befassen sich mit den drei Hauptthemen, die wir genannt haben : Vorbereitung, Spinnerei, Kontrollen.

Als allgemeine Schlußfolgerung soll unterstrichen werden, daß man dank des hervorragenden Geistes der Zusammenarbeit, den die Institute an den Tag legten, die am Programm der Gemeinschaft teilgenommen haben, tatsächlich die Grundlagen für eine entscheidende Weiterentwicklung der Spinntechnologie auf dem Wollsektor schaffen konnte, selbst wenn noch wichtige Probleme zu lösen bleiben, wie z.B. die Übertragung der neuen Ideen auf Maschinen und industrielle Verfahren, die eine Zusammenarbeit zwischen den Instituten und den Textilmaschinen-Herstellern erfordert.

Ein wichtiges Thema für künftige Forschungsarbeiten geht aus den Arbeiten hervor, nämlich die Verwendung neuer Garne, deren Struktur in jedem Fall mehr oder weniger von derjenigen der traditionellen Garne abweicht : das setzt Modifikationen der vor der Spinnerei liegenden Verfahren voraus, sowie die Anpassung der neuen Produkte an die Erfordernisse des Marktes.

VORBEREITUNG DER WIEDERAUFBEREITETEN MATERIALEN FÜR DIE NICHT KONVENTION-NELLE SPINNEREI

Dr. P. ARTZT
Institüt für Textiltechnik

Die Wollfaser ist gegenüber der Baumwollfaser eine sehr teure Spinnfaser. Rohstoffkosten sind bei der Garnherstellung der dominierendste Kostenfaktor. Jeder Prozentsatz erhöhter Rohstoffnutzung schlägt sich somit überproportional in der Kostenrechnung positiv nieder. Zum anderen muß die Forschung nach Wegen suchen; Wollabgänge so aufzubereiten, daß die gewonnene Faser zu möglichst hochwertigen Endprodukten verarbeitet werden kann.

Die Forschungsinstitute CELAC (Belgien), WIRA (UK) und ITF (Deutschland) führten in Kooperation Forschungsprojekte durch, welche die wirtschaftliche Fasernutzung sowie die Prozeßkürzung zum Ziele hatten.

Die bei der Wollverarbeitung hauptsächlich anfallenden Abgänge sind :

1. Krempelflug :
   Hierbei handels es sich um bevorzugt lange Fasern, wie das Stapelschauschild zeigt :
   Der Schmutzanteil beträgt bis zu 80 %.

2. Kämmling :
   Der Kämmling hat eine der Baumwolle ähnliche Faserlängenverteilung. Der Vegetabiliengehalt beträgt bis zu 20 %. Die Vegetabilien sind intensiver mit der Faser verbunden.

Diese Abgänge werden heute ausnahmslos carbonisiert. Dabei handelt es sich um eine chemische Oxydation der zellulosischen Verunreinigungen.

Vorteile des Carbonisierprozesses :

1) sehr gute Reinigungswirkung.
2) hohe Ausbeute.

Nachteile des Carbonisierungsprozesses :

1) hohe Investitionskosten der Anlage,
2) hoher Energiebedarf,
3) hoher Einsatz an Chemiekalien,
4) Umweltprobleme,
5) komplizierte Prozeßsteuerung,
6) negativer Einfluß auf die Faserstoffeigenschaften der Wollfaser (spröde),
7) hoher Kurzfasergehalt der carboniserten Wolle.

Mechanische Verfahren der Reinigung des Wollkämmlings erbrachten Reinigungswirkungen von ca. 50 %. Speziell hierfür entwickelte Verfahren der Baumwollreinigung wurden entsprechend für die Wollkämmlingsreinigung modifiziert.

Meist ist für normal verschmutzte Wollkämmlinge ein Reinigungsgrad der Wollfaser von 50 % ausreichend, da die Faser anschließend beim kardierprozeß und der Verspinnung auf der Rotorspinnmaschine einer

weiteren Reinigung unterliegt. Zum anderen sollten die gereinigten Wollfasern in Mischungen mit Chemiefasern oder Baumwolle verarbeitet werden.
Eine interessante Variante besteht aus der Kombination :
1)   mechanische Vorreinigung,
2)   anschließende Kurzcarbonisierung.

Der eigentliche Carbonisierprozeß für die verbleibenden kleinen
Restschmutzpartikel kann entsprechend verkürzt werden und ist somit
faserschonender. Die mechanische Vorreinigung führt zu einem nissen-
freieren Kardenvlies. Die gereinigten Fasern wurden gemischt mit
Polyester, Acryl und Baumwolle und nach dem Friktionsspinnverfahren
(Dreff 2) sowie dem Rotorspinnverfahren zu Garnen versponnen.
Die Versuch ergaben, daß die Herstellung feinerer Garne nach dem
Friktionsspinnverfahren vorteilhaft unter Einsatz eines Corefilaments
erfolgen sollte.
Es konnten Rotorgarne mit einem Anteil bis 50 % gereinigtem Woll-
kämmling der Feinheit 30 tex auf Rotorspinnmaschinen hergestellt werden.
Der Krempelflug stellt aufgrund seiner Faserlänge den hochwertigsten
Abgang dar. Bei der mechanischen diskontinuierlichen Reinigung kann es
sogar zu einem längeren Mittelstapel der gereinigten Fasermenge kommen da
ein großer Anteil Kurzfasern ausgeschieden wird.
Vergleichsausspinnungen mit carbonisiertem Krempelflug nach dem
Ringspinnverfahren und dem Umwindespinnverfahren ergaben wesentlich
besseres Laufverhalten und Garnwerte mit dem mechanisch gereinigten
Faserstoff einen höherwertigeren Spinnstoff dar, da man diese Faser zu
anderen Produkten verarbeiten kann als die carbonisierte Faser.
Für die Herstellung grober Garne im Bereich Decken, Polsterstoffe
und Teppich ist eine Streckpassage vor der Rotorspinnmaschine nicht notwendig. Es sollte eine 2-tambourige Wollkrempel eingesetzt werden, welche
über eine Vliesquetsche verfügt. Durch Vliesteilung am Auslauf der Karde
erhält man Bänder mit Bandgewichten, welche den Rotorspinnmaschinen
direkt vorgelegt werden können. Der Restfettgehalt der Wolle sollte ca.
0,5 % betragen. Eine Nachavivierung ist dann nicht notwendig. Sie führt
zu Nachteilen hinsichtlich Ablagerungen an den Spinnelementen.
Ein umfangreiches Rohstoffpotential liegt in der Wiederverwendung
der Konfektionsabfälle. Das Problem liegt in der Logistik der Erfassung
nach Farbe und Mischung mit anderen Fasern. Eine umfangreiche Studie gibt
Auskunft über die heutigen Einsatzgebiete, Herkunftsländer und Aufberei-
tungsmethoden dieser Fasern. Die von den Instituten CELAC, ITF Denkendorf
und WIRA in Kooperation durchgeführten Forschungsprojekte ergänzen sich
und zeigen neue Wege der wirtschaftlichen Aufbereitung der Abgänge zur
besseren Wiederverwendung.

NEUE KARDIER-UND SPINNVERFAHREN

J.P. BRUGGEMAN
I.T.F. Nord

Prozesse zur Vorbereitung des Rohmaterials wurden von unseren Kollegen
untersucht; das Kardieren bleibt jedoch die überbrückung zwischen den

Rohmaterialien und dem ersten Florband, welches so sauber wie möglich sein muß (frei von Unreinheiten und Vegetabilien). Es muß auch so einheitlich in dem Titer sein wie möglich und eine gute Homogenität der Zusammensetzung haben.

Beim konventionnellen Wollkardieren liefert die Karde einen Kardenflor, der in Lunten unterteilt ist, welche wiederum in die Woll-Ringspinnmaschine eingeführt werden. Die offenend Spinntechnik hat zur Auslassung des Kondensers an der Kardeneinführung geführt und stattdessen eine Zufuhr von 2 oder 4 Florbändern angenommen, die direkt auf die offenend Spinnmaschine aufgesponnen werden sollen. Dieser für Teppichgarne sehr attraktive Vorgang hat einen Nachteil, der für feinere Garne fast fehlerhaft ist, nämlich die Abwesenheit eines Kontrollsystems für Titer zwischen einem Florband und einem anderen, sowie innerhalb desselben Bandes.

Industrie und Forschung haben sich deshalb in Richtung Baumwollkarde begeben, die es ermöglicht, ein Florband zu erhalten, das mit der offenend Spinnerei kompatibel ist.

Die Ergebnisse, wenngleich sehr ermutigend, haben auch Nachteile dargelegt, wie z.B. :

- einen Überschuß an Vegetabilien,
- schlechte Parallelformation der Fasern,
- eine große Anzahl von Nissen,

Die Labors von CELAC in Belgien, TECNOTESSILE in Italien, WIRA in Grossbritannien und ITF-Nord in Frankreich haben deshalb Maschinen-Hersteller und Leute aus der Industrie miteinbezogen, um eine neue Art von Wollverarbeitungsmaschinen zu untersuchen, die folgende Punkte beachten würden :

- Florband-Sauberkeit,
- Nissenverringerung,
- Materialöffnung,
- Parallelformation der Fasern,
- Florband-Gleichmäßigkeit.

Die Idee, die Öffnungsstärke der Arbeitswalzen einer Wollkarde mit der Kardierstärke für einzelne Fasern und der Säuberungsfähigkeit der Baumwollkarde (wegen ihrer Kardierhauben) zu vereinigen, hat zur Zusammenstellung gemischter Karden geführt, welche hin verschiedenen Formen vorkommen :

- die einzylindrige gemischte Karde mit zwei Arbeitswalzen-Kardierpunkten und einem Haubenteil (SACM-Karde).

- die zweizylindrige Tandem-Karde mit Hauben,

- die zweizylindrige Tandem-Karde; ein Zylinder ist mit Hauben ausgerüstet und der zweite mit 4 Gruppen von Arbeitswalzen,

- die dreizylindrige Karde; zwei Zylinder sind mit Hauben ausgestattet, und zwar mit variabler Bestückung von Nadeln.

Es ist sicher, daß die Leistung eines bestimmten Materials von der Kardiermethode bestimmt wird; in Labors wurden verschiedene Untersuchungen angestellt, und zwar über : Mischungen von Zweitwolle mit Polyester, offene Frischwollen, karbonisierte und unkarbonisierte Kämmlinge, und pratoartige Materialien. Mischkarden mit 2 und besonders 3 Zylindern und abwechselnden Arbeitsgängen von Arbeitswalzen und Hauben produzieren ein Florband, das mit der offenend Spinnerei kompatibel ist.

Die Längendiagramme am Auswurf der Karde zeigen, daß es günstiger wäre, ein Minimum an Streckpassagen zu haben, mit oder ohne automatischem Ausgleich, um die kurz- und langfristige Gleichmäßigkeit zu verbessern. Wenn das Faserdiagramm dies nicht gestattet, dann ist ein automatisches Ausgleichsystem äußerst wünschenswert. Die CBL-Kurven 5 stellen die möglichen Verbesserungen der Florbandqualität dar, welche wiederum eine verbesserte Spinnleistung mit sich bringen würde.

SPINNEREI

Die Anwendung von offenend Spinnerei-Techniken war das oberste Ziel aller an der Arbeit teilnehmenden Institute, und die Reihe der benutzten Anlagen zeigt deutlich den Willen zum Erfolg.

Die folgenden seien erwähnt :

- Dref 2,
- San Giorgo,
- BD 200,
- ITG 300,
- Autocoro,
- RU II.

Es zeigt sich, daß sich reine Wolle nur sehr schwer mit dieser Technik spinnen läßt, vielleicht mit der Ausnahme von DREF. Andererseits lassen sich verschiedene Mischungen mit Synthetikfasern bei 40 tex oder sogar 30 tex erreichen, abhänging von den Faserdiagrammen, wenn einige Vorsichtsmaßnahmen getroffen werden, und zwar :

- Wahl der Vorkratze und ihrer Geschindigkeit
- Wahl der Drallkoeffizienten
- Korrelationswahl der Düse
- Wahl des Durchmessers und Profils des Rotors

Dem Fettstoffgehalt des Florbandes gebührt besondere Beachtung. Ein Level von 0,5 bis 0,8 % wird als Limit angesehen, um eine Wickelung oder Lappung an der Vorkratze zu vermeiden. Außerdem wird die Staubmenge mit einem hohen Anteil an Kutikeln im Rotor durch die mechanische Tätigkeit der Vorkratze hervorgerufen.

Die Überlegenheit einer Klingen-Vorkratze oder eines Selektors bemerkt man im Vergleich mit der Zahn- oder Nadelvorkratze. Die Produktionseinstellung der ITG 300 hat natürlich erneutes Interesse an diesem System geweckt.

Das Problem der Staubentwicklung hat bei Forschungsinstituten dazu geführt, daß man einen Simulator entwarf, dessen Funktion es ist, Mengen frühzeitig zu identifizieren, die möglicherweise Rückstande in den Rotoren hinterlassen.

Offenendig gesponnene Garne unterscheiden sich von denen einer Ringspinnmaschine hauptsächlich in ihrer Struktur, im Griff und ihren mechanischen Eigenschaften, was aber ihren Gebrauch bei Strumpfwaren, Oberbekleidung und Möbeln nicht untersagt. Lediglich Velourstoffe und aufgerauhte Flachware erfordern eine vorsichtige Mischungswahl und Faserdiagramme der Bestandteile.

Wir erwähnen auch die Studie, die von UMIST an Hohlspindeln

durchgeführt wurde; diese führt zur Produktion eines Garns mit interessanten mechanischen Eigenschaften und, nach der Herstellung von Kleidungsstücken, zu Trageeigenschaften, die mit denen konventioneller Produkte identisch sind.

Die in diesem umfassenden Forschungsprogramm erhaltenen Resultate waren nur möglich dank der engen Zusammenarbeit zwischen den verschiedenen Labors innerhalb der EG, welche wiederum als Sprecher der Industrie fungierten, und denen wir auch für ihre Kooperation danken möchten.

PHYSIKALISCHE UND CHEMISCHE ANALYSEN

J. KNOTT
Centexbel

Es ist allgemein bekannt, daß physikalische Parameter wie etwa die Länge und der Durchmesser von Wolle sowie die Vorspinn-Gleichmäßigkeit nicht die einzigen zu beachtenden Faktoren bei der offenend Rotorspinnerei sind.

Die physikalisch chemischen Parameter, die hauptsächlich mit der Oberflächenbeschaffenheit der Faser in Verbindung stehen, spielen eine bedeutende Rolle, besonders bei der Bildung von Staub und der Ablagerung von Staub im Rotor. In ihrer Zusammenarbeit hatten das Deutsche Wollforschungsinstitut Aachen, Institute Textile de France Section Nord und Centexbel-Verviers folgende Forschungsziele :

1) Neue Tests vorzuschlagen oder existierende Tests abzuändern, um die Leistung von Wolle in der offenend Spinnerei vorherzubestimmen,

2) Die Staubbilding beim Verlauf der Garnherstellung zu untersuchen,

3) Den Einfluß folgender Aspekte zu untersuchen :

  - Fettstoffgehalt,
  - die Art des Schmiermittels,
  - Behandlungen, denen die Wolle früher ausgesetzt wurde (Karbonisierung, Färbung).

1. - DIE ENTWICKLUNG SPEZIFISCHER TESTS

1.1. Bestimmung des Staubgehalts in loser Wolle, im Vorgespinst und im Garn.

Staubextraktion vom Material wird durch Hochfrequenzschütteln (30 Hz - Amplitute 5 mm) einer Äthanol-Faser Suspension erreicht. Die Menge wird durch Wiegen bestimmt. Die Kinetik der Staubentlassung ist auch untersucht worden.

1.2. Bestimmung der Spitzigkeit von Wolle

Die Entwicklung einer Technik zur Bestimmung der Prozentzahl oxydierter Faserspitzen, und deshalb mechanisch brüchiger ist ein besonders geeigneter Test im Hinblick auf den hohen mechanischen Stress, dem die Faser während der offenend Spinnerei ausgesetzt ist.

Zwei Methoden wurden untersucht :

1)   Eine Methode, die nur Forschungslabors vorbehalten ist, benutzt ein
      Hochgeschwindigkeits – Abtastelektronenmikroskop. Der Detektor nimmt
      das Licht auf, das durch die fleckigen Teile übertragen wird
      (oxydierte Regionen der Faser).

2)   Eine einfache Methode, die auf der Tatsache beruht, daß die oxydier-
      ten Faserspitzen eine höheren cysteischen Säuregehalt haben, der
      sich übertragen läßt auf eine größere Festsetzfähigkeit bei pH2 von
      einigen gewissen kationischen Färbungen.

Dies ist besonders nützlich für lose Wolle, da es eine schnelle
Anzeige der Schadensausmaße an Faserspitzen ermöglicht.

1.3. Staub-Analyse

Die entwickelten Tests erlauben eine Zerlegung und Bestimmung der
"morphologischen" Zusammensetzung von Staub durch Mikroklassierung
(Faserspitzen und Fragmente, Kutikula und Kortex).

Es war möglich, verschiedene Einflüsse beim offenend Spinnen zu
untersuchen, und zwar durch die oben genannten Tests und auch andere
konventionelle chemische Methoden, wie etwa die Bestimmung von Fettstoff
durch Dichlormethan-Auszug, die Schätzung cysteischer Säure mittels
chemischer Mittel oder mehrfacher interner infrarot Spiegelungen und die
Schätzung von Aminosäuren durch die "Moore und Stein"-Methode, die
Messung der Alkali-Löslichkeit, die Schätzung terminaler Aminogruppen
durch Ninhydrin u.s.w...

2. – ENTWICKLUNG BESCHÄDIGTER TEILE AUF VERSCHIEDENEN STUFEN DER BEARBEI-
   TUNG BIS ZUM SPINNEN

Um die Ursprünge der Unreinheiten, die bei der Garnherstellung für
die Fehler verantwortlich sind, besser entdecken zu können, wollten wir
das Material auf verschiedenen Stufen der Herstellung charakterisieren,
und zwar angefangen bei der losen Faser.

Einen visuellen Eindruck der teilweisen Degradierung von loser Wolle
erhielt man durch die Behandlung des Materials mit Methylenblau. Nach der
Kardierung, dem Kämmen, dem Spinnbandreißen, dem Strecken und Spinnen
wurden das Material und die Staubrückstände analysiert. Wir verglichen
auch die Garne, die auf Offenend-Spinnmaschinen mit Selektor und Vorkratze produziert wurden, mit denen der konventionellen Spinnmaschinen.

Die in Tabelle I dargestellten Resultate zeigen nochmals die
wichtige Rolle der Karde bei der Beseitigung von Staub. Spinnbandreißen
des Materials ist ebenso eine bedeutende Quelle der Staubbildung.

Offenend Spinnmaschinen mit Selektor sind für die Wolle weniger
schädlich als die mit Vorkratze, wie man an dem viel geringeren Kortex/
Kutikula-Verhältnis des erzeugten Staubs ersehen kann.

Wir weisen auch darauf hin, daß der Gebrauch gut karbonisierter
Wolle die Staubablagerung im Rotor vermindert, nicht nur wegen der
Reduzierung von Vegetabilien sondern auch wegen der Reduzierung von
Staub.

Tabelle I

Entwicklung der Staubmenge während
des Spinn-Vorgangs

(* = Durchschnitt von 4 Tests).

Muster	% Staub (*)
Lose Wolle	1,06 – 0,91
Wolle nach der Kardierung	0,55 – 0,46
Wolle vor dem Spinnband-reißen	0,43 – 0,43
Wolle nach dem Spinnband-reißen	0,52 – 0,51
Wolle vor dem Kämmen	0,40 – 0,35
Vorgespinst ins offenend Spinnen	0,50 – 0,51
Vorkratze – offenend Garn	0,76 – 0,76
Selektor – offenend Garn	0,59 – 0,61
Konventionelles Garn	0,53 – 0,51

ZUSAMMENFASSUNG UND SCHLUSSFOLGERUNGEN DES VORSITZENDEN DER SITZUNG

M. BONA
Città Degli Studi

Diese Sitzung, an der rund 70 Delegierte teilnahmen, war ein großer Erfolg. Wie vereinbart wurden drei Vorträge abgehalten, die die Arbeit der acht Institute in fünf Mitgliedsstaaten erläutern. Der Vortrag von Dr. Artzt, Vorbereitung von wiederaubereiteten Stoffen fur nicht konventionelle Spinnerei, stütze sich auf vom Institut für Textiltechnik, Wira und Celac, ausgeführte Arbeiten. Der Vortrag von Herrn Bruggeman über neue Kardier- und Spinntechnologien, beruhte auf den Ergebnissen von ITF-Nord, Centexbel, Celac, Wira, Umist und Tecnotessile. Das Exposé von Dr. Knott über physische und chemische Analysen, berichtete von den Forschungsarbeiten von Centexbel, Deutsches Wollforschungsinstitut und ITF-Nord.

Den Berichten folgte eine lebhafte und konstruktive Diskussion, während der spezifische, technische Aspekte ausführlicher erläutert wurden, u.a. :
- Die Wahl zwischen der konventionellen Carbonisierung und dem neuen mechanischen Reinigungsprozess. Auch wenn die mechanische Verarbeitung zufriedenstellende Ergebnisse erreichen kann, wenn Kämmlinge gehechelt werden, die nur wenig Pflanzenstoffe enthalten, bleibt die Carbonisierung, wenn richtig durchgeführt, die beste Lösung für Abfall, der viel Pflanzenstoffe enthält. Man hat bewiesen, daß man vom Entkletten absehen kann, damit ein Verfilzen während der Neutralisierung vermieden wird. Nach dem Kardieren, haben nicht entklettete Stoffe weniger Fadenverdickungen und sind daher so sauber wie die, die auf traditionelle Weise carbonisiert werden. Die verschiedenen Verfahren — müssen außerdem unter industriellen Bedingungen verglichen werden, und zwar besonders vom Kostenstandpunkt her.
- Die Frage des Staubes ist von großer Bedeutung für eine einwandfreie Arbeitsweise der Turbinen. Mechanische Fehler verursachen mehr oder weniger viel Staub während des gesamten Herstellungsprozesses, je nach Zustand des Rohstoffes und der Arbeitsbedingungen. Es ist genau so wichtig Mittel zum Staubentfernen zu entwickeln als auch dessen Entstehungsweise zu entdecken. Ein interessanter Verweis wurde auf Staub mechanischer Herkunft gemacht, der, obwohl er nur in geringen Mengen vorhanden ist, einen bedeutenden Einfluß wegen seiner scheuernden Wirkung aufweist.
- Hier wurde über die Verwendung von Kardendraht-Öffnungswalzen oder Selektorwalzen im Eingabenmechanismus von OE-Spinnmaschinen vom Typ Rotor diskutiert. Das alte System, bei weitem das meist verbreitete, unterwirft die Fasern einer gröberen Behandlung und während des Hauptberichts wurden Zahlen genannt, die im Laufe der Diskussion bewiesen, daß dies zu einem stärkeren Zerreißen der Fasern führt, obwohl es dafür nur indirekte Beweise gibt, da wirkliche Längenmessungen mit dem Garn noch nicht unternommen wurden.
Es wurde besonders bedauert, daß die einzige Firma, die bisher eine OE-Maschine mit Selektorwalzen herstellt, diese zurückgezogen hat.
Dies scheint das mangelnde Interesse der Maschinenhersteller für den

Wollsektor zu bestätigen und man hofft, daß die Ergebnisse dieses Symposium (dessen Delegierte dringende gebeten werden, keine Mühen zu scheuen, um es den Maschinenbauern mitzuteilen) dazu beitragen werden, die Situation zu berichtigen.

Man hat die Aufmerksamkeit auf die Eigenschaften der mit diesem neuen Garn hergestellten Produkte im Vergleich zu traditionell gestrickte Stoffen- und Geweben gelenkt, unter besonderer Berücksichtigung von umwendegarne (Hohlspindel). Eine Reihe von Mustern, die nach den verschiedenen Techniken hergestellt und von den verschiedenen Instituten geprüft wurden, liefern einen konkreten Beweis für diese Eigenschaften.

- Man erteilte einige Informationen über die zukünftige Forschung, wovon die interessanteste sich mit der Untersuchung von speziellen Lösungen zu gewöhnlichen Problemen beim Wollspinnen auf OE-Maschinen befaßt. Man muß die schwachen Punkte der Faser entdecken (und als Ergebnis dieser Forschung, kann man dies heute sehr genau) und systematisch für Abhilfe sorgen. Das Ziel ist also, die Faser der Maschine anzupassen und nicht einfach die Maschine der Faser. Dies bedeutet eine Bruch mit der Tradition, die durch die höheren Herstellungskosten in bezug auf die Rohstoffkosten entstanden ist.

Um uns schließlich mit allgemeineren Angelegenheiten zu befassen, sollte man die Aufmerksamkeit auf zwei Beiträge lenken, die aus verschiedenen Gründen, von größter Bedeutung, und nicht nur vom technischen Standpunkt her, sind :
- Der von einem spanischen Vertreter, der an diesem Symposium als Beobachter teilnahm, gemachte Diskussionsbeitrag bestätigte einige Punkte der Autoren dieser Berichte. Dies läßt darauf schließen, daß der große Sinn für Zusammenarbeit, der während der gesamten Veranstaltung anwesend war, durch die Ausdehnung der Gemeinschaft noch verstärkt wird.
- Die Bemerkung der Vorsitzenden von Interlaine, höchster amtierender Vertreter der Wollindustrie, der die Institute für ihre Arbeit lobte und deren praktischen Wert hervorhob, den er bei der nächsten allgemeinen Versammlung seiner Organisation erwähnen will.

QUALITÄT VON UNKONFEKTIONIERTEN UND KONFEKTIONIERTEN MASCHENWAREN

J. STRYCKMAN
Centexbel

EINLEITUNG

Die Strickwarenindustrie bildet einen wichtigen Sektor der europäischen Textilindustrie. Mit seinen rund 25.000 Unternehmen – davon allerdings 4/5 in Italien – und 410.000 Beschäftigten stellt dieser Sektor ein Achtel der europäischen Textilproduktion dar. 1984 wurden auf diesem Sektor rund 695.000 t Gespinste und Garne verarbeitet, der Umsatz erreichte nahezu 16 Milliarden ECU. Allerdings exportiert Europa zwar unkonfektionierte Maschenware, die wertmässig das Zweieinhalbfache seiner Importe ausmacht, doch ist bei Fertigartikeln seine Handelsbilanz eindeutig defizitär. Dieses Defizit entspricht in etwa dem Fünffachen des mit unkonfektionierter Ware erzielten Überschusses. Kreativität, Produktivität und Qualität müssen daher als Leitmotiv bei allen Unternehmen des Sektors im Mittelpunkt stehen.

Der Maschenartikel muss funktionnel, zuverlässig und attraktiv sein. Es gilt, Qualität, Einfallsreichtum und Geschmack zum besten Preis zu verkaufen. Der Erfolg hängt von der schnellen Anpassungsfähigkeit an den Markt, insbesondere an den anspruchsvollen Modemarkt ab. Es sind ständige Bemühungen um eine Verbesserung und Erneuerung der Techniken und Produkte notwendig zu einem Zeitpunkt, in dem der Verbraucher hinsichtlich der Gebrauchseigenschaften des Produkts und hinsichtlich der Qualität immer anspruchsvoller wird.

Die Herstellung von Strickwaren erfolgt größtenteils in klein- und mittelständischen Unternehmen, die – um dieser Forderung zu genügen – auf eine externe logistische Hilfe angewiesen sind, die ihnen spezialisierte Institute und Forschungszentren bieten können. Mit dem Ziel einer Verbesserung dieser Unterstützung durch eine Erweiterung ihres Know-how haben sich acht Forschungsinstitute aus sieben Gemeinschaftsländern im Rahmen von zehn Verträgen mit der Kommission verpflichtet, ihre Bemühungen in einem Forschungsprogramm zusammenzufassen und zu koordinieren, das darauf abzielt, den Herstellern die Möglichkeit zu bieten, einen hohen Qualitätsstandard ihrer Produkte zu gewährleisten und das Interesse des Verbrauchers an europäischen Artikeln zu fördern.

Die Herstellung eines attraktiven, funtionellen und zuverlässigen Artikels setzt die Beherrschung einer großen Zahl von Parametern voraus, Parametern im Bereich der Rohstoffe, der Fertigung, der Konditionierung, der Lagerung und des Versands.

Die Erreichung hoher Qualitätsziele bei Strickwaren ist ein vielschichtiges Problem und setzt voraus: zu wissen, welchen Gebrauchseigenschaften der Artikel entsprechen muß; über objektive Messverfahren für diese Gebrauchseigenschaften zu verfügen; die Normwerte zur Beurteilung des Qualitätsniveaus zu kennen; die die Qualität beeinflußenden Parameter und die Auswirkungen einer Änderung dieser Parameter auf die Qualität zu bestimmen.

Hauptziel des Programms war es, den Einfluß der Parameter, der Fertigungsmaterialien und -bedingungen unter Berücksichtigung des Verwendungszwecks der Strickware und der Funktion des Artikels auf die

Qualität der Artikel zu bestimmen.

Die Masshaltigkeit des Artikels beim Tragen und bei der Pflege ist ein wichtiges generelles Kriterium, dem sich alle Hersteller bei allen Kategorien von Strickwaren gegenübersehen. Ihm wurde im Rahmen des Programms besondere Aufmerksamkeit gewidmet. Zu diesem Merkmal kommen noch weitere Qualitätskriterien hinzu, die einen Komplex von Eigenschaften bilden, denen Rechnung getragen wurden muß, um unter Berücksichtigung der Funktion des Artikels das Qualitätskonzept eines Bestimmten Artikels zu erfassen und sein Qualitätsniveau zu ermitteln.

Das Forschungsprogramm führte zur Erarbeitung einer Reihe von Richtlinien hinsichtlich des Materials und seiner Eigenschaften sowie bezüglich der Strickbedingungen und der nachfolgenden Behandlungen, die der Hersteller berücksichtigen muß, um eine optimale Produktqualität gewährleisten zu können. Die Ergebnisse des Programms werden nach dreieinander ergänzenden Hauptschwerpunkten vorgestellt :

- Qualitätsmerkmale konfektionierten Maschenwaren untern dem Blickwinkel des Verbrauchers,

- Qualitätsaspekte für die Herstellung von Maschenwaren,

- Objektive Qualitätsbewertungen.

Die in diesen drei Berichten vorgelegten Ergebnisse bilden einen wertvollen Leitfaden für Strickwarenhersteller, denen daran gelegen ist, die Qualität ihrer Produktion zu beherssehen, zu verbessern und zu stabilisieren.

QUALITÄTSMERKMALE KONFEKTIONIERTER MASCHENWAREN

H.J. SUURMEIJER
T.N.O.

Die Qualitätswahrnehmung von Strickwaren bei Verbrauchern, Händlern und Herstellern wurde untersucht. Konsumenten wurden in einer U.K.-Umfrage nach den Ansichten, Maßen und Werturteilen über ihre Strickwaren gefragt; dies wurde dann auf Befragte in Belgien, Dänemark, Frankreich und den Niederlanden ausgeweitet und etwa 1400 Kleidungsstücke wurden darin eingeschlossen.

Eine Untersuchung der Spezifikationen, nach denen Kleidungsstücke hergestellt werden, zeigt, daß die größeren Wiederverkäufer und die Hersteller von Hausmarken ein großes Maß an Übereinstimmung bei ihren Mindestanforderungen darlegen, aber daß ein Großteil dieser Übereinstimmung mehr scheinbar als wirklich ist, da ihre Testmethoden unterschiedlich sind, mit Ausnahme der von Farbfestigkeit.

Die Einkaufsspezifikationen der kleineren, in der Studie erfaßten, Wiederverkäufer sind völlig unzureichend und wichtige Bestandteile der Hersteller-Richtlinien sind oft zu unverbindlich und werden lediglich mündlich behandelt. Abgesehen von Herstellern abgepasster Maschenware gaben nur wenige die Maschenlänge in ihren Spezifikationen an, obwohl dies der wichtigste Faktor bei der Qualitätskontrolle ist.

Verbesserte Methoden für das Testen der Kleidungsstückdehnung und die Beurteilung der Nähte sind ebenfalls erforderlich. Es gibt auch erhebliche Unterschiede bei den Größenangaben, in Bezug auf Körper- und Kleidungsmaße, sowohl im Neuzustand als auch nach dem ersten Waschen.

In den meisten EG-Ländern gibt es keine allgemein erhältliche anthropometrische Datenliste, welche alle notwendigen Parameter angibt und auch regionale, altersmäßige und sozialschichtige Variationen darstellt.

Für das Design und die Entwicklung bequem sitzender Strickwaren bei der gegenwärtig begrenzten Anzahl von Größen für den Großteil der Kunden sind die publizierten Größentabellen und Methoden zur Größenbestimmung in der EG unzureichend, um von den charakterlichen Eigenschaften der Strickwaren Gebrauch zu machen.

Wegen der unterschiedlichen relativen Körperproportionen von Bevölkerungsgruppen in der EG sollten den Herstellern und Wiederverkäufern ausreichende Größentabellen - abgeleitet von anthropometrischen Studien von Bevölkerungsgruppen in örtlichen oder Heimmärkten - zur Verfügung stehen, um den Großteil der Konsumenten mit gut und bequem sitzenden Strickwaren zu versorgen.

Das Design und die Produktion solcher Kleidungsstücke kann auf dem vorhandenen Wissen über die Eigenschaften gestrickter Strukturen und auf den darauf basierenden Produktionsmethoden beruhen.

Neue Labortests mit speziellen Meßgeräten, vorgeschlagen von TEFO in Göteborg, und spezielle Formen fur Kleidungsstücke, entwickelt von ITF-Maille, und die Inspektion von mehr als 1000 käuflich erhältlichen Unterwäsche-Stücken haben bewiesen, daß die Sitzfähigkeit von Strickwaren in der gegenwärtig limitierten Anzahl von Größen beurteilt werden kann.

Diese Bewertung basiert auf den Tests kritischer Kleidungsstück-Maße bei minimal und/oder maximal annehmbaren Belastungen und dem Vergleich dieser Meßwerte mit den jeweilig größten und kleinsten Körpermaßen, die innerhalb der Größen vorkommen.

T-Shirts wurden als ein Kleidungsstück herausgestellt, das detailliertere Untersuchungen erforderte, da die Rippen- oder Einfachjersey-Strukturen dieser Artikel eventueel durchs Tragen gedehnt werden. Die Werte, die man von einem flach liegenden Kleidungsstück erhält, reichen nicht immer aus, die guten Sitz- und Trageeigenschaften zu beurteilen.

Basierend auf der Erfahrung, die mit flachsitzenden Formen für Socken und Strümpfe etc. gemacht worden war, wurden 26 Flachformen entwickelt, und zwar mit der Oberkörperhöhe, dem Kopf, der Brust, den Schultern und einem Teil der Oberarme. Diese Formen umfassen 95 % der männlichen Erwachsenen in Frankreich.

Die vorgeschlagenen Kriterien für die Beurteilung der Trageeigenschaften und Leistung von T-Shirts sind :

- in der Weite : die Festigkeit, der gute und passende Sitz oder die Flottierung der Artikel wurden untersucht,

- in der Länge : die experimentell definierte erforderliche Mindestlänge wird durch eine Linie auf den Formen angegeben.
Wenn die T-Shirts auf die Formen gebracht werden, wird das leichte Passieren des Kopfes nachgeprüft.
Diese Kriterien sind mit in Labors und der Industrie durchgeführten Exprimenten bestätigt worden. Die dimensionale Leistung von Strickwaren mit unterschiedlichen Festigkeitsfaktoren wurde untersucht, indem man eine Verformung des Kleidungsstückes während des Tragens simulierte.
Die neuen Testmethoden und Meßvorrichtungen können zu folgendem beitragen :

- der Qualitätssicherung und Qualitätskontrolle,

- der Verbesserung von Kleidungsspezifikationen der Wiederverkäufer und Hersteller,

- der Verbesserung von Größenangaben,

- der Verbesserung eines guten und bequemen Sitzes von Strickwaren.

QUALITÄTSASPEKTE FÜR DIE HERSTELLUNG VON MASCHENWAREN

M. BALLAND
I.T.F.-Maille

Die hier beschriebenen Resultate stellen eine Zusammenfassung von Arbeiten dar, die von CENTEXBEL, DANISH TEXTILE INSTITUTE, INSTITUT FÜR TEXTILTECHNIK, und INSTITUT TEXTILE DE FRANCE-MAILLE durchgeführt wurden. Wir werden die einzelnen Punkte anfassen, indem wir versuchen, dem Produktionszyklus - von der Faser bis zum Endprodukt - zu folgen.

Zunächst werden wir jedoch die Kostenstruktur von Strickwaren und die Kosten von Fehlern untersuchen. Im Durschnitt betragen die Materialkosten 56 % beim Flachstrickverfahren und 64 % beim Rundstrickverfahren.

Der Kostenpunkt von Fehlern liegt im Durchschnitt bei 6%, von denen 60 % garnbedingt sind. Garnspleißtechniken ermöglichen eine erhebliche Verbesserung.

Garne für Strickmaschinen müssen gewachst werden. Gängige Paraffin-Waxsorten besitzen Kohlenwasserstoffketten mit 21 bis 28 Atomen und ihre physikalischen und chemischen Eigenschaften werden dadurch bestimmt. Und da die Ablagerungsmengen vom Garn abhängen, ist es unter industriellen Voraussetzungen praktisch unmöglich, einen bekannten einheitlichen Reibungskoeffizienten zu erhalten. Die besten Ergebnisse erhält man mit Werten von 0,4 bis 1 %.

Deshalb scheint es notwendig, die absorbierte Garnmenge auf allen Arten von Strickmaschinen zu beobachten und zu kontrollieren.

Darum wurden verschiedene Prototypen auf einer Flachstrickmaschine untersucht; der neueste Prototyp umfaßt: Einen Garneinführungsdetektor, einen Schlittenpositionsdetektor, eine Zentralsteuerung mit einem "Single Board Computer" und ein Kontrollsystem, das auf Garnspannung reagiert. Erfolgreiche Industrieversuche sind bereits durchgeführt worden und haben es ermöglicht, Längenunterschiede beim eingeführten Garn auf ± 1 % zu verringern. Derzeit wird untersucht, ob es wirtschaftlich ist, eine solche Vorrichtung an allen Maschinen anzubringen. Es ist sicher, daß Vorrichtungen dieser Art weiterhin entwickelt und extrem wichtige Materialeinsparungen mit sich bringen werden.

Die Qualität von Strickwaren aus Wolle bzw. Wolle/Acryl wurde untersucht, und zwar unter folgenden Gesichtspunkten : Berührung, Aussehen, Pill-Bildung und Maschenverwerfung.

Die besten Einstufungen bei der Berührung und beim Aussehen wurden von 100 %-Wollgarnen mit feinsten Fasern erreicht. Außerdem waren die Wollsorten ohne Filzfreibehandlung besser.

Die Woll-Acrylic Mischungen erreichten ähnliche Ergebnisse trotz der Unterschiede in Faserstärke und unabhängig davon, ob die Mischung Antipilling hatte oder nicht.

Nach dem Waschen konnten keine Unterschiede mehr zwischen der besten Wolle und all den Mischungen festgestellt werden.

Was die Pill-Bildung angeht, erzielen die Strickwaren mit den feinsten Wollsorten die ungünstigsten Ergebnisse, besonders beim Flachstrickverfahren. Der Unterschied zwischen behandelter und unbehandelter (gegen Einlaufen) Wolle erscheint nur schwach bei Flachstrick-Waren, wobei die behandelten Garne besser abschneiden. Bei den Mischungen wurden die besten Ergebnisse mit 5,6 dtex Acryl erreicht. Diese sind lediglich den feinen unbehandelten Garnen überlegen.

Was die Maschenverwerfungen bei Jersey betrifft, ist es ganz deutlich gezeigt worden, daß dieser Fehler von äußerst groben Wollfasern an diesen Stellen hervorgerufen wird. Dies erhöht die Biegefestigkeit des Garns und verursacht entweder Verwerfungen oder Welligkeit.

Die Verdrehung schließlich ist ein innerer Fehler bei Jersey-Strickwaren und obgleich gewisse Behandlungen dieses Phänomen verringern können, ist das derzeit einzige praktikable Mittel eine Abwechslung von S-Drall und Z-Drall Garnen.

Dimensionale Variationen bei Strickwaren gehören zu den größten Problemen, die zur Zeit untersucht werden.

Das Entspann-Konzept wird von allen Labors anerkannt und die Methoden zur Erreichung dieses Zustandes sind sich sehr ähnlich. Sie alle benutzen Maschinentrocknung, nachdem entweder Maschinenwäsche-Zyklen (3-5) oder eine statische Behandlung im Wasserbad vorausgegangen sind; dann folgt eine mechanische Behandlung auf einem Vibrationsband.

Die Wichtigkeit des Feuchtigkeitszustandes bei Strickwaren, die aus

343

dem Trockner kommen, wurde deutlich gemacht; dies hat eine Wirkung auf
die Ausmaße von Baumwollstrickwaren und der Trockenvorgang sollte streng
nach einem bekannten Trocknungszustand durchgeführt werden. Diese wird
bei 5 bis 6 % empfohlen.

Die Regeln, die man beim Entspannungszustand erhalten hat, wurden
auch für Baumwolle bestätigt und es wurde bewiesen, daß auch Acryl-Jersey
1:1-Rippe und 2:2-Rippe dem Mundenschen Gesetz folgen, dessen Koeffizien-
ten kalkuliert worden sind.

Wenn Baumwoll-Strickwaren im Verlauf der Herstellung mehrmals
gewaschen werden, wird deutlich, daß der größte Teil der dimensionalen
Veränderungen beim ersten Wachvorgang auftacht, sich dann kaum weiter
verändert, bis es etwa beim zehnten Waschvorgang zur völligen Stabilisie-
rung kommt.

Die Längenschrumpfung der ersten Wäsche geht in derselben Richtung
weiter, während es bei der Breitenverringerung nach dem ersten Waschen zu
einer gewissen Durchhängung kommt.

Die Dimensionen von Strickwaren fluktuieren innerhalb des Appretur-
zyklus. Die relativ hohen Schrumpfungswerte jedoch bei ungefärbtem
Gestrick werden nie beseitigt, weder beim Färben, noch beim Bleichen oder
anderen Naßbehandlungen; auch nicht beim Trocknen oder bei der Trocken-
endbehandlung. Es ist klar, daß jegliches Dehnen oder Ziehen der Strickware zu einem gegebenen Zeitpunkt eine Erhöhung des Schrumpfens beim
Waschen hervorruft.

Was die Stabilität angeht, ist der Trockner von allen Ausrüstungs-
geräten der wichtigste. Die Maschinen mit der besten Leistung sind die
Dauer-Trommeltrockner.

Eine Schwierigkeit bei der Anwendung des Munden'schen Gesetzes auf
Baumwollgestrick ist der Dimensionsunterschied im entspannten Zustand bei
ungefärbten Strickwaren und behandelten Strickwaren. Im Allgemeinen sind
behandelte Waren im entspannten Zustand 5 % länger und 4 % schmaler als
ungefärbte. Datenbänke werden vielleicht die Lösung dieses Problems
ermöglichen.

Die Behandlung von Acryl für Pullover ist auch untersucht worden.
Eine Labormethode zur Messung des "Setzen" des Garns wurde zu diesem
Zweck entwickelt. Zunächst wurde bewiesen, daß eine herkömmliche Preß-
behandlung die Maße einer Strickware nicht setzen konnte, wenn diese,
verglichen mit dem entspannten Zustand, verformt war. Nur 30 % bis 50 %
dieser Verformungen können gesetzt werden. Es wird gezeigt, daß die
Strickflächen vor der Preßbehandlung entspannt und dann auf einer Presse
"gesetzt" werden sollten, und zwar mit einem Luft-Dampf Gemisch bei
Temperaturen, die je nach Art des Acryls variieren. Leacryl bei 75°C.,
Courtelle zwischen 75 und 80°C, Dralon bei 80°C. und Crylor bei 85°C.
Unter 70°C. die Strickwaren werden dann nicht gebügelt.

Versuche mit Testpersonen haben die Laborergebnisse bestätigt.

Deutlich wird einerseits die Wichtigkeit des Ausgangszustands der
Strickware für die Größenentwicklung während des Tragens und Waschens,
und andererseits die Wichtigkeit der Behandlung.

Am stabilsten wird der Stoff mit der besten Setzbehandlung sein, so
nah wie möglich am entspannten Zustand.

In Anbetracht der Neigung zur Längenschrumpfung und zum Durchhängen
in der Breite wäre es jedoch nützlich, vorher Korrekturmaßnahmen von etwa
5 % in den zwei Richtungen vorzunehmen.

Besondere Aufmerksamkeit sollte den Falzen geschenkt werden,
erstens, daß sie nicht zu locker gestrickt werden, und zweitens, daß sie
nicht weit vom entspannten Zustand gefertigt werden.

Unter diesen Voraussetzungen ist es möglich, Acrylpullover mit guten

Trageleistungen zu erhalten.

Schließlich möchten wir noch auf die Existenz von Testmethoden für die Herstellungseigenschaften von Strickwaren hinweisen. Insbesondere die Bedeutung der Nadelstärke und auch des Aufweichverfahrens, indem man die Stärke des Nadeleinstichs in das Gewebe mißt.

Es wurde deutlich, daß die Qualität von Strickwaren im Verlauf der Herstellungskette zunimmt, und die hier gezeigten Ergebnisse können vielleicht zur Qualitätssteigerung beitragen.

OBJEKTIVE QUALITÄTSBEWERTUNGEN

E. FINNIMORE
Deutsches Wollforschungsinstitut

Das allgemeine Thema des hier dargestellten Forschungsprogramms ist die Qualität von gestrickten Stoffen und Strickwaren.

Um wettbewerbsfähig zu bleiben, muß die europäische Textilindustrie ihr Qualitätsniveau aufrechterhalten und ständig verbessern. Aus diesem Grunde sind objektive Qualitätsbeurteilungen notwending, da diese u.a. folgendes ermöglichen : Bessere Qualitätskontrollen in der Fabrik, Aufstellung internationaler Standardwerte und Designs von Stoffen und Kleidungsstücken, die spezifischen Anforderungen gerecht werden. In diesem Vortrag werden verschiedene Aspekte von objektiver Beurteilung besprochen, die auf Arbeiten in drei europäischen Instituten basieren.

Beim Kauf von Strümpfen und Socken kommt es dem Verbaucher besonders darauf an, daß sie gut passen, was wiederum zu einem angenehmeren Tragen führt. Beim ITF Maille in Frankreich machte man eine Studie zur Entwicklung von objektiven Kontrollmöglichkeiten fur die Größe und das Passen von Socken.

Beim Tragen sind Socken in einem Dehnzustand. Dies muß bei objektiven Messungen des Sitzes beachtet werden. Aus praktischen Gründen werden in der Industrie Flachformen bevorzugt. Deshalb mußte eine Methode entwickelt werden, eine flache Oberfläche der äußeren Oberfläche eines dreidimensionalen Volumens (Fuß und Bein) in ihrer Ausdehnung gleichzumachen.

Die nötigen Schritte zur Erreichung der Formen waren folgende : Die Umsetzung eines angepaßten Sockens in eine Starrform, die dann durch Schneiden vom Fuß entfernt wurde; Einschnitte an geeigneten Stellen, um eine Flachform zu erhalten; Umsetzung dieser Form auf eine equivalente Fläche mit durchlaufenden Konturen. Ein Satz von 12 Formen wurde angefertigt, entsprechend den französischen Strumpfgrößen 24-46.

90 Artikel verschiedener Strumpfarten wurden auf diesen Formen und dann von Versuchspersonen getestet.

Bei den 90 getesteten Artikeln kam es nur in drei Fällen zu unterschiedlichen Beurteilungen zwischen dem Objektivtest und den Testpersonen. In diesen drei Fällen konnte gezeigt werden, daß die Fuß- und Beinmaße dieser Personen von den Standardwerten derselben Fußgröße abweichten. Die Formen sind mit Erfolg in 5 Strumpffabriken getestet worden. Die Formen werden nun kommerziell hergestellt und diese Methode ist als ein französisches Standardverfahren eingeführt worden.

Eine wichtige Fehlerquelle in der Strickwarenindustrie ist die schlechte Nähfähigkeit einiger Strickstoffe, die zu Schleifenschäden und Lochbildungen beim Nähen führen kann. Beim Institut für Textiltechnik Denkendorf, BRD, wurde ein objektiver Test für die Nähfähigkeit von Strickwaren entwickelt. Vorstudien zeigten, daß ein bedeutsamer Zusammenhang zwischen der Nadeleinstichstärke und den Schleifenschäden besteht.

Ein Computerunterstütztes System wurde deshalb zur schnellen Aufzeichnung der Einstichstärke entwickelt. Die Vorteile dieser Testmethode sind, daß die Versuche auf Produktionsmaschinen und unter Produktionsbedingungen durchgeführt werden können.

Der Test kann mit Stichproben vor dem Schneiden durchgeführt werden, so daß es möglich ist, das ganze Stück mit Näh-Hilfsmitteln zu behandeln. Die Vorteile einer korrekten Appretur von Strickwaren für eine verbesserte Nähfähigkeit können durch diese Methode deutliche gemacht werden. Bei Stoffen aus 100 % Synthetic-Fasern konnte die Nadelstichstärke mit Hilfe von geeigneten Appreturmitteln um bis zu 93 % vermindert werden. Die Methode ermöglicht also objektive Untersuchungen, um Appreturbehandlungen zu optimieren, die wiederum einen direkten Einfluß auf eine Steigerung der Produktqualität haben.

Der Griff von Strickwaren ist einer der wichtigsten Qualitätsaspekte und kann, gemeinsam mit dem Aussehen der Ware, zum entscheidenden Faktor für den Konsumenten werden, ob er die Ware kauft oder nicht. Dieser Griff ist jedoch eine Eigenschaft, die nur sehr schwer definiert, gemessen und in Zahlen ausgedrückt werden kann.

In den letzten Jahren gab es große Bemühungen, objektive Griffmessungen zu entwickeln. Unter den vielen Vorteilen solcher Messungen seien hier zwei erwähnt : Die Möglichkeit, numerische Spezifikationen zwischen Hersteller und Verbraucher auszutauschen, und die Wirkungen verschiedener Appreturbehandlungen in Zahlen auszudrücken. Die größte Beachtung weltweit hat das KES-F System von Kawabata, Japan, gefunden.

Am Deutschen Wollforschungsinstitut in Aachen wurde die Anwendung dieses Systems zur Messung des Griffes von Strickwaren untersucht. Beachtet werden muß die erhöhte Dehnbarkeit von Strickwaren gegenüber gewebten Materialien.

Gemessen werden die folgenden Eigenschaften : Biegsamkeit, Scherbarkeit, Reißfestigkeit, Kompressionsfähigkeit, Oberflächenbeschaffenheit, und Konstruktion (Gewicht und Dicke). Die Studien befaßten sich zunächst mit der Wirkungsbeurteilung verschiedener Parameter, wie etwa dem Straffheitsfaktor, der Strickart, dem Fasermaterial, der Feinheit und der Appretur bei einer Reihe von Woll- und Woll/Acryl-Strickwaren. Die Wichtigkeit der Dampf-Entspannung zur Erreichung eines vollen weichen Griffes konnte demonstriert werden.

Der zweite Aspekt dieser Studien befaßte sich mit der quantitativen Bewertung der Wirkung, die Aufweichmittel auf Woll- oder Wollgemisch-Strickwaren haben. Mit der Hilfe objektiver Messungen von Steifheitsparametern konnte gezeigt werden, daß es ein optimales Anwendungslevel gab für die Erreichung einer Weichmachwirkung auf schrumpffeste behandelte Wolle.

Diese konzentrierten Effekte wurden auch bei subjektiver Bewertung bemerkt. In Allgemeinen gab es eine ziemliche Übereinstimmung zwischen der objektiven und subjektiven Bewertung von Appreturbehandlungen, aber die Strickart beeinflußte das Urteil der Experten.

Die bei dieser Art von Studie erhaltenen Daten können dem Hersteller helfen, die besten Materialien und Appreturbedingungen zur Erreichung der gewünschten Qualitätsstufe auszuwählen.

ZUSAMMENFASSUNG UND SCHLUSSFOLGERUNGEN DES VORSITZENDEN DER SITZUNG

J. STRYCKMAN
Centexbel

Die auf die Vorträge folgenden Fragen haben zwei wichtige Aspekte
hinsichtlich der Qualität der Strickware hervorgehoben : einerseits die
Dimension sowie die Formbeständigkeit der Strickware, und andererseits
objektive Messungen, welche den Komfort des Artikels festlegen. Die
Diskussion wird im folgenden kurz zusammengefaßt.

Auf die Frage, ob die Dimensionen einer fertigen Strickware anhand
der Faden- und Maschenstrukturparameter festgelegt werden; kann man
antworten, daß diese Parameter gemäß Masche und Fertigstellung der
Strickware miteinander verbunden sind. Die vom International Institute
for Cotton gegründete Datenbank "Starfish" ermöglicht es, die Dimensionen
eines Fertigartikels anhand der Eigenschaften des ungefärbten Fadens
festzulegen. Die Gleichungen sind für Baumwolle in drei Maschenarten
bekannt : Jersey, Rippen 1/1 und Interlock.

Bezüglich der Länge des Fadens, der pro Masche verbraucht wird,
hängen die Maße der Strickware größtenteils von diesem Parameter ab. Es
ist daher äußerst nützlich, daß dieser kontrolliert und beherrscht wird.
Auf der Rundmaschine wird diese Aufgabe vom positiven Fadenzubringer
übernommen. Für die Flachstrickmaschinen existierte keinerlei derartiges
Gerät, aber jetzt zwei neue Vorrichtungen entwickelt und getestet
worden. Bei einem ist das Ziel die Aufrechterhaltung der konstanten
Spannung an einem bestimmten Punkt der Zuführung des Fadens in die
Strickmaschine. Bei dem anderen System wir die Länge des Fadens, der auf
einem festgelegten Teilabschnitt auf jeder Reihe verbraucht wird, durch
einen Mikroprozessor mit einer vorher bestimmten, gewünschten Länge für
die Reihe verglichen. Wenn die beiden Längen nicht übereinstimmen, korri-
giert ein Signal die Fadenzuführung zur Strickmaschine. Beide Systeme
ermöglichen, Strickwaren identischer Länge mit einer Abweichung von 1 %
in der Länge des verbrauchten Fadens zu erhalten.

Die Geschwindigkeit der modernen Fachstrickmaschinen ist kein
Hindernis für das gute Funktionieren der Vorrichtung. Die Beschleunigung
des Aufnehmens könnte in einem der Systeme ein Problem darstellen, aber
die Messung findet im Inneren seiner Fontur statt. Beim anderen System
reagiert der Sensor augenblicklich auf Unterschiede in der Spannung. Die
Wirkung der Fadenbeschleunigung am Anfang einer Reihe wird durch die
gleichzeitige Deaktivierung der Spannungsvorrichtungen aufgehoben.

Das Wachsen des Fadens vor dem Stricken ist ein nicht sehr hoch-
entwickeltes Mittel, um die Länge des verbrauchten Fadens zu regulieren
und bleibt für bestimmte modische Artikel die einzig mögliche Methode.
Jedoch reagieren nicht alle Wachse auf die gleiche Weise. Ihre Zusammenstellung, mit anderen Worten die Länge der Kohlenwasserstoffketten, die
von einem Wachs zum anderen verschieden ist, spielt eine Rolle, da sie
den Schmelzpunkt beeinflußt, der wiederum von der Menge des verwendeten
Wachses abhängt : diese sollte, je nach der Art des Fadens, zwischen 0,4
und 1 % betragen. Es ist also notwendig, das Verhalten des benutzten
Wachses zu kennen und ihn entsprechend der Fadenart anzuwenden.

Die Frage, ob Wachs einen Einfluß darauf hat, wie die Strickware

sich anfühlt, bringt uns zu einem anderen Thema der Diskussion : die
Aspekte "Anfühlen" und "Komfort" der Artikel und ihre objektiven Bewer-
tungskriterien. Zuerst kann man sagen, daß das Wachsen die Art, wie die
Strickware sich anfühlt, verändert, aber diese Frage wurde im Rahmen des
Programms nicht weiter behandelt. Da die objektiven Qualitätsmessungen
noch in ihren Anfängen sind, sind auf diesem Gebiet noch etliche Untersuchungen anzustellen.

Dieses Programm konnte nur die Parameter Strickmasche und Faser
abdecken, und es wurden schon eine große Anzahl an Mustern analysiert.
Zwei Fadentypen wurden untersucht, der eine aus reiner Wolle, der andere
eine Mischung aus Akryl und Wolle. Die objektiven Messungen wurden auf
einer japanischen Maschine Kawabata ausgeführt, die speziell für die
Messung des "Anfühlens" entwickelt wurde. Die Ergebnisse der objektiven
Messungen wurden mit den subjektiven Messungen verglichen, die von Experten ausgeführt wurden, welche man aufgrund ihrer langjährigen Erfahrung
in der Beurteilung von Stoffen oder Strickwaren ausgewählt hatte. Die
Korrelation war insgesamt gesehen recht gut, aber die Experten irren sich
manchmal, wenn der Unterschied zwischen den Mustern nicht auffällig ist.

Man muß allerdings darauf hinweisen, daß das Konzept der "Weichheit"
der Kawabata-Maschine eine Kombination zweier Elemente ist : die Weichheit der Oberfläche und Steifheit. In Europa muß eine Terminologie für
diese Messungen noch ausgearbeitet werden.

Die Erfahrung von Kawabata auf dem Gebiet von Strickwaren ist nicht
so groß wie bei den gewebten Stoffen. Die Parameter müßten den Strickwaren angepaßt werden. Allerdings müßten die subjektiven Messungen nach
und nach einer objektiven Synthese zur Beurteilung der Artikel weichen,
da die Experten nicht immer konsequent sind und sich regelmäßig, selbst
in kurzen Zeitspannen, widersprechen. Für Strickwaren wären Parameter wie
die Hysterese der Abscherung, Dicke oder Gewicht und die Regelmäßigkeit
der Oberfläche besser geeignet. Die Erfahrung wird es zeigen.

Die Bedeutung der Kawabata und der objektiven Messungen liegt auch
darin, daß man den Einfluß der Parameter des Materials und des Herstellungsverfahrens auf das "Anfühlen" leicht beurteilen kann. Dies ist aus
praktischen Gründen schwieriger, um nicht zu sagen unmöglich, wenn man
subjektive Messungen anwendet. Auf die Frage, ob das System Kawabata in
der europäischen Industrie benutzt werden kann, muß man antworten, daß
die Ausrüstung relativ kostspielig ist und die kleinen oder mittleren
Unternehmen kaum einen Umsatz erzielen, der ihre Anschaffung rechtferti-
gen würde.

In Japan sind 150 Ausrüstungen installiert, viele davon in For-
schungseinrichtungen oder Einkaufsbüros, die mit der Industrie zusammen-
arbeiten. Es wäre daher jetzt wünschenswert und notwendig, daß sich die
Textilindustrie für den Bekleidungssektor in der nahen Zukunft diese
Technik zunutze macht. Sie sollte sich dabei auf Gemeinschaftslaborato-
rien stützen, welche so durch ihre Ausrüstung, aber auch die Erfahrung,
die sie hinsichtlich der objektiven Qualitätsmessungen sammeln würden,
dieser Industrie unschätzbare Dienste erweisen könnten.

Das Thema "Qualität der gestrickten Gewebe und Artikel" wurde in
drei Hinsichten untersucht.

Der erste Aspekt : Qualitätskriterien der gestrickten Artikel. Hier
zeigten die von vier Instituten ausgeführten Arbeiten deutlich, daß es
unumgänglich ist, die Qualität, den guten Schnitt, den Komfort und die
Zufriedenheit der Verbraucher besser zu gewährleisten. Dies erfordert
eine einheitliche europäische Methode, um die Maße des menschlichen
Körpers zu messen; diese Methode muß in sämtlichen Ländern der Gemeinschaft angewendet werden und in jedem Land müssen in den Gebieten, wo

Unterschiede im Körperbau der Menschen auftreten, Datenbanken mit diesen Maßen eingerichtet werden. Diese Datenbanken müßten gemäß der etablierten Standardmethode stets auf dem neuesten Stand sein.

Der zweite Aspekt, mit dem sich die vier Institute ebenfalls befaßten, betraf die Qualität bei der Herstellung gestrickter Gewebe und Artikel. Hierbei ging es hauptsächlich um einen sehr kritischen Punkt der Strickwaren : die Formbeständigkeit sowohl während der Herstellung als auch beim Tragen und Waschen durch den Verbraucher. Die Arbeitsergebnisse werden eine Hilfe für eine bessere Kontrolle dieses Problems im Bereich der Artikel aus Wolle, Woll/Akryl-Gemischen und Baumwolle ermöglichen.

Ein besonders interessantes Ergebnis wurd durch die Entwicklung von Vorrichtungen erzielt, die automatisch die Formbeständigkeit der Schnittteile auf Flachstrickmaschinen sicherstellt.

Es ist jedoch nicht zu leugnen, daß auf dem Gebiet der Formbeständigkeit noch viel zu tun ist, sowohl in Bezug auf die Charakteristika des Fadens und der Drehung als auch hinsichtlich des Herstellungsverfahrens einschließlich der Veredlungs-Behandlungen.

Der dritte Punkt schließlich befäßte sich mit einem Aspekt, der in der Zukunft immer mehr an Bedeutung gewinnen wird, um die Qualität gestrickter Gewebe und Artikel zu beurteilen. Es handelt sich um die objektive Qualitätsbewertung. Drei wichtige Merkmale : das Anfühlen, der gute Sitz und die "Konfektionierbarkeit".

Dieses Thema wurde von drei Instituten untersucht, wobei eine neue Methodologie zu ersten Ergebnissen bezüglich der Strickwaren geführt hat, ein Gebiet, das für gewebte Stoffe schon weit entwickelt ist.

Die Anzahl der Parameter, die diese Eigenschaften beeinflussen, ist besonders groß, daher mußte das Programm sich auf Parameter von grundlegender Bedeutung und auf eine begrenzte Anzahl typischer Grundstoffe, Spinnmethoden, Strickwarenstrukturen und Veredlungsverfahren beschränken.

Die Arbeit, die auf diesem Gebiet noch gemacht werden muß, kann auf viele Jahre die Forschungsprogramme der Institute und Forscher versorgen.

Als Vorsitzender dieser Versammlung möchte ich zwei Wünschen Ausdruck verleihen.

Der erste richtet sich an die Institute, die an diesem Thema mitgearbeitet haben. Sie haben eine Menge Resultate erzielt, die nun in die Praxis umgesetzt werden müssen. Dieses Ziel kann nur dann erreicht werden, wenn sämtliche dieser Ergebnisse von der Industrie angewandt werden; daher ist es unumgänglich, daß jedes Institut sich auf seinem Gebiet dafür einsetzt, die erhaltenen Resultate an die Industrie weiterzugeben und sie technisch zu betreuen, um die richtige Anwendung der Informationen in der Praxis zu gewährleisten.

Die Kommission der Europäischen Gemeinschaft bemüht sich um dieses "In-die-Tat-Umsetzen" der Forschungsergebnisse, indem sie alle Anstrengungen unternimmt, den Informationsfluß zu stimulieren und diese Ergebnisse voll zu nutzen. Es hat sich gezeigt, daß der vervielfältigte Effekt einer Zusammenarbeit sehr positiv ist.

Daher ist mein zweiter Wunsch, daß sich in der Zukunft diese europäische Zusammenarbeit zwischen den Instituten, die in diesem Programm in den vergangenen drei Jahren ein so hervorragendes Team gebildet haben, fortsetzen möge.

Diese Zusammenarbeit könnte stimuliert werden, wenn die Kommission auf alle Unterstützungsanträge der großen europäischen Strickwarenindustrie reagieren würde.

Die Erfüllung dieser beiden Wünsche wäre zweifellos ein großer Schritt auf dem Wege zu einer gesicherten Zukunft der europäischen Textilindustrie.

QUALITÄTSVERBESSERUNG VON LEINENERZEUGNISSEN

M. VAN LANCKER
Centexbel

Flachs ist einer der seltenen pflanzlichen Textilrohstoffe, über die die Europäische Gemeinschaft verfügt.

Hier einige Zahlen, um den Flachs allgemein, sowie insbesondere in der Textil-industrie einzuordnen :

- Flachs stellt weniger als 2 % der Textilrohstoffe dar,
- die Produktion von Fasermassen in der EWG (gebrochener Flachs und Flachswerg) belief sich 1984 auf 79.881 T (137.000 T 1962 und 82.100 T 1982),, die Produktion der klassischen Flachsspinnerei betrug 1984 mehr als 33.500 T (80.000 T 1962 und 30.400 T 1982),
- die Handelsbilanz ist ausgesprochen positiv :
  . etwa 145 mil. ECU im Jahre 1984 für Frankreich,
  . etwa 70 mil. ECU im Jahre 1984 für Belgien.

Die seit mehreren jahren bestehende Unausgewogenheit zwischen der Herstellung von Fasermassen und dem Verbrauch der klassischen Spinnerei hat den Handel gezwungen, sich den außerhalb der EWG liegenden Märkten zuzuwenden, und das Gewerbe veranlaßt, neue Absatzmärkte zu suchen.

Seit mehreren Jahrzehnten erfuhr die Verwendung des Flachses (Hanfes) eine ungünstige Entwicklung, und die in der C.I.L.C. (Confédération Internationale du Lin et du Chanvre - Internationaler Bund Flachs und Hanf) vereinten verschiedenen betroffenen Berufe haben ein Aufrichtungs- und Expansionsprogramm erarbeitet, das bedeutende Anstrengungen auf dem Gebiet der Forschung, der Förderung und der Vermarktung vorsah.

Das "EUROLIN" genannte Forschungsprogramm hatte zum Ziel, die Umwandlungsverfahren des Strohs zu Fertigprodukten zu verbessern, um :

- Rohmaterial einzusparen,
- zufriedenstellendere Arbeitsbedingungen zu schaffen,
- die echten Vorzüge des Flachses herauszustellen,
- die Produktionskosten zu verringern,
- über Artikel zu verfügen, die den Bedürfnissen der Verbraucher besser angepaßt sind.

Basierend auf einer Ende der 70er Jahre durchgeführten Analyse der Situation der Flachsindustrie der EWG, auf bestehenden Problemen und Mängeln hinsichtlich der Abschätzung der bereits gemachten Anstrengungen, sowie auf den Ergebnissen vorausgehender Arbeiten, wurde beschlossen, die EWG im Rahmen des zweiten Programmes darum zu bitten, einen Vorschlag zur finanziellen Unterstützung der Textilforschung zu unterbreiten, mit dem Ziel, die Ausführung des in Verwirklichung befindlichen EUROLIN-Projektes zu beschleunigen.

Zahlreiche Faktoren ließen den Augenblick besonders günstig erscheinen, um diesem Programm einen neuen Impuls zu geben :

- die Rückkehr der Verbraucher zu Naturfasern,

- die Rückkehr der Verbraucher zu Naturfasern,
- die Entwicklung von Mischfasern,
- das Erscheinen besonderer Veredelungs-Behandlungen,
- die neuen Spinnereitechniken,
- die dringende Notwendigkeit für die Flachsindustrie, wieder zu investieren,
- das Vorhandensein eines großen Exportmarktes.

Das der EWG vorgestellte Forschungsprogramm wurde von der Wisschenschaftlichen und Technischen Kommission der C.I.L.C. ausgearbeitet und zwar durch einen Forschungs- und Entwicklungs-Ausschuß, das Comité Billaux. Der Ausschuß umfaßt Forschungszentren, die sich in den EWG-Ländern mit Flachs befassen : ATPUL, ITL, ITF-Nord und ITF-Boulogne für Frankreich, Centexbel und OVLT für Belgien, TNO und IBVL für Holland, LIRA für Großbritannien. Textil-Forschung Bielefeld für Deutschland und die Stazione Sperimentale per la Cellulosa fur Italien.

Jedes dieser Zentren ist in einem besonderen Bereich spezialisiert, an den man sich wandte, für das allgemeine Programm war es jedoch erforderlich, eine Koordinierung zwischen den verschiedenen individuellen Bemühungen vorzunehmen. Das Eingreifen der Gemeinschaft hat die Notwendigkeit hervortreten lassen, diese Zusammenarbeit auf hoher Ebene aufrechtzuerhalten. Diese wurde verwirklicht durch die Wissenschaftliche und Technische Kommission der C.I.L.C. und das Comité Billaux. Diese Art des Vorgehens war eine Garantie dafür, daß die Arbeiten auf die tatsächlichen Bedürfnisse der Industrie gerichtet wurden, und daß die wissenschaftlichen und technischen Ergebnisse kurzfristig in der industriellen Praxis bekannt und angewandt wurden.

Eine Aktualisierung der Ziele gestattete, die nachstehend genannten technischen Ziele klar zu definieren. Diese Ziele wurden in Forschungsprogramme übersetzt, die im Zweiten Textil- und Kleidungs-Forschungs- und Entwicklungs-Programm der Gemeinschaft unter dem gemeinsamen Nenner "Qualitätsverbesserung von Leinenerzeugnissen" wiedergegeben wurden.

TECHNISCHE ZIELSETZUNGEN :

Versorgung mit Rohmaterialien und Fasergewinnung
Um eine ständige Versorgung mit passenden Rohmaterialien sicherzustellen, müssen Forschungsbemühungen auf die Verbesserung der Anbautechniken, den Faser-Ertrag, den Wert der Nebenprodukte, die Präsentierung der Materialien und die Reduzierung der Lohnkosten durch Mechanisierung gerichtet werden.

Umwandlung der Faser zu Garn
In diesem Stadium ist es erforderlich, die Faser anzupassen, um entweder Material mit hoher Produktivität oder neue Spinn-Techniken verwenden zu können : OE-Spinnen, Reibungsspinnen, Novacore, u.s.w....

Veredelung
Technische Verbesserungen oder Innovationen der Produkte oder Verfahren müssen experimentell erprobt und entwickelt werden. Weiterhin müssen Antworten auf die allgemeinen Fragen der Flachsveredelung gefunden werden.

Qualifizierung der Produkte
Die Qualifizierung des Flachses in all seinen formen ist erforderlich, um Eigenschaften und Merkmale zu nennen, die es gestatten, die

Eignung des Flachses hinsichtlich verschiedener Verwendungsbereiche zu bestimmen. Die Objektive Qualitätskontrolle muß realisiert und von genau festgelegten Normen und Lastenheften begleitet werden.

Morphologie der Faser

Eine bessere Kenntnis der Morphologie der Flachsfaser und der Auswirkung der Produktions- und Arbeitsbedingungen ist mehr als je zuvor notwendig, um alle potentiellen Mittel der Flachsfaser zu nutzen.

Das "EUROLIN"-Programm, zu dem das Programm "Qualitätsverbesserung von Leinenerzeugnissen" gehörte, wird in den Instituten zugleich mit neuen Förderungsaktivitäten auf internationaler Ebene fortgesetzt.

Durch diese Forschungs- und Förderarbeiten wird sich die Flachsindustrie der EWG einem modernen technologischen und sozialen Kontext vorstellen und wird so der industriellen Strategie der EWG Rechnung tragen.

Die neuen Qualifizierungen werden bereits auf allen Ebenen sichtbar, insbesondere jedoch in Bezug auf neue Absatzmärkte und die Herstellung qualitativ hochwertiger Produkte. Die Flachsindustrie bereitet sich darauf vor, diesen Forderungen zu entsprechen und Produkte vorzustellen, die bei der Kundschaft der Weltmärkte gefragt sind.

PRODUKTION DER LEINENFASER

W.W. FOSTER
LIRA

Die Tabelle zeigt eine Auflistung der Aktivitäten in fünf Forschungszentren innerhalb des EUROLIN-Programms, abgezielt auf die Verbesserung
der Gesamternte und der Flafaserqualität sowie die Senkung der Verarbei-
tungskosten.

AKTIVITÄTEN	Forschungszentren				
	ITL	LIRA	Wagen-ingen	ATPUL	Centex-bel
1. Anwendungen beim Anbau					
– Chem.Sprays gegen Umlegung	x				
– Austrocknungsspray (chem.) Wasserflachs	x				
– Austrocknungsspray (chem.) Rasenflachs			x		
– Schnitt/Wasserflachs			x		
– Halm-Hechelungs-Techniken	x				
– Verseuchende Materialien				x	x
2. Anwendungen in der Fabrik					
– Verbesserung bei der Strang-ausbreitung				x	
– Florbandproduktions-Systeme				x	
– Florband-Verarbeitung :					
. Enzyme		x			
. Chemikalien		x		x	
– Kurzfaser-Systeme				x	
– "Lin-Totale"				x	

1. – ANWENDUNGEN BEIM ANBAU
    Während der letzten fünf Jahre wurde die Benutzung chemischer Sprays
zur Verhinderung des Umlegens entwickelt und der Prozeß hat sich nunmehr
voll etabliert. Etwa 10 % der gesamten Flachsernte in Frankreich wurde
1984 behandelt. Die Nachteile dieser Chemikalien, die normalerweise zur
Wachstumsverlangsamung benutzt werden, sind eine geringere Halmhöhe und
ein kleinerer Saatertrag. Die verkümmerten Stämme sind auch dicker als
normal. Diese Sprays sollten deshalb nur im Notfall gebraucht werden. Die
drei als sicher angesehenen Produkte sind Cerone, Atheverse und Terpal.
Alle Arten reagieren auf die Chemikalien, und zwar auf allen Stufen des
Wachstums. Netzmittel verstärken ihre Wirkungen.
Die Pionierarbeit beim Gebrauch von herbiziden Sprays zur Austrocknung
und Röstung von Flachs im Feld hat sich weiterentwickelt. Das Verfahren
wurde auch erweitert, um Trocknung und normale Tauröste zu beschleu-

nigen. Wegen des außergewöhnlich trockenen Wetters in den letzten zwei Jahren sind die Ergebnisse noch begrenzt und es bleibt noch viel zu lernen.

Maschinen wurden entwickelt zur Anhebung taugerösteten Flachses, bevor er in Ballen gepreßt wird, und zwar zur Aushechelung von Unkraut, Beseitigung von Steinen und Trockenförderung. Etwa 200 Maschinen sind bereits in Betrieb. Ein Zusatzgerät mit ähnlicher Technik kann auch an Ballenpressen angebracht werden, wobei 85 % der Steine ausgeschüttelt werden können, bevor der Flachs in die Ballen geht.

Gearbeitet wurde auch an der Entwicklung von Verfahren, die den Gebrauch spezieller Flachsmaschinen, wie z.B. "Abziehern", überflüssig machen und stattdessen normale Landwirtschaftsmaschinen benutzen. Im Prinzip würde der Halm geschnitten, nicht gezogen, und der daraus resultierende Flachs würde taugeröstet und getrocknet, in Ballen gepreßt und später gehechelt, um saubere kurzstapelige Fasern zu produzieren. Erste Eindrücke sind, daß der Faserpreis wettbewerbsfähig sein wird mit dem, der bei konventionelleren Landwirtschaftsverfahren erzielt wird, und daß die Faser für den Gebrauch in Mischungen sehr geeignet ist.

Verfahren wurden auch entwickelt, die eine Gegenwart von Polypropylen in Flachs aufzeigen, jedoch müssen die entsprechenden Mittel zur Beseitigung dieser Verunreinigung erst noch entwickelt werden. Ausgedehnte Arbeit wurde auch bei der Identifizierung von vielen verschiedenen Unkraut-Fragmenten verrichtet, welche im Flachs-Werkgarn vorkommen und sich später im Endprodukt als Verschmutzung darstellen. Die Anzeichen sind, daß es weitaus besser ist, Unkraut von der Flachsernte zu eliminieren, als später zu versuchen, es zu entfernen oder zu verdecken.

2. - ANWENDUNGEN IN DER FABRIK

Nützliche Abänderungen beim Design von Strangausbreitungsabschnitten wurden entwickelt. Andere Entwicklungen : Untersuchung eines Systems zur besseren wirtschaftlichen Verarbeitung, abgezielt auf höhere Produktionserträge und höhere Qualität. Eine Vorrichtung für wiederausgebreitetes Werkgarn zur Produktion von Material in Florbandform. Und ein Ausrüstungselement für Florbandformation von "Lin Totale". Einige dieser Entwicklungen sind äußerst vielversprechend.

Echte Fortschritte wurden erzielt bei der Aufwertung niederqualitativer Arten von Flachsfasern durch chemische und biochemische Behandlungen, insbesondere bei der Entwicklung von Enzymen zur Aufwertung der Florbandqualität.

NEUE VORBEREITUNGS- UND SPINNTECHNIKEN

M. VAN LANCKER
Centexbel

Die hierunterliegende Tabelle zeigt die Aktivitäten der vier Forschungszentren innerhalb des EUROLIN-Programms. Die Arbeiten wurden unternommen, um technische und wirtschaftliche Verbesserungen bei Faser-Verarbeitungsverfahren in Endprodukten zu erzielen, und zwar unter folgenden Aspekten :

- Wirtschaftlichkeit bei Rohmaterialien,
- Entwicklung der wahren Flachsqualitäten,
- Reduzierung der Produktionskosten,
- Herstellung von geeigneteren Waren für den Verbraucherbedarf.

Aktivitäten	Forschungszentren			
	Biele-feld	Centex bel	I.T.F.	LIRA
1. Aufwertung von Rohmateria- lien mit niederer Qualität	x	x		
2. Durch neue Produktions- routinen hervorgerufene Probleme :				
- Optimierung des Spinn- bandreißens		x		
- Vorbereitung und Trenn- fähigkeit der Faser			x	x
- Vorgarn-Behandlung, Faserdissoziation			x	
- Wirtschaftliche Aspekte einer neuen Produktions- anlage			x	x
3. Neue Spinnverfahren :				
- Rotorspinnerei	x			
- Friktionsspinnverfahren		x		
- Novacore			x	
- Drehlos-Spinnerei				x
4. Veredlung von Produkten und Verfahren.			x	x

In jedem Jahr ist ein Teil der Ernte unbrauchbar für die konventionelle Flachsspinnerei; der bei der Spinnvorbereitung produzierte Abfall
könnte wiedergewonnen werden. Spezifische Verfahrenszyklen konnten durch
Forschungsarbeit identifiziert und entwickelt werden. Neue, auf diesen
Materialien beruhende Produkte sind untersucht und industrielle Versuche
begonnen worden.

Die Spinnbandreiß-Konversion des langen Fadens für seine Verwendung
auf einem Alternativzyklus, genannt "Hochproduktivitäts-System", und die
Spinnbandreiß-Konversion als Mittel zur Korrektur des Faserlängen-
Diagramms sind untersucht worden. Ein Reißkonverter-Prototyp wurde von
der Firma Piersen et Digneffe gebaut. Die Entwicklung eines neuen Reiß-
konverters, basierend auf den Erkenntnissen der Studie und den Bedürfnissen der Flachsfaser angepaßt, wurde von Ateliers Houget-Duesberg-
Bosson durchgeführt.

Systematische Forschung auf dem Gebiet der Vorbereitung, Trenn-
barkeit und Vorgarnbehandlung vor dem Spinnen hat Aufschlüsse gegeben,

die es den Spinnereien ermöglichen, die mechanische und chemische Arbeit
an der Faser zu optimieren. Ein durch Forschung entworfenes und
entwickeltes "Tenomètre"-Gerät ermöglicht ein perfektes Erkennen der
Faser-Dissoziation nach chemischer Einwirkung. Der Flachshandel hat die
Nützlichkeit des Geräts offiziell anerkannt und wird momentan mit diesem
Kontrollgerät ausgestattet.

Im Hinblick auf die Notwendigkeit, Maschinen zu erneuern, wurden die
wirtschaftlichen Aspekte einer neuen Produktionsanlage gemeinsam mit
einem Textilmaschinenhersteller untersucht. Obwohl die Produktion
erheblich verbessert würde, bleibt die Tatsache bestehen daß die Amor-
tisationsphase die möglichen Käufer noch zögern läßt.

Die neuen Spinntechniken, wie z.B. offenend Spinnerei, Umspinnerei
und Drehlosspinnerei, sind untersucht worden. In einigen Fällen wird an
den Maschinen eine Abänderung vorgenommen, z.B. für Novacore, und in
anderen Fällen werden der Flachs und die Nebenprodukte modifiziert, um
den neuen Verfahrenstechniken gerecht zu werden. Für kurze Flachsfasern
wurde durch Forschung eine spezielle Spinnmaschine entwickelt. Dieses
Verfahren wird gegenwärtig in verschiedenen Mitgliedstaaten in der
Industrie benutzt.

Das Studium und die Entwicklung neuer Garne führten zur Modifizie-
rung von Produkten und Verfahren und zur Schaffung neuer Produkte, sowohl
bei Strick- als auch bei Webwaren. Neue Produkte in verschiedensten
Bereichen : Sportkleidung, Oberbekleidung, Möbel- und Freizeitstoffen.
Einige dieser Produkte werden bereits in den meisten Mitgliedstaaten
kommerziell vertrieben.

DER BEREICH DER LEINENVEREDLUNG

M. SOTTON
I.T.F.

Die Hauptaktivitäten der vier Forschungszentren innerhalb des
EUROLIN-Programms sind in der Tabelle unten zusammengefaßt. Die einzelnen
Projekte wurden auf mehreren, sich ergänzenden Forschungsebenen durchgeführt, um erstens durch bessere Materialkenntnisse allgemeine Fragen
der Flachsveredlung zu beantworten und zweitens, um technische Verbesserungen und Erneuerungen bei Produktion und Verfahrensweisen in den
verschiedenen spezifischen Leinensektoren der Mitgliedstaaten zu
erzielen.

Das EUROLIN-Forschungsprogramm konnte eine Lücke bei dem Grundwissen
über die Flachsfaser schließen. Diese Leistung ist bereits zur Quelle
neuer Fortschritte und Verbesserungen geworden. Die ursprüngliche, natür-
liche Flachsfaser scheint ein echter "zelluloser Kevlar" zu sein, besonders anisotropisch und kristallisiert, mit einer geringen Zugänglichkeit
zu Reagenzien, und besonders sensibel gegenüber Kompression und Scher-
kräften, die erhebliche, strukturelle Veränderungen hervorrufen. Das Auf-
tauchen solcher Veränderungen ist mitverantwortlich für die geringe
Widerstandskraft des Flachses gegen Verschleiß. Diese Merkmale lassen
sich aber vielleicht ausnutzen (mechanische Ausrüstung) und sind Grund
genug, das Material mit der nötigen Alkali-Verhandlung zu versehen; dies

Aktivitäten	Forschungszentren			
	Centex bel	I.T.F.	LIRA	Stat. exper.
1. Verbesserte Kenntnisse über die vertrauliche Struktur der Faser, Stand der Technik		x		
2. Bedeutung von Alkali-Vorbehandlung bei der Ausrüstung von Stoffen	x		x	x
3. Veredlungserneuerungen : – von Kleidungstoffen				x
– von Möbelstoffen : –– Stühle, Wandbehänge, Haushaltsleinen	x			
–– Kontrolle der Entzündbarkeit gepolsterter Stühle			x	
4. Möglichkeiten von : – Kurzbad-Behandlungen (Schaum)			x	
– mechanischen Veredlungsprozessen		x		

homogenisiert die einzelnen Fasern und macht sie gleichzeitig zugängiger zu Reagenzien.

Diese Alkali-Vorbehandlungen wurden in allen Mitgliedstaaten untersucht, da es sich gezeigt hat, daß industrielle Weiterverarbeiter unzureichende Kenntnisse über diese Behandlung haben, und daß es unbedingt notwendig ist, Flachs ganz anders als Baumwolle anzugehen. In den verschiedenen Ländern haben diese Projekte zur Einsetzung präziser Empfehlungen für die Handelspraxis geführt, die jedoch von Land zu Land recht unterschiedlich sind, je nach Art des beabsichtigten Produkts (Kleidung, Hemden, Möbelstoffe....)

Die Veredlung von Bekleidungstextilien wurde in Verbindung mit Veredlungsbetrieben untersucht. Ein Resultat dieser Arbeit ist es, daß eine zur Erreichung von Griff- und Dimensionsstäbilität durchgeführte Alkali-Vorbehandlung ein entscheidender Faktor für die Qualität des Endproduktes ist. Sie verdeckt alle ursprünglichen Unterschiede bei den Rohmaterialen, beim Garn und bei den Stoffen. Eine der größeren technischen Leistungen - geteilt mit dem Handel - ist die Einführung einer Spezifikation für Bekleidungsstoffe, die auf eine Qualitätsbeschreibung abzielt.

Etwas Ähnliches wurde für Flachsfasern unternommen, die für Möbel, besonders Stühle und Wandbedeckungen vorgesehen sind.

Vom Fortschritt bei der Anwendung von Anti-Schmutz Appreturen kann auch berichtet werden : Empfehlung der effektivsten Mittel (Fluorkarbon) und Anwendungstechnik (Klotzwalzen eher als Dampfphase oder Schaum). Bei den Stuhl-Polsterstoffen fand man einen eleganten Kompromiß, indem man die Abnutzungs-Widerstandsfähigkeit erhöhte und das Aussehen und die Zusammensetzungs-Charakteristiken modifizierte.

Eine wichtige Studie wurde auch mit dem Ziel unternommen, Leinenstoffe zu produzieren, welche die neuen, gesetzlichen Anforderungen bezüglich der Entzündbarkeit von stoffgepolsterten Stühlen erfüllen, und zwar sowohl im Haushaltsbereich als auch auf dem öffentlichen Bereich können Flachs/Nylon-Stoffe (80/20) geeignet sein, vorausgesetzt sie besitzen eine Unterschicht aus nicht-brennbarem Material, eine industrielle Versuchsreihe ist bereits mit dieser Flachs/Nylon Mischung begonnen worden.

Die Leistung "knitterfester" Stoffe, die mit der Schaummethode behandelt wurden, ist ähnlich wie die von konventionell behandelten Stoffen; außerdem sind die Trocknungskosten erheblich niedriger.

358

ZUSAMMENFASSUNG UND SCHLUSSFOLGERUNGEN DES VORSITZENDEN DER SITZUNG

M. VAN LANCKER
CENTEXBEL

Die Vorträge haben mehrere Fragen hervorgehoben, die eine bestimmte Reihe
von Aspekten erläutern, welche sowohl für die Flachs- als auch allgemeine
Textilindustrie von Bedeutung sind :
- alternative Roestverfahren;
- die Verwendung neuer Herstellungstechniken;
- Produktklassifizierung,
- die morphologische Struktur von Flachs.
In bezug auf die Frage, wie lange es dauern wird, bis alternative
Roestverfahren (chemisch und enzymatisch) als wirklich effizient bezeich-
net werden können, sollte darauf hingewiesen werden, daß nach dem heutigen Stand der Dinge, die neuen Verfahren nicht dazu geeignet sind, das
Problem wegen nachteilhafter Witterungsverhältnisse zu lösen. Jedoch
bedeuten die chemischen Roestverfahren, daß Flachs auch in Gegenden mit
allgemein nachteilhaften Bedingungen angebaut werden kann. Bis heute vor-
liegende Resultate weisen darauf hin, daß es in naher Zukunft möglich
sein wird, adäquates Roesten durch chemische Mittel zu erreichen und
daher qualitativ hochwertige Textilfasern bei schlechtem Wetter zu erzeu-
gen und Profite durch solche Verfahren bei normalen Umständen zu ernten.
Die Verwendung neuer Herstellungsverfahren, besonders das Reißspinnen von langen Stapeln und das Spinnen von kurzen Fasern erweckten
lebhaftes Interesse. Die Anzahl der Reißzonen bei Reißspinnen hing weitgehend von der maximalen annehmbaren Länge und dem besagten Rohmaterial
ab. Tests haben ergeben, daß es vorteilhaft wäre, mit einer dritten Zone
zu arbeiten.
Die für kurze Fasern verwendeten Verfahren entsprechen meist denen
für Baumwolle. Wird jedoch Flachs mit solchen Verfahren verwendet, muß
man der Wahl der zu appretierenden Stoffe und dem Herstellungsverfahren
des Faserprodukts besondere Aufmerksamkeit schenken. Unter bestimmten
Umständen kann man für die Verarbeitung kurzer Fasern, die Maschinen für
kardierte Wolle verwenden. Jedoch wurden nur wenige Industrielle Tests
ausgeführt.
Eine Objektive Bewertung wäre unerlässlich für die nicht auf Flachs
basierenden Textilsektoren, die hauptsächlich Flachs zusammen mit anderen
Fasern verwenden.
Seit mehreren Jahren werden nun Untersuchungen durchgeführt, mit dem
ziel Untersuchungs- und Arbeitsmethoden zur Klassifizierung von Stoffen
und Produkten zu entwickeln. Zur Zeit gibt es drei Testverfahren, wovon
zwei die Simulierung mit einbeziehen, die ausgeführt werden können,
obwohl man ihnen nicht volles Vertrauen schenken kann. Eines davon
ermöglicht eine Bewertung des Hechelns der Ernte, die zweite eine mechanische Teilung des gehechelten, zu bestimmenden Materials, wohingegen das
dritte eine Messung der Faserfeinheit durch Anwendung der Luftströmtech-
nik ermöglicht. Diese Techniken können jedoch nicht mit organoleptischen
Expertenanalysen gleichgestellt werden. Die Produkte (d.h. Garne, Textil-
erzeugnisse, usw.) werden normalerweise aufgrund der Standardtests

spezifische Tests verwendet werden (z.B. Messen der Spannkraft der Fasern
oder ihres Polymerisationsgrades).

Die Kenntnisförderung der morphologischen Flachsfaserstruktur
erweckte besonderes Interesse. Zwei wichtige Aspekte ragen heraus :
erstens der Reaktionsunterschied von Flachsfaser und Baumwollfaser auf
alkole Verfahren und zweitens die mechanischen Eigenschaften der Grund-
faser, die denen des Kevlar ähneln.

Werden Flachsfasern mit Ätznatron (Merzerisation oder Kaustifizie-
rung) behandelt, bilden die besonders hohe kristalline Struktur und
mehrere andere Faktoren eine Barriere. Im Gegensatz zu Baumwolle, bei der
die Wirkung von Ätznatron homogen verläuft, dringt das Ätznatron sehr
ungleich durch die Flachsfasern ein und hindruch. Die Gesamte Flachsfaser
entwickelt sich langsam zu merzerisierter Zellulose. Zu diesem Zeitpunkt
ist die Wirkung die gleich wei bei Baumwollfasern, d.h. erhöhte Durchlässigkeit für Chemikalien und gesteigerte Stabilität. Ein bedeutender
Aspekt bei der Ätznatronbehandlung von Flachsfasern ist die Tatsache, daß
die Verbindungen und Verlagerungen in der Faserstruktur beibehalten
werden. Verwendet man flüssiges Ammoniak als Merzerisationsmittel erreicht man verschiedene Resultate. In diesem Fall wird die Faserstruktur
langsamer und nicht so drastisch verändert wie bei der Verwendung von
Ätznatron.

Hinsichtlich der mechanischen, typischen Eigenschaften der Neufaser,
könnte man sie möglicherweise bei zusammengesetzten Stoffen in einer
organischen oder mineralen Matrize verwenden. Das Problem beim Vorschlag, Flachs als Ersatzmittel zur Verstärkung von Fasern wie Kevlar bei
Verbundwerkstoffe zu verwenden, ist darauf zurückzuführen, daß die
Flachsfaser nicht die gleichen innerlichen mechanischen Eingenschaften
besitzt, es sei denn, die bestehenden Extraktionstechniken werden verändert. Es gibt ebenfalls keine auf Flachs basierende Faser, die die
gleiche Hitzebeständigkeit wie Kevlar aufweist. Dies wird die mögliche
Verwendung von Flachs bei Verbundwerkstoffe für spezifische nichtther-
mische Anwendungen beschränken. Außerdem findet man Flachs in seiner
natürlichen Form in einer Höchstlänge von 6 cm, wobei andere Verstär-
kungsfasern wie Kevlar dem Benutzer als Endlosfaser angeboten werden.

Im Hinblick auf die Frage nach der wirtschaftlichen Position von
Flachs in der Zukunft, kann man antworten, daß es schon heute zu begrüßen
wären, wenn die Forschung und die Industrie ernsthaft diese Möglichkeit
untersuchen würden. Dies wird in absehbarer Zukunft von vorrangiger
Bedeutung sein.

ALLGEMEINE SCHLUSSFOLGERUNGEN DES SYMPOSIUM

D. FINLAY MAXWELL
Vorsitzender des Wissenschaftlichen Forschungsausschusses von Comitextil

Was die wissenschaftlichen Leistungen angeht, hat dieses Symposium Ihnen eine Vorschau, vier ausführliche Sitzungen, vier Erklärungsvorträge und eine Erläuterung dieser Erklärungsvorträge geboten. Nun muß die Nachricht ja sehr deutlich sein. Wir haben auf jedem der behandelten Gebiete beachtliche Erfolge verzeichnet. Wir haben fraglos bewiesen, daß mit der Hilfe der Kommission unsere sich ergänzenden Forschungseinheiten über die Landesgrenzen hinaus erfolgreiche Forschungsarbeit leisten können.

Warum gemeinsame forschen ? Warum nich einfach diese Arbeit auf nationaler Ebene ausführen ? Die zunehmende Geschwindigkeit des Technologieaustauschs hat zur Folge, daß die Textilforschung immer mehr Disziplinen einbeziehen muß.

Ich möchte hier ein ganz hypothetisches Beispiel nennen, und zwar das Färben oder die Oberflächenbehandlung von Leinen oder eines beliebigen Stoffs. Man muß die Wirkung zahlreicher Chemikalien studieren, wahrscheinlich auch eine Reihe von Hitzebehandlungen, einschließlich der Verwendung von Radiowellen, damit ein Hitzegradient nach innen oder anders eine Mikrowelle bzw. Infrarotwelle nach außen kann, um durch Schaum hinzugefügte Zusatzstoffe zu entfernen. Es besteht noch immer die Möglichkeit, Alpha- oder Betastrahlen oder sogar Laserstrahlen zu verwenden. Wir können ebenfalls auf die Kawabata-Untersuchungstechniken zurückgreifen, um diese Behandlungen zu optimieren.

Kann man von irgendjemand in unserer Organisation erwarten, daß er genügend Erfahrung auf all diesen Gebieten besitzt ? Wenn nein, wie soll man dann eine solche Forschungsarbeit unternehmen . Eine der großen Stärken der gemeinsamen Forschung liegt in der Fähigkeit eine ständig anwachsende Reihe von Disziplinen zu vereinen – wobei jede spezialisierte Einheit ein hohes Kompetenzniveau erreicht.

Solche Programme besitzen einen zusätzlichen Wert, weil sie viel öfter konstruktiver Bewertung und Selbstkritik ausgesetzt sind. Von dem oder jenem Delegierten habe ich erfahren, daß ein Teil der vorgestellten Arbeiten bereits anderswo ausgeführt wurden. Dies wird immer so sein, besonders da, wo nationale Programme unabhängig von einander arbeiten. Hier spielen gemeinsame Programme eine wichtige Rolle, da sie kostspielige Doppelarbeit verhindern oder beträchtlich vermindern.

Im Großen und Ganzen, bin ich der Auffassung, daß dies die Rolle von Gemeinschaftsprogrammen bei ausgedehnten Projekten auf breiterer Basis liegt. Nationale Programme sind vielleicht besser auf die Entwicklung einer besonderen Fachkenntnis innerhalb dieses Fragenkomplexes ausgerichtet. Diese beiden Entwicklungsrichtungen können sich ergänzen.

Wir haben über die Erfolge, die wir bei diesem Symposium verzeichnet haben diskutiert. Was wollen wir nun mit diesen Ergebnissen tun ?

Zuerst müssen wir dafür sorgen, daß diese Ergebnisse soviel wie möglich bekanntgegeben und verbreitet werden. Der erste Schritt der Kommission wird sein, innerhalb von 6 Monaten, allen Abgeordneten eine vollständige Reihe von Referaten auszuhändigen. Dann liegt es an uns, das

Wissen soweit wie möglich über unsere Organisationen und Vereinigungen zu verbreiten.

Wir müssen uns ganz entschlossen bemühen, die durchgeführten Arbeiten in die Praxis umzusetzten. Unvollständige Arbeiten bleiben jedoch noch übrig. Und wir sollten ihnen jedoch die nötige Mühe schenken, um sie zu vollenden. Um diese Arbeit zu überwachen und zu leisten, würde ich ein von Comitextil unterstütztes "Durchführungsprogramm" begrüßen.

Ich bin mir der zunehmenden Arbeitsbelastung bewußt, der unsere Techniker ausgesetzt sind, doch muß ich sagen, daß dies schon seit einiger Zeit deutlich ist. Von meinem persönlichen Standpunkt aus, habe ich festgestellt, daß meine Anwesenheit bei Versammlungen, ein-oder zweimal im Monat, als Amateur-Vorsitzender, mit Sitz in Hudderfield, völlig unzureichend für Ihre zukünftigen Anforderungen war. Deshalb habe ich vor kurzem den Platz geräumt für einen Nachfolger, der die nötigen Veränderungen durchführen wird.

Dies Änderungen erfordern eine Neubewertung eines Großteils unserer Zeit und Mittel. Ich glaube, daß dies jedoch möglich ist und eine entsprechende Organisation aufgebaut werden kann, damit sichergestellt wird, daß Programme auf einer kontinuierlichen Grundlage erarbeitet werden. Wir brauchen Programme, die der Gemeinschaft entsprechen - Programme von höchster technischer Qualität, die durch unsere gemeinsamen verfügbaren Mittel an den Tag gebracht und durch sie ausgeführt werden.

RESUMES EN LANGUE FRANCAISE

Physiologie et construction du vêtement

Nouvelles technologies de filature dans l'industrie
lainière

Qualité des étoffes et des articles tricotés

Valorisation du lin

Conclusions générales du symposium

364

PHYSIOLOGIE ET CONSTRUCTION DU VETEMENT

R. JEFFRIES
Shirley Institute

L'objet de la présente réunion est de décrire les travaux de recherche qui ont été effectués dans deux domaines précis : l'aspect physiologique du vêtement et le degré de bien-être qu'il procure, et la relation
entre la qualité du tissu et la réalisation du vêtement d'une part et les
procédés de fabrications utilisés, le style et l'aspect esthétique du
vêtement d'autre part. Ces deux domaines d'études figuraient dans le
second programme de recherche pour le secteur textile de la CEE, en
raison de l'importance primordiale qu'ils revêtent tant pour l'industrie
textile que pour les fabricants et les détaillants de ce secteur.
L'amélioration du degré de confort que présentent tissus et vêtements
constitue indéniablement l'une des préoccupations essentielles des fabricants et des détaillants. La mise en oeuvre de divers projets y relatifs
est dès lors particulièrement opportune. Les autres grands axes de préoccupation actuels sont d'une part l'élaboration de vêtements stylés, de
qualité et de bonne coupe et l'automatisation de la confection d'autre
part; la qualité, la classification et la confection des tissus et partant des vêtements jouent dans chacun des ces domaines un rôle essentiel.
    Cinq projets ont ainsi été commentés. L'état d'esprit qui a présidé
à la mise en oeuvre de ces projets a été de produire des résultats et des
conclusions présentant un intérêt direct et immédiat pour l'industrie
textile et le secteur de la confection des pays de la CEE et partant de
leur fournir des outils adéquats qui leur permettront d'élaborer des produits de meilleure qualité et de mettre au point des procédés de fabrications plus performants.
    Le premier compte-rendu dresse l'état des recherches effectuées dans
le domaine très complexe du confort vestimentaire et des facteurs qui
contribuent à le déterminer, c.à.d. la nature et les propriétés des vêtements et des tissus qui entrent dans leur confection. Deux types de confort vestimentaire ont été examinés : le confort "thermophysiologique",
qui est relatif au maintien d'un équilibre thermique et hygrométrique
indispensable pour garantir au cours une température de fonctionnement
adéquate, et le confort "sensoriel", qui est relatif aux propriétés
tactiles des vêtements en contact direct avec la peau et à leurs effets.
    Trois projets traitent du confort thermophysiologique. L'un d'eux
s'est attaché à fournir des données tangibles quant au confort thermophysiologique des vêtements et à établir des directives et des méthodes
d'évaluation. Des modèles de comportement de la peau humaine ont ainsi pu
être dégagés et les résultats des mesures thermiques et hygrométriques
ont été insérés dans une formule offrant la faculté d'établir une "cote
de confort".
    Cette cote de comfort peut parfaitement remplir la fonction de
l'essai sur mannequin, très coûteux. Un autre projet avait pour thème
l'amélioration du confort thermophysiologique des vêtements imperméables. L'étude portait sur les types les plus divers de tissus imperméables et perméables à la vapeur d'eau. Une attention toute particulière a

été prêtée à la combinaison de la perméabilité et de la ventilation. Ce projet a permis de mettre au point une technique d'étude des effets de la ventilation et de présenter des directives pour la création de vêtements imperméables. Un troisième projet avait pour objet d'étudier les effets d'ordre physiologique (augmentation de la température du corps et transpiration) produits par les vêtements de laine et de coton ou réalises à partir de mélanges de laine ou de coton.

Le quatrième projet traite du confort sensoriel des vêtements en contact direct avec le corps. Il s'est avéré nécessaire d'éffectuer d'abord des essais sur mannequin complets et précis, en vue d'identifier les sensations et les facteurs physiques qui déterminent le confort sensoriel des vêtements en contact direct avec le corps. Une série de procédures d'essai simples, tant objectives que subjectives, ont pu être élaborés à la lumière des résultats obtenus au terme des essais sur mannequin. Ces procédures permettent de mesures des facteurs tels que l'irritation, le picotement, le chatouillement, la pression, la démangeaison, le contact froid ou humide avec la peau, les effets des fibres rebelles et de l'électricité statique.

Le second compte-rendu traite de la détermination et de la classification des propriétés physiques des tissus dans le cadre de la création et partant à l'amélioration des procédés de fabrication et de l'esthétique des vêtements. La résistance des tissus aux tractions multidirectionnelles inhérentes à la fabrication a été étudiée. Ce type d'étude requiert l'utilisation de techniques de pointe, qui ne sont pas nécessairement adaptées aux besoins d'exploitation des fabricants.

Les fabricants ont participé à la vérification des résultats des essais. La second partie de ce projet avait pour thème l'étude de l'interaction tissus/machine au cours de la fabrication; les effets de l'ajustement automatique des machines sur la qualité des tissus ont également été étudiés. Deux domaines spécifiques ont fait l'objet d'expérimentations : l'alimentation du tissu en couture (afin d'identifier les propriétés des tissus et lescaractéristiques des machines affectant le défilement du tissu dans la machine) et le pressage intégré (afin de déterminer le juste milieu entre vitesse de production, conditions de pressage et dimensions du vêtement.

EVALUATION DU CONFORT DES ETOFFES ET DES VÊTEMENTS

K.H. UMBACH
Bekleidungsphysiologisches Institut Hohenstein E.V.

ASPECT THERMOPHYSIOLOGIQUE DU CONFORT AU PORTER

De nos jours, afin d'être compétitifs, à côté de bonnes propriétés d'ordre mécanique/technologique, d'être faciles à laver ou à nettoyer, les vêtements doivent posséder un confort au porter thermophysiologique élevé. Toutefois, de bonnes caractéristiques au porter ne peuvent être incorporées en tant qu'aspect planifié du développement du vêtement que s'il est possible de mesurer et d'évaluer quantitativement le confort au porter physiologique. Le but de ce projet fut d'établir un système d'analyse à cette fin, capable d'être mis en oeuvre par l'industrie dans le développement des produits.

Ainsi que ce système d'analyse comprend plusieurs étapes. Pour l'évaluation des propriétés thermophysiologiques au porter des textiles, il est fait usage d'un modèle de thermorégulation de la peau humaine (modèle de la peau). Ce modèles a permis de développer des méthodes d'essai par lesquelles les propriétés des textiles, quant à l'isolation thermique aussi bien qu'à la transmission ou à la rétention de l'humidité, peuvent être déterminées quantitativement, à la fois sous les conditions de porter "normales" et lors de transpiration profuse et prolongée du porteur, et être exprimées en données spécifiques au tissu. Dans ce projet, des formules de prédiction furent établies dans lesquelles ces données peuvent être utilisées. Elles fournissent une grille permettant l'évaluation du confort thermophysiologique au porter que l'on peut attendre d'un article textile. Les formules furent mises au point séparement pour les divers groupes de marchandises tels que, par exemple, sous-vêtements, chemises, pantalons etc. Des essais au porter avec participants ont montré que les grades de confort au porter, calculés pour un article textile grâce à ce modèle prévisionnel, sont en accord avec la sensation de confort ressentie de façon subjective par la personne durant le port de l'article.

Cette évaluation des propriétés des textiles relatives au confort au porter reposant sur des méthodes d'essai en laboratoire rationnelles, le projet de recherche a, comme résultat direct, d'offrir maintenant, à l'industrie textile, la possibilité de planifier objectivement le confort au porter en tant que trait de contexture, déjà dans la première phase d'un nouveau développement. Ceci, est confirmé par le fait que les expériences fondamentales exécutées dans le projet ont mené à des directives de contexture par lesquelles certains éléments de confection (vêtements d'athlétisme, de pluie, vêtements thermiques) peuvent être conçus avec des propriétés physiologiques optimales.

A l'étape 2 du système d'analyse, les éléments de vêtements confectionnés sont étudiés à l'aide d'un mannequin mobile, léquel représente un modèle de thermorégulation du corps human.

Il fournit des données thermophysiologiques spécifiques (résistance à la transmission de chaleur et de vapeur) de tout le système d'habillement, y compris les couches d'air contenues dans le micro-climat et adhérant à la surface externe des vêtements.

Plusieurs centaines d'essais au porter, contrôlés, avec participants dans la chambre climatique, (figure 5) inclus dans le système d'analyse de la figure 1 comme étape 3 et dans lesquels à la fois les valeurs objectives des fonctions corporelles et les évaluations subjectives faites par les participants aux essais furent obtenues, aboutirent à la création d'un modèle prévisionnel pour vêtements.

Avec ce modèle qui comprend les données vestimentaires mesurées sur le mannequin, il est possible, en termes universellement compris et en accord avec l'expérience pratique, de décrire le confort au porter de vêtements sous n'importe quelle variante désirée de conditions contraignantes de climat et d'activité du porteur.

Ainsi, il est maintenant non seulement possible de définir la gamme de températures ou le champs d'utilisation dans lequel le porteur, à un niveau d'activité défini, ne gèle ni ne transpire de façon excessive, sur la base de quelques essais de laboratoire, mais il est également possible de calculer directement les variations dans le temps de diverses fonctions du corps humain et de l'humidité dans le microclimat lors du port de vêtements dans des conditions externes spécifiques de climat et d'activité.

Les études entreprises dans ce projet ont montré que ces valeurs relatives aux fonctions corporelles et au microclimat commandent directement les sensations de confort thermophysiologique de la personne. Avec les seuls vêtements textiles, il a été possible de mettre au point des formules bio-physiques qui fournissent directement un grade de confort au porter pour le système complet d'habillement. Dans l'industrie de la confection, pour le développement de produits à venir, le besoin en essais au porter, prolongés et coûteux, a été largement éliminé.

A leur place, comme résultat du projet de recherche, et sur la base de quelques tests de laboratoire, une évaluation de l'état satisfaisant des vêtements du point de vue physiologique est maintenant possible.

2. - CONFORT AU PORTER SENSORIEL CUTANE

L'objectif de ce projet était de déterminer les paramètres de contexture spécifiques qui commandent le confort au porter sensoriel cutané des textiles, causé par leur contact avec la peau. De simple méthodes d'essai en laboratoire furent mises au point permettant de déterminer les divers aspects du confort sensoriel cutané. Une recherche à travers la littérature technique, un questionnaire dans lequel environ 1000 personnes furent interrogées sur leurs habitudes vestimentaires et des essais au porter avec des participants portant 22 échantillons textiles confectionnés en T-Shirts, montrèrent que les facteurs influents suivants contrôlent le confort au porter sensoriel :

2.1. Ajustement localisé

La pression exercée par des vêtements reposant trop près de la peau peuvent masquer toutes les autres sensations de confort. Il faut vérifié que les pressions sur diverses parties du corps peuvent tout juste être tolérées. Des appareils d'essai furent développés grace auxquels la pression de contact d'articles d'habillement sur le corps peuvent être déterminée quantitativement.

2.2. Irriations localisées causées par les étiquettes de vêtements.

Les arrêtes et coins aigus rencontrés en particulier sur les étiquettes thermocollées sont susceptibles de causer des sensations de gêne considérables.

2.3. Picotements

Les textiles contenant de la laine sont ressentis comme "provoquant des picotements" lorsque 3,5 % des fibres sont d'un diamètre dépassant 30um ou lorsque 0,6 % des fibre sont d'un diamètre excédant 40um.

2.4. Chatouillement

Les textiles à surface "poilue" chatouillent de façon désagréable en particulier la peau moite de transpiration. Une photographie des pointes de fibres en saillie à la surface du textile permet de déterminer si un textile aura tendance à chatouiller.

2.5. Chute de fibres

Les fibres se détachent de l'article textile pendant le port et celles qui adhèrent à la peau entraînent des sensations de confort négatives. L'utilisation d'appareils dans lesquels l'article est secoué d'une manière spécifique permet d'exprimer quantitativement la propensité des fibres à la chute par un "indice de chute des fibres" en comparaison de normes photographiques.

2.6. Tendance à coller à la peau (des textiles humides)

Un appareil d'essai fut mis au point dans lequel la force nécessaire pour arracher une bande textile d'un cylindre en plexiglass reproduit la sensation désagréable causée par les textiles qui collent à la peau. Les essais au porter entrepris ont montré une bonne corrélation entre les valeurs d'adhésion et les évaluations subjectives des participants aux tests.

2.7. Sensation initiale de froid

Bien que lorsqu'il revêt des articles d'habillement, la sensation de froid éprouvée éventuellement par le porteur ne dure que quelques secondes, celle-ci aura peut-être une influence significative et durable sur son impression sensorielle de confort. En utilisant un appareil d'essai développé spécialement, on est en mesure de déterminer quantitativement cette sensation initiale de froid causée par les textiles.

2.8. Tissus qui grattent la peau

Mesurer la friction de cohésion entre un article textile et un sujet d'essai permet de juger si oui ou non une matière textile donnera la sensation de gratter la peau.

2.9. Effets d'élasticité statique

Des décharges électrostatiques, alors que le porteur ôte les articles d'habillement ou lors de mouvements corporels, ainsi que l'adhésion électrostatique des textiles à la peau, sont susceptibles d'entraîner un confort sensoriel cutané médiocre. La mesure, dans des conditions spécifiques, de l'importance de la charge se produisant quand on enlève des articles d'habillement, permet d'évaluer si la performance de l'article textile est défectueuse.

Pour résumer, le projet de recherche a non seulement impliqué le développement de méthodes permettant la mesure du confort au porter sensoriel cutané, mais, à partir des connaissances fondamentales obtenues, il a été possible d'établir des directives permettant mieux qu'auparavant à l'industrie, de fabriquer, avec des données objectives, des articles d'habillement à porter à même la peau, dotés de bonnes propriétés au porter sensoriel cutané.

3. - AMELIORATION DU CONFORT AU PORTER DE VÊTEMENTS PROTECTEURS

Une grande proportion de vêtements protecteurs a pour but de protéger le porteur des effets du temps humide. Du point de vue physiologique, ce type de vêtement présente le problème d'associer simultanément l'imperméabilité à la transmission adéquate des vapeurs d'humidité. Si cette dernière propriété est absente, il en résulte un confort au porter médiocre qui réduit l'acceptabilité des vêtements de protection.

Afin d'être en mesure de réaliser un examen critique au point de vue physiologique, des textiles protecteurs, disponibles aujourd'hui, contre les intempéries, des méthodes d'essai spécifiques furent tout d'abord développés, dans ce projet de recherche, pour la détermination, dans des conditions pratiques, de l'imperméabilité des textiles et de leur perméabilité à l'air et à l'eau.

Une série d'expériences accomplies en utilisant ces méthodes d'essai ont montré que certains tissus stratifiés à membrane intégrante, des revêtements spéciaux microporeux et des tissus à microfibres, sont capables de satisfaire simultanément aux exigences technologiques et physiologiques imposées. La combinaison de ces diverses technologies à révélé des possibilités de développer des propriétés supérieures aux structures disponibles actuellement. Si de plus, ces textiles sont associés à une conception des vêtements appropriée permettant, dans le microclimat crée par les mouvements corporels du porteur ou le mouvement de l'air ambiant, la convection et la ventilation, l'objectif d'optimisation du vêtement protecteur contre les intempéries est alors atteint. Des possibilités de mettre ceci en pratique furent démontrées dans le projet de recherche, par des essais au porter, conduits avec des participants, et pour lesquels une méthode spéciale fut développée pour la mesure quantitative du taux de ventilation dans l'habillement. Les essais menèrent aussi à un modèle mathématique décrivant les processus de transmission de chaleur et d'humidité physiques dans les vêtements et grace auquel il est possible de déterminer quel est l'effet de la perméabilité des textiles à la vapeur d'humidité, alliée à la ventilation incorporée dans la conception des habits, sur les propriétés de confort au porter des vêtements protecteurs dans des conditions climatiques spécifiques.

OPTIMISATION DE LA CONCEPTION ET DE LA REALISATION DES VETEMENTS

J. DESCHAMPS
CETIH

La partie du programme qui est étudiée ici concerne la prise en compte des caractéristiques physiques des tissus lors de la conception des vêtements de manières à obtenir :

- une conception des articles optimisée éliminant au maximum les retouches,
- une fabrication plus facile à réaliser,
- l'obtention de vêtements de forme confortable et d'esthétique satisfaisante.

1. - CLASSEMENT DES ETOFFES EN FONCTION DE LEURS PROPRIETES PHYSIQUES

Les modes de classification des tissus le plus souvent utilisés, encore de nos jours, se réfèrent à des catégories d'usage. Ils sont essentiellement subjectifs et en relation directe avec les informations visuelles ou tactiles recueillies par les observateurs lorsqu'ils voient et palpent le tissu bien qu'ils soient parfois pondérés par des caractéristiques mesurables élémentaires comme le poids du tissu.

La modernisation des Industries de l'Habillement, l'automatisation des processus de fabrication et l'utilisation à plus ou moins long terme d'outils informatiques pour la Conception et la Fabrication Assistées par Ordinateur, rendent nécessaires un classement plus objectif des tissus en fonction de critères mesurables.

Afin d'être facilement utilisable par la majorité des industriels de la confection, le système de classement objectif retenu ne doit prendre en compte qu'un nombre de critères mesurables restreints, la méthode de classement optimale pouvant être celle qui détermine des classes de tissus par référence à une ou plusieurs propriétés convenablement choisies.

Un recensement des propriétés des étoffes actuellement prises en compte plus ou moins intuitivement par les industriels lors de la conception et de la fabrication des vêtements et qui mériteraient d'être systématiquement examinées, a été effectué.

Les propriétés les plus significativement influentes sur la facilité de fabrication des produits et l'obtention de vêtements de qualité, de confort et d'esthétique conformes aux attentes du concepteur ont été recherchées.

L'étendue maximale de variations, toutes catégories de vêtements confondues, a été déterminée pour chacune des propriétés susceptibles d'être influentes dans le cas des tissus d'usage courant.

Les variabilités de chacune de ces propriétés pour un même tissu, au long d'une même pièce, et entre deux pièces différentes ont été recherchées, ceci afin de déterminer les variations réellement significatives.

La mise en évidence des déformations qui se produisent dans la structure des tissus du fait de leurs transformations en vêtements s'est révélée nécessaire.

2. - MISE EN EVIDENCE DES CONTRAINTES IMPOSEES AUX TISSUS ET DES DEFORMATIONS QU'ILS SUBISSENT LORSQU'ILS SONT TRANSFORMES EN VETEMENTS.

L'étude photographique des modifications qui interviennent dans la structure des tissus au niveau des principales coutures d'assemblages (coutures de côté des vêtements, coutures d'épaule, empiècements, couture de montage de manche) et des pinces, a permis de mettre en évidence les propriétés des étoffes qui peuvent avoir une influence prépondérante sur la "confectionnabilité" des tissus.

Les capacités de cisaillement et de courbures multidirectionnelles des tissus paraissent devoir être les propriétés ayant une influence majeure, la mise au point de méthodes de mesure simples permettant aux industriels d'appréhender ces propriétés est donc nécessaire.

3. - RECHERCHE DE METHODES PERMETTANT LA MESURE PAR LES INDUSTRIELS DES CAPACITES DE CISAILLEMENT ET DE COURBURES MULTIDIRECTIONNELLES DES ETOFFES.

Les capacités des tissus de ce cisailler et de se courber simultanément dans plusieurs directions semblant être prépondérantes dans leurs aptitude à être facilement confectionnés en vêtements, une mesure globale de l'ensemble de ces paramètres a paru souhaitable.

Des méthodes expérimentales ont été mises au point pour réaliser ce type de mesure.

Il est apparu opportun de voir au niveau de conceptions et de fabrications réelles en entreprise de confection, l'influence de ces caractéristiques des étoffes. Une vérification "in situ" a donc été réalisée.

4. - EVALUATION ET VERIFICATION "IN SITU" DE L'IMPORTANCE DES PROPRIETES DES TISSUS, RETENUES POUR LEUR CLASSEMENT

La réalisation pratique de vêtements (jupes - robes - manteaux - vestes), à partir d'un cahier des charges et de tissus imposés, par des industriels de l'habillement selon leurs processus de conception et de fabrication habituels, a été effectuée pour vérifier le bien fondé des hypothèses émises.

Les appréciations ont été faites quant aux résultats obtenus (volume obtenu - confort et esthétique de l'article), par des professionnels de l'habillement et des représentants des consommateurs.

5. - CONCLUSION

Cette étude ne constitue qu'une première approche du problème de l'incidence des propriétés physiques des étoffes sur la conception des vêtements.

La conception à coup sûr de la forme recherchée tant du point de vue esthétique que du point de vue confort, n'est pas encore réalisée, sa venue serait accélérée par une meilleure maîtrise de la régularité des propriétés des tissus fournis aux industriels de la confection.

RESUME ET CONCLUSIONS PAR LE PRESIDENT DE LA SESSION

R. JEFFRIES
Shirley Institute

L'objet de la session sur la physiologie et la construction du vêtement était double :
- évaluation du confort des etoffes et des vêtements,
- optimisation de la conception et de la réalisation des vêtements.

Une première conclusion de la session s'impose : le choix de ces deux domaines d'étude était judicieux, étant donné qu'ils présentent un intérêt direct pour l'industrie textile et l'industrie de la confection de la Communauté si elles entendent rester créatives et compétitives.

Le confort des textiles et des vêtements, dans tous ses aspects, relève du domaine du fonctionnel qui est une préoccupation essentielle des fabricants et détaillants de tissus et de vêtements et qui gagnera en importance pour plusieurs raisons. Premièrement, on demande de plus en plus des vêtements protecteurs de qualité et donc, les exigences de confort seront plus élevées; deuxièmement, les goûts du consommateur se portent de plus en plus sur des vêtements de sport qui sont confortables, particulièrement en cas de transpiration abondante; troisièmement, le consommateur demande de plus en plus des vêtements de tous les jours dont le confort est grand.

Dans son rapport sur l'étude du confort, le Dr. Umbach nous a fait part des recherches menées au "Hohenstein Institute", au "Shirley Institute" et au "TNO Institut for Perception" traitant respectivement du confort thermo-physiologique, du confort sensoriel et du confort des vêtements de pluie semi-perméables ("ventilés"). Ce projet a fourni une source d'informations utiles à l'industrie pour concevoir des vêtements plus confortables dans diverses situations, ce qui a permis de réduire le nombre d'essais, longs et coûteux, sur mannequin.

Toutefois, nous sommes arrivés à la conclusion qu'en dépit des excellents progrès réalisés, de nombreuses améliorations doivent encore être apportées dans ce domaine.

Premièrement, une connaissance scientifique plus approfondie de tous les aspects du confort vestimentaire. Une partie de cette étude doit absolument se donner pour objectif une meilleure compréhension des aspects physiologiques, psychologiques et quasi-médicaux du problème, par example, bien que l'on dispose d'abondantes informations sur la sensibilité de la peau au toucher, on ne connaît que fort peu la sensibilité d'une peau qui ne touche que fort peu le tissu du vêtement porté. Il faut acquérir davantage d'informations sur l'ensemble du concept psychologique de confort. Ainsi, par confort, entend-t-on plutôt l'absence d'inconfort ? Ne pourrait-il exister une définition plus positive du confort du vêtements ? Ne pourrait-on pas, par example, concevoir des tissus et des vêtements plus à même de créer un sentiment de bien-être ? Peut-être n'est-ce-qu'une élucubration ....

Deuxièmement, il faut fabriquer des tissus encore plus confortables, des tissus qui susciteront un sentiment de bien-être plutôt qu'un manque d'inconfort (cfr supra).

Troisièmement, il faut réactualiser la conception du vêtement en vue d'une utilisation optimale des propriétés et du potentiel de ces nouveaux tissus confortables. Il sera peut-être nécessaire de trouver de nouvelles façons de ventiler les vêtements afin d'optimiser toutes les propriétés de confort du vêtement.

L'intervention du Dr. Umbach a donné lieu à une discussion intéressante où l'on a souligné l'importance des travaux sur le confort pour l'industrie textile et l'industrie de la confection. Les recherches effectuées sur ce projet sont pour l'industrie une source sérieuse d'informations sur le confort du vêtement au porter. Le Dr. Manni nous a présenté les travaux de Lanerossi sur le confort, qui a été évalué par le biais de mesures physiologiques en comparant des tissus pure laine et des mélanges de laine, et du coton et du lin. Une attention particulière a été accordée à la nature technique et scientifique du vêtement protecteur qui laisse passer la vapeur d'humidité mais pas l'eau (à l'état liquide).

En réponse à une question supplémentaire, le Dr Umbach a déclaré qu'une autre étude à "Hohenstein Institute" avait analysé l'influence de l'apprêt du tissu sur le confort, mais cette étude n'avait pas été mentionnée au Symposium car elle ne faisait pas partie du projet (en matière d'apprêt, tout est fonction de sa nature hydrophilique ou hydrophobique). On s'intéressa beaucoup à la discussion sur la nature des techniques utilisées pour l'évaluation des facteurs de confort; ces derniers sont au coeur même du problème du confort.

Au cours de la deuxième partie de la session, M. Deschamps a exposé les travaux du CETIH sous-traités par le British Clothing Centre sur l'optimisation de la conception et de la réalisation des vêtements. Nous en sommes arrivés aux deux conclusions suivantes.

Premièrement, ce sujet est d'une importance primordiale pour l'avenir de l'industrie textile et de l'industrie de la confection. La relation entre les propriétés des tissus et la conception et la réalisation des vêtements doit être mieux comprise si nous voulons assurer la compétitivité et le bien-être de l'industrie. Une meilleure connaissance de l'interaction textile/vêtement est indispensable pour remplacer ou du moins compléter les techniques déjà existantes dans le domaine de la conception et de la réalisation des vêtements. Ces techniques se basent principalement sur l'habileté personnelle et l'expérience acquises au cours des ans, et, bien qu'elles se soient révélées satisfaisantes jusqu'à présent, elles ne suffiront désormais plus à elles seules. Les nouvelles données sur la relation textile/vêtement permettront certes d'accroître la qualité et la productivité obtenues avec les méthodes actuelles, mais ce qui est plus important encore pour l'avenir, elles signifieront un pas important vers l'automatisation de la réalisation des vêtements et le recours à des systèmes robotisés. Cette étude présentée par M. Deschamps analyse les propriétés physiques des tissus en rapport avec l'optimisation de la conception des vêtements fonctionnels et confortables bien stylés et esthétiques en vue d'une meilleure utilisation de l'automatisation et de la robotique dans le processus de réalisation des vêtements (c.à.d. une meilleure compréhension des interactions tissus/machine au cours de la fabrication).

Il faut tirer une deuxième conclusion du rapport de M. Deschamps. Cette étude sur l'interaction tissu/vêtement n'en est encore qu'à ses débuts. Des enquêtes sur le projet ont apporté une contribution substantielle mais de nombreux travaux restent encore à faire sur ce problème fort complexe si l'on veut disposer d'une bonne base de connaissances, indispensables pour l'avenir. Une collaboration étroite entre les scientifiques, les technologues et l'industrie est primordiale si l'on veut réussir dans cette tâche. M. Deschamps a obtenu une excellente collaboration de l'industrie; c'est un fait encourageant. Il faut donc non seulement la poursuivre mais aussi l'intensifier.

Les deux sujets abordés à la session, la Physiologie du vêtement et la Construction, sont d'une importance capitale pour l'industrie textile et l'industrie de la confection de la Communauté si elles entendent rester créatives et compétitives.

NOUVELLES TECHNOLOGIES DE FILATURE DANS L'INDUSTRIE LAINIERE

M. BONA
Città Degli Studi

Depuis des nombreuses années, la filature du coton a subi une innovation technologique importante, grâce à la mise au point de techniques
non conventionnelles, en substitution partielle du système classique du
continu à filer; parmi ces techniques, les principales font partie de la
catégorie des systèmes à bout libéré, d'abord du type à rotor mais récemment également – du moins en perspective – à friction ou à jet d'air.

Il est normal que ces importants développements aient concerné avant
tout le secteur du coton et certaines fibres synthétiques à la fois pour
des raisons techniques objectives qui rendent plus aisé le traitement de
ces fibres plus courtes et plus résistantes que la laine, et à cause du
fait que les investissements de Recherche et Développement des construc-
tures mécanico-textiles se sont avant tout adressés à un marché de rem-
placement potentiel qui est au moins dix fois plus grand que le marché
lainier.

Pour également favoriser l'adoption de nouvelles technologies de
filature pour la laine, dans le but de réduire ou limiter les coûts
croissants de fabrication, il était donc nécessaire d'effecteur un effort
de recherche, en partant des besoins réels de l'industrie et avec le
support de celle-ci; cet effort devait porter sur le comportement de la
fibre soumise aux nouveaux procédés non conventionnels et préciser les
conditions mécaniques et textiles pour leur application éventuelle de
manière optimale. Cette recherche, qui a intéressé 8 Instituts apparte-
nant à 5 Pays membres de la CEE, avait déjà débuté sous l'initiative de
certains d'entre eux, mais elle a été supportée de manière décisive, et
surtout coordonnée par le lancement du Second Programme de Recherche
Textile Communautaire et par ses financements.

La première phase du programme a évidemment consisté dans l'exécution d'essais expérimentaux sur les différents modèles de métiers à bout
libéré à rotor disponibles sur le marché, sans négliger toutefois
d'autres solutions non conventionnelles, telles que la filature à friction ou à broche creuse.

Dans cette phase, le programme de recherche communautaire a fourni
des nombreuses indications de grand intérêt, à la fois en peigné et en
cardé, dont l'importance pratique est soulignée par le fait que, grâce à
la collaboration de plusieurs Instituts, il a été possible d'essayer des
nombreux types de machines.

Toutefois, on s'est progressivement rendu compte que la nature
technique du problème devait être déplacée en amont de la filature
proprement dite, en intéressant la préparation initiale des fibres et le
cardage.

En ce qui concerne la préparation, le programme a permis de proposer
quelques modifications intéressantes des procédés conventionnels employés
pour les fibres vierges, comme par exemple, une épuration mécanique qui
peut s'accompagner, en le simplifiant, du carbonisage classique; une
attention particulière a été évidemment réservée, à l'emploi de matières

premières de récupération, qui revêt une grande importance économique pour les secteur du cardé européen et pose les problèmes techniques les plus difficiles en vue des opérations ultérieures.

Dans la perspective de l'emploi futur des nouvelles technologies de filature, que ce soit en partant de matières vierges ou, surtout, de récupération, un des résultats les plus importants du programme a sans doute été la conception, la mise au point comme prototype et l'expérimentation pratique de cardes de nouvelles conception, qui réunissent les avantages des systèmes traditionnels lainier et cotonnier, dans le but de produire un ruban qui soit en même temps homogène, bien ouvert et bien propre, pour l'alimentation du métier à filer à bout libéré. Il est souhaitable que les constructeurs, à la fois des métiers à filer et des cardes conventionnelles, établissent les contacts utiles avec les Instituts de Recherche, afin de développer des machines industrielles basées sur les nouvelles conceptions, qui pourraient du reste également conduire à une évolution intéressante dans le domaine de la filature conventionnelle.

Un dernier aspect du programme de recherche a concerné l'examen du comportement physico-chimique de la laine pendant la fabrication, une condition indispensable pour pouvoir utiliser les technologies nouvelles d'une façon optimale, en évitant les inconvénients qui, comme par exemple la formation excessive de poussières et de dépôts dans les rotors, en empêcheraient une exploitation économique.

Parmi les caractéristiques de la laine qui la rendent plus difficile à filer avec les nouvelles technologies par rapport au coton, par exemple, il faut signaler sa plus grande fragilité et la nécessité d'employer des additifs qui doivent être soigneusement choisis et dosés, si l'on veut filer sur un métier à bout libéré.

Il faut par conséquent tout particulièrement considérer comme un résultat important du programme de recherche la mise au point de méthodes de laboratoire qui pourront servir en pratique pour le contrôle du processus de fabrication, ainsi que pour la sélection des laines les mieux adaptées à l'emploi avec les technologies nouvelles.

Les trois rapports techniques qui vont suivre, fournissent les résultats du programme de recherche. Ils auront trait aux trois thèmes principaux que l'on vient de citer : préparation, filature, contrôles.

Comme conclusion générale, il faut souligner que, grâce à l'excellent esprit de coopération manifesté par les Instituts qui ont participé au programme communautaire, on a réellement établi les bases pour une évolution déterminante de la technologie de la filature dans le secteur lainier. Même si, des problèmes importants restent encore à résoudre, tel que le transfert des idées nouvelles en des machines et des procédés industriels qui demande la coopération entre Instituts et firmes mécanico-textiles.

Un thème important pour des recherches futures se dégage des travaux à savoir l'utilisation des fils nouveaux, dont la structure est de toute façon plus ou moins différente de celle traditionnelle : cela implique en effet des modifications aux procédés en aval de la filature, ainsi que l'adaptation des nouveaux produits aux exigences du marché.

PRÉPARATION DES MATIÈRES RÉCUPÉRÉES POUR LA FILATURE DE TYPE NON-CONVENTIONEL.

P. ARTZT
Institüt für Textiltechnik

Par rapport à la fibre de coton, la fibre de laine est une matière première très chère. Dans la fabrication des fils, le coût des matières premières représente l'élément prédominant du coût. En conséquence, chaque augmentation du pourcentage d'utilisation de la matière première se reflète positivement, en proportion accrue, dans le calcul des coûts. La recherche doit donc essayer de découvrir des moyens de préparer les déchets de laine afin que les fibres récupérées soient transformées en produits finaux de valeur maximale.

Les Instituts de recherche CELAC (Belgique), WIRA (Royaume-Uni) et ITF (Allemagne) ont collaboré à des projets de recherche dont le but était l'utilisation économique des fibres et l'abrègement du traitement.

Les déchets principaux résultant de la transformation de la laine sont :

1. Les déchets des rubans de carde :
    ceux-ci consistent essentiellement en fibres longues de laine : la teneur en salissures va jusq'à 80 %.

2. La blousse :
    La blousse a une distribution de longueurs de fibres, semblable à celle du coton. La teneur en matières végétales atteint 20 %. Les particules végétales sont intimement mêlées aux fibres.

Ces déchets sont carbonisés sans exception aujourd'hui. Ceci implique l'oxydation des impuretés cellulosiques.

Avantages de la carbonisation :

1) action de nettoyage éfficace,
2) rendement élevé.

Inconvénients de la carbonisation :

1) l'installation nécessite des capitaux importants,
2) gros besoins en énergie,
3) utilisation importante de produits chimiques,
4) problèmes liés à l'environnement,
5) contrôle d'un processus complexe,
6) influence négative sur les propriétés de la fibre de laine (fibre rendue cassante),
7) forte teneur en fibres courtes de la laine carbonisée.

Les techniques mécaniques de nettoyage de la blousse de laine aboutirent à une efficacité de nettoyage d'environ 50 %. Des techniques développées tout particulièrement pour le nettoyage du coton furent modifiées de façon appropriés pour nettoyer des blousses de laine.

En règle générale, pour ce qui concerne les blousses de laine dont la teneur en salissures est d'un niveau normal, une efficacité de

nettoyage de 50 % suffit, la fibre étant, par la suite, soumise à de
nouveaux nettoyages, lors du cardage et du filage dans la machine à filer
à rotor. Les fibres de laine nettoyées pourraient également être
utilisées dans des mélanges avec des fibres synthétiques ou du coton.
Une variante intéressante découle de la combinaison :

1)   pre-nettoyage mécanique suivi,

2)   d'un procédé de carbonisation de courte durée.

Le procédé réel de carbonisation, pour les petites particules
d'impuretés résiduelles qui demeurent, peut-être raccourci, selon les
cas, et est ainsi moins endommagenat pour la fibre. Le procédé de pré-
nettoyage mécanique produit un voile de carde sans boutons.
Les fibres nettoyées furent mélangées à des fibres de polyester,
acrylique et coton et transformées en filés sur le système de filature à
friction (Dref 2) et le système de filature à rotor.
Les essais montrèrent, qu'en utilisant un fil à âme, la fabrication
de filés par la technique de filature à friction présente des avantages.
Sur machines à filer rotor, il fut possible de produire des filés
rotor de 30 tex à teneur atteignant 50 % de blousses de laine nettoyée.
En raison de la longueur de la fibre, les déchets des rubans de
cardes qui ont le plus de valeur. Avec un nettoyage mécanique périodique,
une proportion accrue de fibres courtes étant éliminées, il est même
possible d'obtenir, dans la fibre nettoyée, une longueur de fibre moyenne
plus grande.
Lors d'essais de filature comparatifs, en filature à anneaux et en
filature des fils à âme avec des débourrures carbonisées des performances
sensiblement meilleures et des filés de valeur supérieure furent obtenus
avec les fibres nettoyés par procédé mécanique. La fibre nettoyées
mécaniquement représente ainsi pour la filature une matière première de
qualité, ces fibres étant susceptibles d'être transformées en produits
différents de ceux réalisés à partir de fibres carbonisées.
Dans la fabrication de filés grossiers pour couvertures, tissus
d'ameublement et tapis, le passage d'étirage avant la filature rotor
n'est pas nécessaire. Une carde à laine à deux tambours devrait être
utilisée avec un briseur de voile.
A la sortie de la carde, par la division du voile, des bandelettes
sont obtenues; leur poids en ruban est tel qu'elles peuvent être fournies
directement aux machines à filer rotor. La teneur de la laine en matières
grasses résiduelles devrait être d'environ 0,5 %. Un nouvel ensimage
n'est donc pas nécessaire. Il entraîne des problèmes de dépôts sur les
pièces de la machine à filer.
Il existe de vastes possibilités pour la réutilisation des déchets
provenant de ḷa confection. Le problème réside dans la logistique du
triage selon les couleurs et les mélanges de fibres. Une étude couvrant
divers aspects fournit des renseignements sur les champs d'application en
cours, les pays d'origine et les méthodes de transformation de ces
fibres.
Les projets de recherche conjoints réalisés par les instituts CELAC,
ITF Denkendorf et WIRA sont complémentaires et ils indiquent de nouvelles
voies dans la préparation des déchets en vue d'une réutilisation perfec-
tionnée.

NOUVELLES TECHNIQUES DE CARDAGE ET DE FILATURE

J.P. BRUGGEMAN
I.T.F.-Nord

Le cardage reste l'opération qui va conduire à partir des matières premières préparées suivant les procédés étudiés par nos collègues, à un premier ruban qui doit être le plus propre possible (exempt d'impureté et de matières végétales) le plus régulier possible en titre et présentant une bonne homogénéité de composition.

En filature type cardée traditionnelle, la carde délivre un voile découpé en lanières qui alimentent le continu à filer. la technique Open-End a fait germer l'idée de supprimer le continu diviseur à la sortie de la carde, et de le remplacer par une sortie en 2 ou 4 rubans destinés a être filés directement sur la machine à Fibres libérées. Ce procédé, très séduisant en fil tapis, présente un inconvénient quasi rédhibitoire pour les fils plus fins, à savoir l'absence de tout système de régulation de titre d'un ruban à l'autre et à l'intérieur même du ruban.

L'industrie et la recherche se sont alors orientées vers la carde type coton, permettant d'obtenir un ruban de titre compatible avec la filature Open-End.

Les résultats obtenus, bien que très encourageants, ont laissé transparaitre des lacunes telles que :

- une présence trop importante d'impuretés végétales,
- un mauvais parallélisme des fibres
- une grande quantité de boutons.

Les laboratoires du CELAC en Belgique, de TECNOTESSILE en Italie, de la WIRA en Grande Bretagne, et d'I.T.F.-Nord en France ont alors mis à contribution constructeurs et industriels pour étudier un nouveau type de matériel pour la laine qui devrait prendre en compte :

- la propreté du ruban,
- la diminution des boutons,
- l'ouvraison de la matière,
- le parallélisme des fibres,
- la régularité du ruban.

L'idée d'allier la capacité d'ouvraison des travailleurs débourreurs de la carde laine, à celle du cardage fibre à fibre et du nettoyage de la carde coton grâce aux chapeaux a conduit à l'élaboration ou à l'utilisation de cardes mixtes qui se présentent sous plusieurs formes :

- la carde mixte 1 tambour comprenant 2 points cardants travailleurs-débourreurs et une section de chapeaux (carde SACM).

- La carde tandem à 2 tambours munis de chapeaux.

- La carde tandem à 2 tambours dont d'un équipé de 4 groupes de travailleurs-débourreurs et le deuxième équipé de chapeaux.

- La carde à 3 tambours dont 2 garnis de chapeaux à population de pointes variable .

Il est sûr que le comportement d'une matière donnée va dépendre du mode de cardage mais il résulte des différentes études réalisées par les laboratoires tant sur des mélanges laine de récupération et polyester, que sur des laines vierges Open-Top ou blousses carbonisées ou non ou encore des matières du type Prato, que les cardes mixtes à 2 et surtout à 3 tambours alternant l'action des travailleurs-débourreurs et des chapeaux conduisent à un ruban compatible avec la filature Open-End.

En fonction des diagrammes de longueur en sortie de carde il sera préférable de prévoir un minimum de passages d'étirage régulés ou non pour améliorer la régularité à court et long terme. Si le diagramme ne le permet pas, un système d'auto-régulation sur carde est très souhaitable. La lecture des courbes CBL traduit les améliorations possibles de la qualité du ruban qui entraineront un meilleur comportement en filature.

FILATURE

L'utilisation de la technique Open-End a été l'objectif principal de l'ensemble des Instituts ayant participé aux travaux, et la gamme des matériels utilisés est très significative de la volonté d'aboutir.

On citera :

- la Dref 2,
- la San Giorgo,
- la BD 200,
- l'ITG 300,
- l'Autocoro,
- la RU II.

Il se dégage que la pure laine est très difficilement filable par cette technique sauf peut-être avec la DREF. Par contre, en mélange avec des fibres synthétiques en proportions variables, une gamme de numéros pouvant atteindre le 40 Tex voire le 33 Tex en fonction des diagrammes est parfaitement réalisable moyennant quelques précautions comme :

- choix du briseur et de sa vitesse
- choix du coefficient de torsion
- corrélativement choix de la buse
- choix du Ø de turbine et de son profil.

En particulier, le taux de gras sur ruban nécessite une surveillance. Un taux compris entre 0,5 et 0,8 % est à considérer comme limite pour éviter les enroulements au briseur. Par ailleurs, le taux de poussière à forte proportion de cuticules dans les turbines est surtout provoqué par l'attaque mécanique du briseur.

On signalera la supériorité du briseur à lamelle ou sélecteur par rapport aux briseurs à dents ou à aiguilles. La suspension de la fabrication de l'ITG 300 a bien entendu remis en cause l'intérêt du système.

Le problème posé par les poussières a amené la recherche à concevoir un simulateur devant, à terme, permettre de classer les lots susceptibles d'encrasser les turbines.

Enfin, les fils Open-End diffèrent de ceux obtenus sur continu à filer, surtout de par la structure, le toucher et les propriétés mécaniques, ce qui n'empêche pas leur utilisation en bonneterie, en vêtement, en ameublement. Seuls les velours coupés et les tissus plats grattés nécessitant un choix judicieux du mélange et des diagrammes des constituants.

Signalons l'étude particulière menée par l'UMIST sur les filatures à broches creuses qui conduit à l'obtention de fils aux caractéristiques mécaniques très intéressantes et dont les comportements au porter après confection sont absolument indentiques aux produits traditionnels.

Les résultats obtenus dans le cadre de ce vaste programme de recherche n'ont pu être obtenu que grâce à une coopération étroite entre les divers laboratoires de la Communauté qui eux-mêmes étaient les porte-paroles des industriels concernés par les études et que nous remercions, au passage pour leur collaboration.

ANALYSES PHYSIQUES EN CHIMIQUES

J. KNOTT
Centexbel

Il est bien connu que les paramètres physiques tels que la longueur et le diamètre de la laine, la régularité de la mèche ne sont pas les seuls facteurs à prendre en considération en filature rotor open-end.

Les paramètres physico-chimiques liés principalement à l'état de surface de la fibre interviennent d'une façon notable principalement par la formation de poussières et leur dépôt dans le rotor. Les objectifs de la recherche effectuée en collaboration avec le Deutsches Wollforschungs-institut Aachen, l'Institut Textile de France, Section Nord et Centexbel-Verviers, étaient :

1) de présenter de nouveaux tests ou d'adapter des tests existants, afin d'essayer de prévoir le comportement des laines lors de la filature open-end,
2) d'étudier la formation de poussières lors du processus de fabrication du fil,
3) d'étudier l'influence :
   - du taux de gras,
   - de la nature de l'ensimage,
   - et des traitements subis préalablement par la laine (carbonisage, teinture), sur la filabilité.

1. – MISE AU POINT DE TESTS SPECIFIQUES

1.1. Détermination de la teneur en poussières, sur la laine en bourre, la mèche et le fil.

L'extraction de la poussière des matières est réalisée par agitation à haute fréquence (30 Hz – amplitude 5 mm) d'une suspension éthanolique de fibres. La quantité de matière est déterminée par gravimétrie. La cinétique de la libération des poussières a également été étudiée.

1.2. Détermination du caractère tippy de la laine.

La mise au point d'une technique pour la détermination du pourcentage d'extrémités de fibres oxydées, donc mécaniquement fragiles est un test particulièrement approprié, vu les sollicatations mécaniques importantes subies par la fibres lors de la filature open-end.

Deux méthodes ont été étudiées :

1) une méthode uniquement réservée aux laboratoires de recherche utilise un microscope photométrique à balayage rapide. le détecteur enregistre la lumière transmise par les parties teintes (régions oxydées de la fibre).

2) Une méthode simple de détection basée sur le fait que les pointes de fibres oxydées possèdent une teneur plus élevée en acide cystéique, ce qui se traduit par une capacité plus grande à fixer sélectivement à pH 2, certains colorants cationiques sélectionnés.
Ce test est surtout intéressant pour la laine en bourre, car il permet de donner très rapidement une indication sur le degré d'altération des pointes de fibres.

1.3. Analyse des poussières.

Les tests mis au point permettent de séparer par microtamisage et de déterminer, en particulier, la composition "morphologique" des poussières (pointes et fragments de fibres, cuticule et cortex).
A l'aide de ces méthodes et d'autres méthodes chimiques classiques, telles que la détermination du gras extractible au dichlorométhane, le dosage de l'acide cystéique par voie chimique ou par infra-rouge en réflexion interne multiple et le dosage des acides aminés par la méthode de Moore et Stein, la mesure de la solubilité alcaline, le dosage des groupements aminés terminaux par la ninhydrine, etc..., différentes influences dans le processus de filature O.E. ont pu être étudiés.

2. - EVOLUTION DES PARTIES ALTEREES AUX DIFFERENTS STADES DE LA PREPARATION JUSQU'A LA FILATURE

Afin de mieux cerner l'origine des impuretés responsables des défauts de fabrication du fil par la technique Open-End, nous avons voulu caractériser la matière aux différents stades de fabrication en partant de la matière en bourre.
La visualisation de l'altération partielle de la laine en bourre a été réalisée en traitant la matière par le bleu de méthylène. Après cardage, peignage, craquage, étirage et filature, la matière et les poussières résiduelles ont été analysées. Nous avons également comparé les fils obtenus sur métiers open-end à sélecteur et à briseur, ainsi que celui obtenu par filature classique.
Les différents résultats repris dans le tableau 1, montrent de nouveau le rôle important de la carde en ce qui concerne l'élimination des poussières. Le craquage de la matière est une source non négligeable de formation supplémentaire de poussières.
Les métiers open-end à sélecteur sont moins dégradants pour la laine que les métiers open-end à briseur, comme l'indique un rapport cortex/cuticule beaucoup plus faible dans les poussières produites dans ce type de machine.
Signalons que, l'utilisation de laines bien carbonisées atténue fortement l'encrassement de la turbine, vu non seulement la diminution des matières végétales, mais également des poussières.

Tableau 1

Evolution du taux de poussières lors
du processus de filature
(* = moyenne de 4 déterminations)

Echantillon	% de poussières (*)
Laine en bourre	1,06 — 0,91
Laine sortie carde	0,55 — 0,46
Laine avant craquage	0,43 — 0,43
Laine après craquage	0,52 — 0,51
Laine avant peignage	0,40 — 0,35
Mèche entrée Open-End	0,50 — 0,51
Fil Open-End défibreur	0,76 — 0,76
Fil Open-End sélecteur	0,59 — 0,61
Fil Classique	0,53 — 0,51

384

RESUME ET CONCLUSIONS PAR LE PRESIDENT DE LA SESSION

M. BONA
Città Degli Studi

Quelques 70 délégués participaient à la session. Ce fut un réel succes.
Comme convenu, trois rapports ont été rendus résumant les travaux exécutés par huit instituts dans cinq pays membres. L'exposé du Dr. Artzt
intitulé "Préparation des matières récupérées pour la filature de type
non conventionnel" était basé sur des travaux exécutés à l'Institut für
Textiltechnik, par Wira et Celac. L'exposé de M. Bruggeman, "Nouvelles
techniques de cardage et de filature", reprenait les résultats d'ITF-
Nord, Centexbel, Celac, Wira, Umist et Tecnotessile. Quant à l'exposé du
Dr. Knott, "Analysis physiques et chimiques", il mentionnait les recherches de Centexbel, du Deutsches Wollforschungsinstitut et d'ITF-Nord.
  Ces exposés furent suivi d'une discussion animée et positive au
cours de laquelle des aspects techniques spécifiques firent l'objet d'une
analyse détaillée, à savoir :
- Le choix entre le carbonisage classique et les procédés mécaniques
  d'épuration a été examiné en détail. Le traitement mécanique peut
  certes donner des résultats satisfaisants pour carder les blouses
  contenant un faible taux de matière végétale, mais le carbonisage,
  quand il est bien fait, apparaît encore comme la meilleure solution
  pour les déchets contenant un taux élevé de matière végétale. Il a été
  démontré que le battage peut être omis pour éviter le feutrage pendant
  la neutralisation.
  Après le cardage, les fibres qui n'ont pas été feutrées ont peu
  d'enchevêtrements et sont virtuellement aussi pures que celles
  traitées par le carbonisage classique. Les différents procédés devront
  également être comparés dans un contexte industriel, et plus
  particulièrement au point de vue des coûts.
- Les poussières constituent un problème crucial pour le fonctionnement
  des turbines. Des défauts mécaniques entraînent la production de poussières tout au long du processus de fabrication, et cela à un degré
  variable, suivant l'état de la matière première et des conditions de
  travail. Il est aussi important de trouver le moyen d'éliminer les
  poussières que de découvrir comment elles sont produites. On a, à
  dessein, fait référence aux poussières d'origine minérale qui, quoique
  présentes en faible quantité, peuvent avoir une influence significative, étant donné leur action abrasive.
- Il s'agissait d'une discussion portant sur l'utilisation des rouleaux
  d'entrée du fil pour garniture de carde ou des rouleaux de sélection
  dans le mécanisme d'alimentation des machines à bout libéré du type à
  rotor. L'ancien système, de loin le plus couramment utilisé, soumet
  les fibres à un traitement plus grossier et de nombreux chiffres ont
  été cités dans le rapport principal. Chiffres à l'appui, il est res-
  sorti de la discussion que ce traitement entraînait un taux élevé de
  rupture des fibres, quoique l'on dispose à ce sujet de preuves indi-
  rectes, étant donné que la longueur n'a pas encore été effectivement
  mesurée sur le fil continu.
  Fait regrettable, la seule entreprise qui avait produit jusqu'à

présent une machine à bout libéré avec des rouleaux de sélection vient
de la retirer du marché.

Cela semblerait confirmer le manque d'intérêt des fabricants de machines du secteur lainier, et il est à espérer que les résultats de ce
symposium (dont les délégués sont priés de prendre connaissance)
aideront à rectifier la présente situation.

- Une attention particulière fut accordée aux propriétés des produits
fabriqués avec les nouveaux fils, comparativement aux vêtements et
tissus tricotés de manière traditionnelle avec une référence particulière aux fils guipés (broche creuse).
Une série d'échantillons, fabriqués avec les différentes techniques
expérimentées par les divers instituts, fournissent des preuves
concrètes de ces propriétés.

- Quelques informations furent données sur les futures recherches, la
plus intéressante étant l'étude des additifs spécialement conçus pour
éliminer les problèmes couramment rencontrés lors du tissage de la
laine sur des machines à bout libéré. Ce qu'il faut faire, c'est identifier les points faibles de la fibre. (Ce qu'il est désormais possible de faire avec précision, grâce aux recherches) et y remédier
systématiquement. Ainsi, l'objectif est d'adapter la fibre à la
machine, et pas vraiment la machine à la fibre. Cela constitue une
rupture avec la tradition; elle n'est en fait que le résultat des
coûts de production proportionnellement plus élevés que ceux des
fibres vierges.

Et enfin, pour en venir à des préoccupations plus générales, il
faudrait attirer l'attention sur deux contributions qui, pour des raisons
diverses, revêtent plus d'importance que les simples problèmes d'ordre
technique.

- La contribution apportée à la discussion par le représentant espagnol,
qui assistait au symposium en qualité d'observateur, vint confirmer
quelques points mentionnés par les auteurs des rapports. Cela montre
que cet esprit de coopération présent tout au long de ce programme
aura d'autres prolongements avec l'élargissement de la Communauté.

- Le Président d'Interlaine, instance suprême de l'industrie lainière, a
félicité les instituts pour leurs travaux dont il a souligné la valeur
pratique. Il a d'ailleurs l'intention de les mentionner à la prochaine
assemblée générale de son organisation.

QUALITE DES ETOFFES ET DES ARTICLES TRICOTES

J. STRYCKMAN
Centexbel

INTRODUCTION

L'industrie de la mialle est un secteur important de l'industrie textile européenne. Avec ses quelque 25.000 entreprises, dont il est vrai quatre cinquième en Italie, avec ses 410.000 personnes qu'il met au travail, ce secteur représente un huitième de la production textile européenne. En 1984, il aconsommé environ 695.000 tonnes de fils et filés; son chiffre d'affaires a atteint près de 16 milliards d'ECU. Cependant si l'Europe exporte des étoffes à mailles qui, en valeur, représente deux fois et demi ce qu'elle importe, en articles finis sa balance commerciale est très nettement déficitaire. Ce mali représente en effet cinq fois la valeur du boni sur les étoffes. C'est pourquoi, créativité, productivité, qualité, doivent présider au leitmotiv de toutes les entreprises du secteur.

L'article en maille doit être fonctionnel, faible et attrayant. Le secteur doit vendre de la qualité, de l'imagination et du goût au meilleur prix. Son succès viendra de sa capacité d'adaptation rapide aux marchés et en particulier au marché exigeant de la mode. Il doit faireun effort permanent d'amélioration et de renouvellement de ses techniques et de ses produits au moment même où le consommateur devient plus exigeant sur l'adaptation du produit à l'usage et sur l'assurance de la qualité.

La production des articles tricotés est réalisée dans sa grande majorité dans des entreprises de moyenne et petite dimension qui, pour répondre à cet impératif, doivent pouvoir s'appuyer sur une aide logistique extérieure qu'elles peuvent trouver dans des instituts et centres de recherche spécialisés. C'est pour améliorer cet appui en élargissant leur know-how qui huit instituts de recherche de sept pays de la Communauté, dans le cadre de dix contrats avec la Commission, se sont engagés en conjugeant et coordonnant leurs efforts, dans un programme de recherche qui vise à pouvoir fournir au producteur les atouts susceptibles d'assurer la qualité de leurs produits et de favoriser l'attrait du consommateur vers les articles européens.

Produire un article attrayant, fonctionnel et fiable nécessite la maitrise d'un nombre particulièrement important de paramètres, paramètres de matières premières, paramètres de fabrication, paramètres de conditionnement, de stockage et d'expédition.

Maîtriser la qualité des articles de bonneterie est cependant un problème complexe et implique de savoir à quelles propriétés l'article doit satisfaire; de disposer de méthodes de mesures objective de ces propriétés, de connaître les valeurs normatives pour apprécier le niveau de qualité, de déterminer les paramètres qui influencent la qualité et l'incidence d'une variation des paramètres sur la qualité.

L'objectif essentiel du programme a consisté à déterminer l'influence des paramètres, matières et conditions de fabrication, sur la qualité des articles compte tenu de la destination du tricot et de la fonction de l'article.

Le maintien des dimensions de l'article au porter et à l'entretien est un critère important et général avec lequel tous les fabricants sont confrontés pour toutes les catégories d'articles tricotés. Il a particulièrement retenu l'attention dans le programme. A cette caractéristique, viennent s'ajouter d'autres critères de qualité qui forment un ensemble de propriétés qu'il faut prendre en considération pour cerner le concept qualité d'un article déterminé et pour évaluer son niveau de qualité, tenant compte de la fonction qu'il doit remplir.

Le programme de recherche a abouti à l'élaboration d'un ensemble de directives quant à la matière et ses caractéristiques, aux conditions de tricotage et aux traitements subséquents dont le producteur doit tenir compte pour garantir l'obtention de la meilleure qualité de ses produits. Les résultats du programme sont présentés suivant trois axes complémentaires, à savoir :

- critères de qualité des articles tricotés, sous l'angle du consommateur,

- aspects qualité dans la fabrication des étoffes et des articles tricotés,

- évaluation objective de la qualité.

L'ensemble des résultats, présentés dans ces trois rapports, constituent un guide particulièrement précieux pour le bonnetier qui a le souci de maîtriser, d'améliorer et de stabiliser la qualité de sa production.

CRITERES DE QUALITE DES ARTICLES TRICOTES

H.J. SUURMEIJER
T.N.O

Comment les consommateurs, détaillants et fabricants perçoivent la qualité des vêtements tricotés fut examiné. Les vues des porteurs, les mesures et une évaluation de l'ajustement des tricots qu'ils possédaient furent obtenus dans une enquête menée au Royaume-Uni, étendue par la suite à des porteurs en Belgique, au Danemark, en France et aux Pays-Bas, couvrant environ 1400 vêtements.

L'examen des cahiers des charges d'après lesquels les vêtements sont fabriqués démontre que les grands détaillants et les fabricants ayant leur propre marque sont d'accord, en règle générale, sur les normes de performance minimale mais, étant donné que, sauf pour la solidité des couleurs, leurs méthodes d'essais diffèrent, cet accord est en grande partie plus apparent que réel.

A l'achat, les cahiers des charges des petits détaillants étudiés sont totalement inadéquats, et, souvent, des parties importantes des cahiers des charges des fabricants sont négligées et communiquées verbalement. A l'exception des fabricants de vêtements dimensionnés, peu d'entre eux incluaient, dans leurs cahiers des charges, la longueur de la maille, alors que, dans le contrôle de qualité, c'est le paramètre-clé.

Des besoins se manifestent clairement concernant l'amélioration des méthodes d'essai à l'extension des vêtements et l'évaluation des coutures. Il existe également des différences considérables dans la détermination des tailles, reliant mensurations et mesures des vêtements "à la vente" et après lavage.

Dans la plupart des pays de la CEE, il n'existe pas de séries de données anthropométriques généralement disponibles, indiquant tous les paramètres nécessaires et montrant les variations régionales ou celles liées à l'âge et à la classe sociale.

Lorsqu'on conçoit et développe des vêtements tricotés "allant bien" dans le nombre de tailles restreint à l'heure actuelle pour une majorité de consommateurs, les tableaux de tailles publiés dans la CEE et les méthodes d'appellation des tailles ne permettent pas d'utiliser les propriétés spécifiques des tricots.

En raison des différences entre les proportions corporelles relatives des groupes de population de la CEE, des tableaux d'appellations de tailles adéquats, dérivés d'études anthropométriques de groupes de population dans les marchés locaux ou nationaux, devraient être à la disposition des fabricants et détaillants, pour qu'ils puissent fournir à une grande majorité de consommateurs des vêtements tricotés confortables, bien à leur taille.

La conception et la fabrication de tels vêtements peuvent se baser sur les connaissances dont on dispose quant aux structures tricotées et sur les méthodes de production fondées sur ce savoir.

De nouveaux essais de laboratoire utilisant des dispositifs de mesure spéciaux, proposés par TEFO à Goteborg, des formes spéciales pour vêtements, développées par ITF Maille ainsi que l'inspection de plus de 1000 vêtements de dessous, disponibles dans le commerce, montrent qu'il est possible de juger la capacité "du bien aller" des vêtements tricotés dans le nombre restreint de tailles actuel.

Ce jugement peut se fonder sur les essais des mesures critiques de vêtements soumis à des charges acceptables minimales ou maximales et sur

la comparaison des résultats de ces mesures avec les mensurations corres-
pondantes, des plus petites aux plus grandes, rencontrées à l'intérieur
des tailles.

Les T-Shirts furent identifiées comme un type de vêtement demandant
une étude plus détaillée, les structures de ces tissus en jersey, à côtes
ou uni, étant susceptibles d'extension à l'usage. Les mesures prises à
partir des vêtements étendus à plat ne suffisent pas toujours lorsqu'il
s'agit de juger leurs caractéristiques au porter et déterminer s'ils vont
bien.

En se basant sur l'expérience obtenue avec des formes plates pour
chaussettes et bas etc. 26 formes plates furent développées, pour repré-
senter la hauteur du tronc, la tête, la poitrine, les épaules et une
partie des hauts de bras. Ces formes recouvrent 95 % de la population
adulte male de France.

Les critères proposés pour juger les caractéristiques et la performance au porter des T-Shirts sont les suivants :

- dans le sens de la largeur : on examine si les vêtements serrent,
  s'ils sont bien ajustés, proches du corps, ou s'ils flottent.
- dans le sens de la longueur : la longueur minimale requise, définie
  expérimentalement, est indiquée par une ligne sur les formes.
  Les vêtements étant placés sur les formes, il est vérifié qu'il y a
  passage aisé de la tête.
  Ces critères ont été vérifiés lors d'expériences menées en laboratoire et dans l'industrie. La performance dimensionnelle des tissus
  tricotés, serrant le corps à divers degrés, a été examinée en
  simulant la déformation des vêtements pendant l'usage.

Les nouvelles méthodes d'essai et dispositifs de mesure sont capables de contribuer à :

- l'assurance de qualité et le contrôle de qualité,
- l'amélioration des cahiers des charges pour vêtements des détaill-
  lants et fabricants,
- l'amélioration relative aux appellations de tailles,
- mieux adapter les tricots à la taille des porteurs.

ASPECTS QUALITE DANS LA FABRICATION DES ETOFFES ET DES ARTICLES TRICOTES

M. BALLAND
I.T.F.-Maille

Les résultats présentés ici sont une synthèse des travaux effectués
à CENTEXBEL, au DANISH TEXTILE INSTITUTE, à l'INSTITUT FÜR TEXTILTECHNIK
et à l'INSTITUT TEXTILE DE FRANCE-MAILLE.

Nous aborderons les sujets en essayant de suivre le cycle de production de la fibre à l'article fini.

Nous commencerons, cependant, par examiner la structure des coûts
des étoffes et le coût des défauts. Les coûts matières sont en moyenne de

6 % dont 60 % sont dus au fil. Les techniques d'épissures permettent une
nette amélioration.

Les fils destinés à la bonneterie doivent être paraffinés. Les
paraffines actuelles possèdent des châines hydrocarbonées de 21 à 38
atomes et leurs propriétés physiques et chimiques en dépendent. Comme de
plus, les quantités déposées dépendent du fil, il est pratiquement impossible, industriellement, d'obtenir un coefficient de frottement connu et
régulier. Les meilleurs résultats sont obtenus avec des taux allant de
0,4 % à 1 %.

Il apparaît donc nécessaire de contrôler et d'asservir la longueur
de fil absorbée sur tous types de métiers.

C'est pourquoi plusieurs prototypes ont été étudiés sur un métier à
tricoter rectiligne; le dernier prototype comprend : un détecteur de
débit du fil, un détecteur de position du chariot, une unité centrale
constituée par un "single board computer" et un système d'asservissement
qui réagit sur la tension du fil.

L'expérimentation industrielle a été un succès et permet d'abaisser
les fluctuations de longueur de fil absorbée à plus ou moins 1 %. La
rentabilité de l'implantation du dispositif sur chaque machine est en
cours d'étude. Il est certain que des dispositifs de ce type vont se
développer et permettre des économies de matières extrêmement importantes.

La qualité des tricots, sur laine ou mélange laine/acrylique, a été
étudiée du point de vue du toucher, de l'aspect, du boulochage et de la
distorsion des mailles.

Pour le toucher et l'aspect, ce sont les fils 100 % laine avec les
fibres les plus fines qui ont obtenu les meilleurs notes. De plus, les
laines non traitées infeutrables étaient les meilleures.

Les mélanges laine/acrylique ont donné des résultats semblables
malgré les différences de finesse de fibre et que le mélange soit anti-
boulochage ou non.

Après lavage, on n'a plus constaté de différence entre la meilleure
laine et l'ensemble des mélanges.

En ce qui concerne le boulochage, ce sont les tricots avec les
laines les plus fines qui donnent les moins bons résultats et ceci d'une
manière plus nette sur les tricots rectilignes. La différence entre laine
traitée antifeutrage et non traitée n'apparaît faiblement qu'en rectiligne où les laines traitées donnent de meilleurs résultats. Parmi les
mélanges, le meilleur résultat a été obtenu avec de l'acrylique 5,6
dtex. Ils ne sont supérieurs qu'aux laines fines non traitées.

En ce qui concerne la distorsion des mailles en jersey (Maschenver-
werfungen), il à été parfaitement mis en évidence que ce défaut provient
de la présence de très grosses fibres de laine à ces endroits. Ceci
augmente la rigidité à la flexion du fil et provoque, soit le défaut de
distorsion, soit le défaut d'ondulation.

Enfin, le vrillage est un défaut inhérent aux tricots à base jersey
et, si certains apprêts peuvent en diminuer l'effet, le seul moyen pratique consiste actuellement à alterner les fils de torsion S et de torsion
Z.

Les variations dimensionelles des tricots sont l'un des problèmes
majeurs actuellement à l'étude.

La notion d'état relaxé est admise par tous les laboratoires et les
méthodes pour y parvenir sont très voisines. Elles consistent toutes à
utiliser le séchage en tumbler, précédé soit d'un certain nombre de

lavages en machine (3 ou 5), soit d'un traitement statique en bain aqueux suivi d'une action mécanique sur un tapis vibrant.

On a montré l'importance du taux d'humidité dans l'étoffe, à la sortie du tumber, sur les dimensions des tricots de coton et, en toute rigueur, il faudrait sécher jusqu'à un taux d'humidité connu. Il est préconisé de 5 à 6 %.

Les lois obtenues, à l'état relaxé, ont été confirmées pour le coton et il a été prouvé que les tricots acrylique en jersey, côte 1 et 1 et côte 2 et 2 suivent aussi les lois de Munden dont les coefficients ont été calculés.

Lorsque l'on effectue des lavages successifs sur des tricots coton prélevés en fabrication, il ressort clairement que la majorité des variations dimensionnelles apparaît au premier lavage puis celles-ci évoluent peu, pour finir par se stabiliser vers le 10ème lavage.

En long, le retrait obtenu au premier lavage se poursuit dans le même sens alors qu'en large, après le retrait du premier lavage, on assiste à des avachissements.

Les dimensions des tricots varient tout au long du processus de finition. Cependant, les retraits au lavage relativement élevés constatés sur les tricots écrus ne sont jamais effacés ni durant la teinture, le blanchiment ou d'autre traitement au mouillé, ni au séchage, ni au finissage. Il est clair que toute extension du tricot à un stade donné provoque un accroissement du retrait au lavage.

Parmi l'ensemble des machines de finition, la plus importante du point de vue de la stabilité est le séchoir. Les machines les plus performantes sont les Tumbler à la continue.

Une difficulté d'utilisation des lois de Munden pour les tricots de coton est la différence de dimensions à l'état relaxé des tricots écrus et des tricots traités. En général, à l'état relaxé, les tricots traités sont de 5 % plus longs et de 4 % moins larges que les tricots écrus. Des banques de données permettront, peut-être, de résoudre ce problème.

Le traitement des acryliques pour pull-overs a été étudié. Pour cela une méthode de laboratoire de la mesure de la "fixation" des fils a été mise au point. Il a d'abord été prouvé qu'un traitement classique sur presse ne permettrait pas de fixer les dimensions d'un tricot si celui-ci était déformé par rapport à l'état relaxé. Seuls 30 à 50 % de ces déformations peuvent être fixés. Il en ressort qu'avant d'être traités sur presse, les panneaux devraient être relaxés puis "fixés" sur presse par un mélange air + vapeur à des températeurs variant selon les types d'acryliques. Le Léacryl à 75 °C, la Courtelle entre 75 et 80°C, le Dralon à 80°C et le Crylor à 85°C. Notons qu'en dessous de 70°C, les tricots ne sont plus repassés.

Les essais de porter confirment les résultats de laboratoire.

Ils mettent en évidence d'une part l'importance de l'état initial du tricot dans l'évolution des dimensions au porter-lavage et d'autre part l'importance du traitement.

Le plus stable sera le tricot qui aura été le mieux fixé le plus proche de l'état relaxé.

Compte tenu, malgré tout, de la tendance au retrait sens longueur et avachissement sens largeur, on aura intérêt à prévoir des corrections inverses, de l'ordre de 5 % dans les deux sens.

Une attention particulière devra être portée aux bord-côte et, en particulier, d'une part de ne pas les trictor trop lâche et d'autre part, de ne pas les traiter loin de leur état relaxé.

Dans ces conditions, il est possible d'obtenir des pull-overs en acrylique ayant un bon comportement au porter.

Pour terminer, notons enfin qu'il existe des moyens de contrôle de la confectionnabilité des tricots. En particulier, il a été mis en évidence d'une part l'importance de la grosseur réelle de l'aiguille et d'autre part l'importance de l'adoucissage, ceci par mesure de la force de pénétration de l'aiguille dans le tricot.

On constate que c'est tout au long de la chaîne de fabrication qui se construit la qualité des tricots; les résultats présentés ici peuvent contribuer à l'amélioration de cette qualité.

EVALUATION OBJECTIVE DE LA QUALITE

E. FINNIMORE
Deutsches Wollforschungsinstitut

Le thème général du programme de recherche présenté, dans cette séance, est la qualité des tissus et articles tricotés.

Afin que l'industrie textile européenne demeure compétitive, il faut maintenir et améliorer constamment le niveau de la qualité. Pour atteindre ce but, des évaluations objectives de la qualité sont nécessaires, puisqu'elles permettent par exemple, un meilleur contrôle de qualité dans l'usine, l'établissement de normes internationales et la conception de tissus et vêtements répondant à des besoins spécifiques. Dans cette communication seront décrits plusieurs aspects de l'évaluation objective basée sur des travaux menés dans trois Instituts européens.

Pour le consommateur, un aspect important de la qualité lors de l'achat de chaussettes est qu'elles aillent bien, ce qui mène aussi à un confort au porter accru. A l'ITF Maille de France, une étude fut entreprise afin de mettre au point un moyen de contrôle objectif de la taille des chaussettes (demi-bas) et de leur aptitude à bien aller.

Lorsqu'elles sont portées, les chaussettes sont en état d'extension. Une mesure objective de leur aptitude à bien aller doit en tenir compte. Les formes utilisées pour l'étalage ne conviennent pas à cette fin. Pour des raisons d'ordre pratique, il est préférable, dans l'industrie, d'avoir des formes plates. Ainsi, il fut nécessaire de mettre au point une méthode qui permettrait à une surface plate d'être transformée en une superficie équivalente à la surface externe d'un volume tri-dimensionnel (pied et jambe).

Les diverses étapes nécessaires à l'obtention des formes furent: transformation d'une chaussette enfilée en une forme rigide qui fut enlevée de la jambe par découpage; insertion d'incisions en des points appropriés afin d'obtenir une forme plate; transformation de celle-ci en une forme au contour continu et dont la superficie est équivalente. Un jeu de 12 formes fut préparé correspondant aux tailles de la bonneterie française 24 à 46.

90 articles, regroupant différents types de chaussettes furent testés sur ces formes et ensuite par des porteurs.

Sur les 90 échantillons mis à'essai, il n'y eut désaccord entre le test objectif et l'essai au porter que pour trois seulement. Il fut possible de montrer que, dans ces trois cas, les dimensions du pied et de la jambe s'écartaient des valeurs normales dans cette pointure. Les

formes ont été essayées avec succès dans cinq usines de bonneterie. Les
formes sont fabriquées d'une manière commerciale et la méthode a été
introduite en tant que norme française. Il est à espérer, que sur cette
base, une norme européenne pourra être établie.

Une source importante de défauts dans l'industrie de la Maille est
la confectionnabilité médiocre de certains tricots. Celle-ci est suscep-
tible d'entraîner des dégâts par formation de boucles et de trous durant
la couture.

A l'Institut für Textiltechnik Denkendorf, en Allemagne de l'Ouest,
un essai objectif de la confectionnabilité des tricots fut mis au point.
Des études préliminaires montrèrent qu'il existe une très bonne corrélation entre la force de pénétration de l'aiguille et le défaut par formation de boucle.

Un système assisté par ordinateur a en conséquence été conçu pour
l'enregistrement rapide de la force de pénétration de l'aiguille. Les
avantages présentés par cette méthode d'essai résident dans le fait
qu'elle peut être mise en oeuvre sur des machines de production, dans les
conditions de production.

L'essai peut se faire sur des échantillons représentatifs avant la
découpe, offrant ainsi la possibilité de traiter toute la pièce avec des
agents auxiliaires de couture. Il est facile de démontrer, par cette
méthode, les avantages que l'apprêt correct des tricots présente pour
améliorer leur aptitude à la couture. Avec des tissus fabriqués à 100 %
de fibres synthétiques, la force de pénétration de l'aiguille pourrait
être réduite jusqu'à concurrence de 93 % par des apprêts appropriés.
Cette méthode permet donc de mener des études objectives pour optimiser
des traitements d'apprêtage ayant une influence directe sur l'accroissement de la qualité du produit.

Le toucher des tissus est l'un des aspects de la qualité le plus
important, et, avec l'impact visuel, il est souvent, dans la décision du
consommateur, le facteur déterminant s'il va acheter ou pas. Toutefois,
le toucher est une caractéristique particulièrement difficile à définir,
quantifier et mesurer.

Au cours des années récentes, des efforts accrus ont été faits pour
développer des mesures objectives du toucher. Parmi les nombreux avantages de telles mesures, on peut mentionner : la possibilité qu'ont ache-
teurs et fournisseurs d'échanger des spécifications numériques, ainsi que
la quantification des effets des différents traitements d'apprêtage.
L'approche qui a attiré le plus d'attention dans le monde entier est le
système venu du Japon : KES-F de Kawabata.

Au Deutsches Wollforschungsinstitut à Aachen, Allemagne, on a étudié
l'application de ce système à la mesure du toucher des tricots. Il faut
tenir compte de la plus grande extensibilité des tricots comparés aux
tissés.

Les mesures comprennent les caractéristiques de ployage, cisaillement, limite élastique à la traction, compression, surface et contexture
(poids et épaisseur). Les études se proposaient tout d'abord d'évaluer
l'effet, sur les paramètres objectifs du toucher d'une série de tricots
en laine et laine/acrylique, de divers paramètres tels que l'épaisseur,
le type de tricot, la matière textile, la finesse et l'apprêt. L'importance de la relaxation à la vapeur, pour l'obtention d'un toucher volumi-
neux et doux, fut démontrée.

Le second aspect de ces études se préoccupait de l'évaluation quantitative de l'effet des produits adoucissants sur les tricots de laine et
mélanges de laine. Par la mesure objective des paramètres de raideur, on

peut démontrer qu'il y a un niveau d'application optimum pour l'obtention d'effets adoucissants sur de la laine traitée pour résistance au rétrécissement.

Ces effets de concentration furent notés également au cours des évaluations subjectives. Un bon accord fut en général observé entre les évaluations objectives et subjectives des traitements d'apprêt, mais le type de tricot utilisé influença le jugement des experts.

Les données quantitatives obtenues dans ce type d'étude peuvent aider le fabricant à choisir les matériaux et les conditions d'apprêt les meilleurs pour obtenir le niveau de qualité voulu.

SOMMAIRE ET CONCLUSIONS PAR LE PRESIDENT DE LA SESSION

J. STRYCKMAN
Centexbel

Des questions qu'inspirèrent les exposés se dégagent deux aspects importants de la qualité des tricots : la dimension et la stabilité dimensionnelle des tricots d'une part, le confort des articles caractérisé par des mesures objectives d'autre part. Nous commenterons ce débat succinctement ci-après.

A la question de savoir si les dimensions stabilisées d'un tricot achevé peuvent être prévues à partir des paramètres du fil et du tricotage, l'on peut dire qu'il existe des corrélations entre ces paramètres en fonction du liage et du type d'achèvement du tricot. La banque de données "Starfish" élaborée par l'International Institute for Cotton permet de prévoir la dimension d'un article fini à partir des caractéristiques du tricot écru. Les équations sont connues pour le coton dans trois liages : le jersey, la cote 1/1 et l'interlock.

Quant à la longueur de fil absorbée par maille, les dimensions du tricot dépendent pour une très grande part de ce paramètre. Il est donc hautement utile que celle-ci soit contrôlée et maîtrisée. Ceci s'opère sur métier circulaire par les fournisseurs positifs dont ils sont équipés. Aucun dispositif n'existait pour les métiers rectilignes. Deux dispositifs nouveaux ont été développés et testés. Dans l'un, on s'efforce de maintenir la tension constante en un point déterminé d'alimentation du fil sur métier. Dans l'autre système, la longueur du fil absorbée sur un tronçon prédéterminé à chaque rangée est comparée par microprocesseur à une longueur prédéterminée souhaitée pour la rangée. Si les deux longueurs sont différentes, un signal de correction agit sur l'alimentation du fil au métier. Les deux systèmes permettent d'obtenir des panneaux de longueur identique avec une tolérance de 1 % sur la longueur de fil absorbée.

La vitesse des métiers rectilignes modernes ne constitue pas un obstacle au bon fonctionnement du dispositif. La mise en vitesse du capteur pourrait constituer un problème dans l'un des sytèmes, mais la mesure s'effectue à l'intérieur de la fonture de celle-ci. Dans l'autre système, le sensor réagit immédiatement aux différences de tension. L'effet des accélérations du fil en début de rangée est annulé par la suppression de la réaction sur les organes de tension à ces moments-là.

Le paraffinage du fil avant tricotage est un moyen moins sophistiqué pour réguler la longueur de fil absorbée et, pour certains articles fantaisie, cette pratique reste le seul moyen.

Toutes les paraffines ne réagissent cependent pas de la même façon. Leur composition joue un rôle, en l'occurence la longueur des chaînes hydrocarbonées différente d'une paraffine à l'autre. Elles ont une incidence sur le point de fusion de base, celui-ci étant lui-même corrélé à la quantité de parafine emportée. La quantité de parafine emportée doit se situer entre 0,4 et 1 % en fonction de la nature du fil. Il est donc nécessaire de connaître le comportement de la paraffine que l'on utilise et de modifier le parafinnage en fonction de la nature du fil.

A la question de savoir si la paraffine a une influence sur le toucher des tricots, l'on passe à l'autre thème des débats : les aspects "toucher" et "confort" des articles et leurs mesures objectives. En premier lieu, l'on peut dire que la paraffine modifie le toucher des tricots, mais cette question n'a pas été approfondie dans le cadre du programme. Les mesures objectives de la qualité étant à ses débuts, des investigations restent à faire dans ce domaine.

Dans le programme, seuls les paramètres tricot et fibre ont pu être cernés. Ceux-ci ont déjà donné lieu à un nombre important d'analyse d'échantillons. Deux types de fil ont été examinés, l'un en pure laine, l'autre était en mélange laine-acrylique. Les mesures objectives ont été réalisés sur l'appareil japonais Kawabata spécialement conçu pour la mesure du "toucher".

Les résultats des mesures objectives ont été comparés aux mesures subjectives effectués par des experts choisis pour leur longue expérience dans l'appréciation des tissus ou tricots. La corrélation est grossièrement assez bonne, mais les experts se trompent parfois lorsque la différence entre les échantillons n'est pas flagrante.

Il faut cependant préciser que le concept "douceur" du Kawabata est une combinaison de deux éléments : douceur de surface et raideur. Une terminologie appropriée pour ces mesures devrait être élaborée en Europe.

L'expérience du Kawabata dans le domaine des tricots est moins étendue que dans les tissus chaine et trame. Les paramètres devraient être adaptés au tricot. Il n'empêche que les mesures subjectives doivent progressivement faire place à une synthèse objective qualifiant l'article, parce que les experts ne sont pas toujours conséquents et se contredisent régulièrement même sur une courte période. Pour les tricots, des paramètres tels que l'hystérésis de cissaillement, l'épaisseur ou le poids, la régularité de surface seraient plus appropriés. L'expérience devrait le montrer.

L'intérêt du Kawabata et des mesures objectives résident également dans le fait que l'on peut plus facilement apprécier l'incidence des paramètres de la matière et du processus de fabrication sur le "toucher". Ce qui est plus difficile, pour ne pas dire impossible, pour des raisons pratiques, pas des méthodes subjectives.

Quant à savoir si le système Kawabata est utilisable dans l'industrie européenne, il faut répondre que l'équipement est relativement coûteux et que les entreprises de taille moyenne ou petite n'en auront pas un rendement qui justifie son acquisition.

Au Japan, 150 équipements sont installés, beaucoup sont utilisés par des instituts de recherche ou des offices d'achat qui collaborent avec l'industrie. A cette question, la réponse est donc qu'il serait souhaitable, maintenant, et nécessaire, dans un proche avenir, que l'industrie textile pour l'habillement y fasse appel et s'appuie à cet effet sur des laboratoires collectifs qui pourront ainsi par l'équipement mais aussi par l'expérience qu'ils accumuleront sur les mesures objectives de qualité rendre d'énormes services à cette industrie.

Le thème "Qualité des étoffes et articles tricotés" s'est développé suivant trois axes.

Dans le premier axe : critères de qualité des articles tricotés, les travaux réalisés par quatre instituts, on mis en évidence qu'il était indispensable de mieux assurer la qualité, le bien-aller et le confort et la satisfaction des consommateurs, de disposer, à cet effet, d'une méthode uniforme européenne pour mesurer les dimensions du corps humain, d'appliquer cette méthode dans tous les pays de la Communauté et, dans chaque pays, dans les régions où peuvent apparaître des différences de conformation des personnes, d'élaborer des banques de données de ces mensurations. Ces banques de données devraient être tenues à jour dans le temps suivant la méthode standard établie.

Le deuxième axe, également traité par quatre instituts, a porté sur les aspects qualité dans la fabrication des étoffes et des articles tricotés. Il y fut essentiellement question d'un aspect très critique des tricots : la stabilité dimensionnelle, tant lors de la fabrication qu'au porter et à l'entretien par le consommateur. Les résultats des travaux aideront à mieux maîtriser ce phénomène dans le domaine des articles en laine, en mélange laine-acrylique et en coton.

Un résultat particulièrement intéressant a été obtenu par le développement de dispositifs assurant de façon automatique la constance des dimensions des panneaux sur métiers rectilignes.

Il n'en reste pas moins vrai que dans le domaine de la stabilité dimensionnelle beaucoup reste à faire tant au niveau des caractéristiques des fils et filés, que du processus de production,en cela compris les traitements d'ennoblissement.

Enfin, le troisième axe fut consacré à un aspect qui revêtera à l'avenir une importance croissante pour juger de la qualité des étoffes et articles tricotés. Il s'agit de l'évaluation objective de la qualité. Trois importantes caractéristiques : le toucher, le bien-aller et la "confectionabilité".

Dans ce thème auquel ont collaboré trois instituts, une méthodologie récente a conduit aux premiers résultats sur tricots, là où une large expérience existe déjà pour les tissus chaîne et trame.

Le nombre de paramètres qui influencent ces propriétés est particulièrement étendu, aussi le programme a dû se limiter aux paramètres qui ont paru essentiels et sur un nombre limité de types de matières premières, de techniques de filature, de structures de tricots et des processus d'ennoblissement.

Le travail qui reste à fournir dans ce domaine peut encore pour de longues années alimenter les programmes de recherche des instituts et des chercheurs.

Le président de cette session conclut par deux souhaits.

Le premier s'adresse aux instituts qui ont collaboré à ce thème de recherche. Ils ont accumulé une masse de résultats qui doivent être valorisés. Ce but ne sera atteint qui se l'ensemble de ces résultats sont appliqués par l'industrie; aussi est-il indispensable que chaque institut dans sa sphère s'attache à transférer l'acquis dans l'industrie et apporte son appui logistique pour assurer la bonne application des informations par les praticiens.

Pour ce follow-up à la recherche, la Commission des Communautés européennes déploie des efforts tout particuliers pour donner une impulsion à ce transfert et valorise les résultats de la recherche.

La masse des résultats acquis est important, mais le nombre de paramètres est incommensurable et la diversité des produits laisse un large champ ouvert à de futures investigations. Il a été démontré que l'effet multiplicateur par la coopération est bénéfique.

Aussi, le deuxième voeu serait de voir se poursuivre, à l'avenir, cette collaboration européenne entre les instituts qui ont formé un excellent team dans ce programme au cours de ces trois dernières années. Cette collaboration serait encouragée, si la Commission européenne pouvait répondre favorablement à toute demande de moyens d'accompagnement par l'importante industrie européenne de la maille.

La réalisation de ces voeux constitue sans conteste un moyen indispensable pour assurer un avenir à l'industrie textile européenne.

VALORISATION DU LIN

M. VAN LANCKER
Centexbel

Le lin est une des rares matières premières textiles végétales dont dispose la Communauté Européenne.

Quelques chiffres pour situer le lin en général et dans l'industrie textile en particulier :

- le lin représente moins de 2 % des matières textiles,
- la production dans la CEE de filasses (lin teillé et étoupes) était de 79.881 T en 1984 (137.000 T en 1962 et 82.100 T en 1982), la production de la filature de lin classique de plus de 33.500 T en 1984 (80.000 T en 1962 et 30.400 T en 1982),
- la balance commerciale est largement positive :
  . env. 145 milj. ECU en 1984 pour la France,
  . env. 70 mil. ECU en 1984 pour la Belgique.

Le déséquilibre existant depuis plusieurs années entre la production de filasses et la consommation de la filature classique a contraint le négoce à se tourner vers les marchés extérieurs à la CEE et la profession à rechercher des débouchés nouveaux.

Depuis plusieurs décennies l'utilisation du lin avait connu une évolution défavorable et les différentes professions concernées, réunies dans la C.I.L.C. (Confédération Internationale du Lin et du Chanvre) ont mis au point un programme de redressement et d'expansion qui prévoyait des efforts importants dans les domaines de la recherche, de la promotion et de la commercialisation.

La programme de recherche, appelé "EUROLIN" avait comme objectif d'améliorer les procédés de transformation de la paille aux produits finis de façon à :

- économiser la matière première,
- créer des conditions de travail plus satisfaisantes,
- mettre en valeur les qualités intrinsèques du lin,
- diminuer les coûts de production,
- disposer d'articles plus adaptés aux besoins des consommateurs.

Basés sur une analyse de la situation de l'industrie linière de la CEE fin des années 70, des problèmes et lacunes existants de l'évaluation des efforts déjà entrepris, et des résultats des travaux antérieurs, il a été décidé de solliciter la CEE dans le cadre du 2ème programme de faire une proposition de support financier de recherche textile dans le but d'accélerer l'exécution du projet EUROLIN en cours de réalisation.

De nombreux facteurs rendaient le moment particulièrement choisi pour donner une nouvelle impulsion à ce programme :

- le retour des consommateurs vers les fibres naturelles,
- le développement des mélanges,
- l'apparition de traitements d'ennoblissement particuliers,

399

- les nouvelles techniques de filature,
- l'urgence pour l'industrie linière de réinvestir,
- l'existence d'un large marché à l'exportation.

Le programme de recherche présenté à la CEE a été mis au point par la Commission Scientifique et Technique de la C.I.L.C. à travers un comité de recherche et de développement, le Comité Billaux. Ce comité comprend les centres de recherches qui s'occupent du lin dans les pays de la CEE : ATPUL, ITL, ITF-Nord et ITF-Boulogne pour la France, Centexbel, et OVLT pour la Belgique, TNO et IBVL pour la Hollande, LIRA pour la Grande Bretagne, Textil-Forschung Bielefeld pour l'Allemagne et la Stazione Sperimentale per la Cellulosa pour l'Italie.

Chacun de ces centres est spécialisé dans un domaine particulier auquel il a été fait appel mais dans ses conditions pour le programme général, il était nécessaire d'assurer une coordination entre les différents efforts individuels. L'intervention communautaire a accentué le besoin de maintenir cette coopération à un haut niveau. Celle-ci s'est réalisée au travers de la Commission Scientifique et Technique de la CILC et du Comité Billaux. Cette façon de procéder était une garantie pour que les travaux soient orientés vers les besoins réels de l'industrie et que les résultats scientifiques et techniques soient connus et appliqués rapidemment dans la pratique industrielle.

Une actualisation des objectifs a permis de définir clairement les objectifs techniques énumérés ci-après. Ces objectifs ont été traduits en programmes de recherche repris dans le second programme communautaire de recherche et développement textile et habillement sous le dénominateur commun "Valorisation du lin".

OBJECTIFS TECHNIQUES :

Approvisionnement en matières première et extraction des fibres

Pour assurer la permanence d'un approvisionnement en matière première convenable, des efforts de recherche doivent portés sur l'amélioration des techniques de cultures, du rendement de fibres, de la valeur des sous-produits, de la présentation des matières et la réduction de la main-d'oeuvre par la mécanisation.

Transformation de la fibre en fil.

A ce stade il est nécessaire d'adapter la fibre pour pouvoir utiliser soit du matériel à haute productivité, soit les nouvelles techniques de filature : filature OE, filature à friction, Novacore, etc...

Ennoblissement

Des améliorations techniques ou innovations dans les produits et procédés sont à expérimenter et à développer, par ailleurs, il est nécessaire d'apporter des réponses aux questions générales posées par l'ennoblissement du lin.

Qualification des produits

La qualification du lin sous toutes ses formes est nécessaire afin de préciser des caractéristiques permettant de déterminer l'aptitude du lin à divers usages. Le contrôle objectif de la qualité doit être réalisé et accompagné de normes et de cahier des charges bien définis.

Morphologie de la fibre

Une meilleure connaissance de la morphologie de la fibre de lin et de l'incidence des conditions de production et de travail est plus que jamais nécessaire afin d'exploiter toutes les resources potentielles de la fibre de lin.

Le programme "EUROLIN", dont le programme "Valorisation du lin" était une partie intégrale se poursuit dans les instituts conjointement avec de nouvelles activités promotionnelles sur le plan international.

Par ces travaux de recherche et de promotion, l'industrie du lin de la CEE se présentera avec un contexte technologique et social industriel moderne et répondra ainsi à la stratégie industrielle de la CEE.

Les qualifications nouvelles se font déjà jour à tous les niveaux et plus spécialement au niveau des nouveaux débouchés et de la production de produits de haute qualité. L'industrie linière se prépare à y répondre et à présenter des produits recherchés par la clientèle des marchés mondiaux.

PRODUCTION DE LA FIBRE DE LIN

W.W. FOSTER
LIRA

Les activités entreprises dans cinq centres de recherche, dans le cadre du programme EUROLIN, sont énumérées dans le tableau ci-dessous. Ces travaux ont pour objectif une amélioration du rendement global de la fibre de lin et de sa qualité, ainsi qu'une réduction des coûts de transformation.

ACTIVITES	Centre de recherche				
	ITL	LIRA	Wagen-ingen	ATPUL	Centex-bel
1. Applications sur le terrain					
- vaporisation des produits chimiques anti-verse	x				
- vaporisation de prod.chim. séchage/rouissage sur pré	x				
- vaporisation de prod.chim. séchage/rouissage sur tige			x		
- coupe/rouissage sur pré				x	
- tech.de démêlage des pailles	x				
- matériaux de contamination				x	x
2. Applications en usine					
- amélior.du brisage-teillage				x	
- systèmes de prod. des rubans				x	
- traitements des rubans :					
. enzymes		x			
. produits chimiques		x		x	
- systèmes à fibres courtes				x	
- lin-totale				x	

1. - APPLICATIONS SUR LE TERRAIN

Au cours des 5 dernières années, l'usage des pulvérisations chimiques pour empêcher la verse s'est développé et le procédé est maintenant bien établi à l'échelle commerciale. En 1984, environ 10 % du total de la récolte de lin français furent traités. Normalement utilisés pour ralentir la croissance, ces produits chimiques entraînent les inconvénients suivants : hauteur inférieure de la paille et rendement moindre de la graine. les tiges rabougries sont aussi plus épaisses que les tiges normales. Les pulvérisations ne devraient donc être utilisées qu'en dernier recours. Les trois produits considérés comme inoffensifs à l'usage sont : Cerone, Atheverse et Terpal. Toutes les variétés du lin, à toutes les phases de leur croissance, sont sensibles à ces produits dont les effets sont accentués par des mouillants.

Les travaux pionniers sur l'usage de vaporisations pour le séchage et le rouissage du lin poussant dans les champs se sont poursuivis. Le procédé a été aussi mis en oeuvre pour accélérer le séchage et le rouissage sur pré normal. Au cours des deux dernières années, le temps ayant été exceptionnellement sec, les résultats demeurent limités et il reste beaucoup à apprendre.

Des machines furent développées pour la levée du lin roui sur pré avant sa mise en balle, éliminant ainsi les mauvaises herbes, enlevant les pierres et favorisant le séchage. Quelque 200 machines sont déjà en usage. Il est possible, également, d'adapter aux ramasseuses – presses un accessoire utilisant un procédé semblable, faisant sortir ainsi jusqu'à 85 % des pierres avant que le lin ne soit mis en balle.

Des travaux furent consacrés, d'autre part, au développement de techniques qui supprimeraient l'usage de machines linières spécialisées telles que les arracheuses et permettraient par contre d'employer les machines agricoles ordinaires. En bref, la paille serait coupée, par arrachée et le lin enchevêtré obtenu serait roui sur pré et séché, mais en balle, demêlé par la suite pour produire des fibres courtes et propres. D'ores et déjà, il apparaît que, par rapport aux fibres produites par des procédés agricoles plus conventionnels, la fibre sera à un prix concurrentiel et se prêtera parfaitement aux mélanges.

Des techniques furent développées pour déceler la présence de reste de ficelles de polypropylène dans le lin, bien qu'il reste à développer un moyen d'ôter ce contaminant. Des travaux approfondis ont été menés afin d'identifier la grande diversité de fragments de végétaux se trouvant dans l'étoupe et qui apparaissent comme contaminants dans le tissus final. Tout semble indiquer qu'il vaut bien mieux éliminer les mauvaises herbes des récoltes de lin que d'essayer, ensuite, d'enlever ou de déguiser les résidus.

2. - APPLICATIONS EN USINE.

D'utiles modifications ont été développées dans la conception d'équipements de teillage y comrpis des avant trains de cardes et sections de batteurs plus performantes. D'autres développements comprennent l'étude de systèmes pour une transformation meilleure et plus économique visant des rendements de produits accrus et une qualité supérieure : un dispositif à étoupe re-teillée pour la production de matériaux sous forme de rubans et un agent d'apprêt pour la formation de rubans du système lin-totale. Certains de ces développements sont très prometteurs.

De véritables progrès furent réalisés dans l'ennoblissement des fibres de lin de basse qualité au moyen de traitement chimiques et biochimiques, tout particulièrement par l'usage d'enzymes afin d'élever la qualité du ruban.

NOUVELLES TECHNOLOGIES DE PREPARATION ET DE FILATURE

M. VAN LANCKER
Centexbel

Le tableau, ci-après, reprend les activités de quatre centres de recherche dans le cadre du programme EUROLIN. Ce travail a été entrepris afin de réaliser une amélioration technique et économique des procédés de transformation de la fibre en produits finis de façon à :

- économiser la matière première,
- mettre en valeur les qualités intrinsèques du lin,
- diminuer les coûts de production,
- disposer d'articles plus adaptés aux besoins des consommateurs.

Activités	Centres de recherche			
	Biele-feld	Centex bel	I.T.F.	LIRA
1. Révalorisation des matières de basse qualité	x	x		
2. Problèmes posés par les nouveaux schémas de fabrication :				
- optimisation du craquage		x		
- préparation et divisibilité de la fibre			x	x
- traitement de la mèche, dissociation de la fibre			x	
- aspect économique d'une nouvelle ligne de fabrication			x	x
3. Nouveaux procédés de filature :				
- filature à rotor	x			
- filature à friction		x		
- Novacore			x	
- filature sans torsion				x
4. Adaptation des produits et procédés.			x	x

Chaque année, une partie de la récolte est inutilisable pour la filature de lin classique, en outre des déchets formés lors de la préparation de filature pourraient être récupérés. Les travaux de recherche ont permis de déterminer et de mettre au point des circuits de transformation spécifiques. Des produits nouveaux ont été étudies à base de ces matières et des expériences industrielles lancées.

Le craquage du long brin pour sa transformation dans un circuit alternatif, appelé "système de haute productivité", ainsi que le craquage comme moyen de rectification du diagramme de longueur ont été étudiés. Un premier prototype de craqueuse a été réalisé par la firme Pierson et Digneffe. Le développement d'une nouvelle craqueuse sur base des résultats de l'étude et adapté aux spécifications de la fibre de lin a été repris par les ateliers Houget-Duesberg-Bosson.

La recherche systématique dans le domaine de la préparation, de la divisibilité et du traitement de la mèche avant filature a abouti à des conclusions permettant au filateurs d'optimiser le travail mécanique et chimique de la fibre et d'orienter les investissements futurs. Un appareil "Ténomètre" conçu et mis au point par la recherche permet de cerner parfaitement la dissociation de la fibre après action chimique. La Profession linière a reconnu formellement son utilité et se dote actuellement de ce dispositif de contrôle.

Dans l'optique d'un besoin de renouveler le parc de matériel de filature, les aspects économiques d'une nouvelle ligne de fabrication ont été étudiés avec un constructeur de matériel textile. Si la production s'en trouverait très sensiblement améliorée, il n'en demeure pas moins que la durée d'amortissement fait hésiter la Profession.

Les nouveaux procédés de filature comme par exemple les procédés OE, core-yarn et fil sans torsion ont été étudiés. Dans certains cas une transformation particulière du matériel est réalisée, par exemple pour le Novacore, dans d'autres cas la fibre de lin et les sous-produits sont adaptés aux nouvelles techniques de transformation. Pour le lin fibres courtes, une filière particulière a été mise au point par la recherche. Ce procédé est aujourd'hui industrialisé dans plusieurs Etats membres.

L'étude et la mise au point de fils nouveaux ont conduit la recherche à adapter des produits et procédés et réaliser des nouveaux produits, tant en bonneterie qu'en tissage, dans des produits aussi variés que le sportwear, l'habillement, l'ameublement et le loisir. Certains de ces produits sont maintenant déjà commercialisés dans la plupart des Etats membres.

LE DOMAINE DE L'ENNOBLISSEMENT DU LIN

M. SOTTON
I.T.F.

Les principales activités développées par quatre centres de recherches dans le cadre du programme EUROLIN sont exposées de manière synoptique dans le tableau ci-dessous. Ces différents travaux ont été entrepris sur plusieurs niveaux complémentaires de recherche de manière à apporter, d'une part, à travers une meilleure connaissance de la matière, des réponses aux questions générales posées par l'ennoblissement du lin, d'autre part des améliorations techniques ou innovations dans les produits et procédés et ceci dans les différents secteurs spécifiques liniers des pays membres.

Ce programme de recherche EUROLIN a permis de combler un déficit de connaissance fondamental sur la fibre "lin". Cet acquis est d'ores et déjà source de progrès et d'innovations. La fibre de lin native apparaît tel un véritable "Kevlar cellulosique", particulièrement anisotrope et cristallisé, peu accessible aux réactifs, particulièrement fragile aux sollications en compression et en cisaillement qui créent de profondes dislocations structurales. La présence de telles dislocations rend compte notamment de la faiblesse des tissus de lin à l'abrasion. Ces événements peuvent être exploités (apprêts mécaniques) et suffisent à justifier les prétraitements alcalins requis pour l'ennoblissement des étoffes et qui, tout en homogénéisant les fibres élémentaires, les rendent plus accessibles aux réactifs.

Activités	Centres de recherche			
	Centex bel	I.T.F.	LIRA	Stat. exper.
1. Meilleure connaissance de la structure intime de la fibre, condition de progrès technologiques		x		
2. Importance des prétraite- ments alcalins pour l'ennoblissement des étoffes	x		x	x
3. Innovations dans l'enno- blissement : - des textiles d'habille- ment				x
- des textiles d'ameuble- ment : — sièges, revêtement mural, linge de maison	x			
— la maîtrise de l'in- flammabilité des sièges rembourrés			x	
4. Potentialités : - des traitements en bains courts (mousses)			x	
- des apprêts mécaniques		x		

Ces prétraitements alcalins ont été étudiés dans tous les pays membres, car il s'est avéré que les industriels transformateurs avaient une connaissance insuffisante de ce traitement et qu'il convenait absolu- ment de considérer, dans cette approche, le lin différemment que le coton. Dans les divers pays, ces travaux ont abouti à l'établissement de recommandations précises à l'usage des professionnels et qui d'ailleurs diffèrent assez sensiblement d'un pays à l'autre selon le type de pro- duits visé (vêtement, chemiserie, ameublement ...).

L'ennoblissement des textiles d'habillement a été étudié en étroite relation avec les entreprises de finissage. Il ressort de ce travail que le prétraitement alcalin, effectué pour l'obtention du toucher et de la stabilité dimensionnelle, s'avère déterminant pour la qualité finale. Il estompe toutes les différences initiales de matières, fils, tissus. Une des retombées techniques majeures, partagée avec la profession, est l'établissement d'un cahier des charges pour les tissus d'habillement en vue d'un étiquetage qualitatif.

Une démarche de qualification analogue a été entreprise pour les tissus en lin destinés à l'ameublement, notamment les tissus de sièges et revêtements de murs.

Des progrès sont à souligner sur l'application des apprêts anti-
tache, recommandation des produits les plus efficaces (fluorocarbonés),
de la technique d'application (cylindres lécheurs plutôt que vaporisage
ou apprêt mousse). En ce qui concerne le tissu pour siège, un compromis
élégant a pu être trouvé entre l'augmentation de résistance à l'abrasion
(apprêts vinylique ou silicone) et les modifications d'aspect et de la
confectionnabilité.

Une importance étude a été entreprise dans le but de produire des
étoffes en lin susceptibles de satisfaire aux exigences de la nouvelle
législation sur la résistance à l'allumage des sièges rembourrés revêtus
de textile et qui touche aussi bien les aspects domestiques que contrac-
tuels. Il a été montré que les applications domestiques peuvent être
couvertes sans problème avec des tissus en lin ou lin-fibre chimique.

Quant aux applications contractuelles, des tissus lin/nylon (80/20)
peuvent convenir à condition qu'ils soient revêtus sur l'envers d'une
couche de matière ignifuge : une expérience industrielle est lancée sur
ce mélange lin/nylon.

Les performances de tissus traités "infroissables" en phase mousse,
sont similaires à celles des tissus traités selon la méthode classique :
de plus, les coûts de séchage sont considérablement réduits.

SOMMAIRE ET CONCLUSIONS PAR LE PRESIDENT DE LA SESSION

M. VAN LANCKER
CENTEXBEL

Suite aux exposés de nombreuses questions ont été posées. La nature des questions permet de dégager plusieurs aspects qui ont attiré l'attention de l'industrie textile (industrie textile du lin et industrie textile en générale) :
- le processus alternatifs de rouissage,
- l'utilisation des nouveaux schémas de fabrication,
- la qualification du lin,
- la structure morphologique du lin.

A la question de savoir combien de temps il faudra avant que les processus de rouissage alternatifs (rouissage chimique et enzymatique) pourront être considérés comme vraiment efficaces, il faut noter que lors de conditions climatiques très défavorables les nouveaux procédés ne sont pas, à l'état actuel, à même de régler complètement le problème. Les procédés de rouissage chimique permettent toutefois d'introduire la culture du lin, là où normalement les conditions sont en moyenne défavorables. Les résultats à ce jour laissent entrevoir que dans un avenir proche il sera possible à partir de procédés de rouissage chimique, de réaliser un rouissage normal et d'obtenir une fibre de lin textile de qualité lors de conditions climatiques défavorables et de tirer profit de ces processus lors de conditions normales

L'utilisation des nouveaux schémas de fabrication a suscité un intérêt tout particulier, principalement le craquage du long brin et le circuit "fibre courte". Il semble que pour le craquage du long brin le nombre de zones de craquage est fortement conditionné par la longueur limite admise et la matière première. Les essais ont indiqué un certain avantage de travailler avec une troisième zone.

Les circuits pour fibres courtes sont généralement proche du circuit traditionnel coton. Toutefois l'utilisation du lin dans cette filière implique une attention toute particulière dans le choix des matériaux à affiner et dans l'élaboration du produit fibre. Le matériel type laine cardé peut sous certaines conditions être utilisé pour le travail de la fibre courte. Les essais industriels sont restés limités.

Pour les secteurs textiles hors liniers, utilisant la fibre de lin principalement en mélange avec d'autres fibres, une qualification objective de la matière semble indispensable. Depuis plusieurs années la recherche s'est efforcée de mettre au point des méthodes d'analyse et de travail afin d'essayer de qualifier les matières et les produits. A ce jour on maîtrise, avec qu'en même quelques réserves, trois points, deux tests de simulation, un qui permet d'apprécier le rendement au peignage, le deuxième permet de déterminer la divisibilité mécanique de la matière peignée et un troisième test qui permet de mesurer la finesse de la fibre par une méthode air-flow. Toutefois on est obligé de reconnaître qu'on ne fait pas aussi bien ce que font les experts (analyse organoleptique). La qualification des produits (fils, tissus, etc.) se fait généralement à partir de tests normalisés. Dans certains cas particuliers des tests spécifiques ou adaptés au lin sont toutefois nécessaires (par exemple mesure de la résistance des fibres, mesure du degré de polymérisation).

Une meilleure connaissance de la structure morphologique de la fibre de lin a suscité un intérêt tout particulier. Deux aspects importants se dégagent, primo la différence entre la réaction de la fibre de lin et de coton pour des traitements alcalins et secundo les propriétés mécaniques de la fibre élémentaire de lin à l'état natif, qui sont similaires aux propriétés du Kevlar.

Lors d'un traitement de la fibre de lin avec un solution de soude (caustification ou mercerisage), il y a des barrières d'accès de par la structure fortement cristalline de la fibre de lin et un ensemble d'artefacts. La pénétration de la soude dans la fibre et au travers de la fibre de lin est très hétérogène en comparaison avec la fibre de coton où l'action de la soude est très homogène. Progressivement toute la fibre de lin sera transformée en cellulose mercerisée. A ce state on trouve les mêmes effets que pour la fibre de coton, augmentation de l'accessibilité pour les produits chimiques et augmentation de la stabilité. Un point important lors du traitement à la soude des fibres de lin est que les articulations et les dislocations dans la structure fine de la fibre vont persister. L'utilisation de l'ammoniaque liquide comme agent de mercerisage conduit à des résultats différents. La transformation de la structure fibreuse est plus lente et moins prononcée comparativement à l'action de la soude.

Les caractéristiques mécaniques particulières de la fibre native pourraient laisser entrevoir son utilisation dans des matériaux composites avec matrice organique ou minérale. La difficulté de proposer du lin en substitution à des fibres de renfort comme par example le Kevlar dans des applications composites, réside dans le fait que la fibre de lin n'aura les propriétés mécaniques intrinsèques similaires à celle du Kevlar qu'à condition de l'extraire dans des conditions différentes de celles qui existent actuellement dans l'industrie du lin. Il faut ménager l'extraction, la faire autrement. Par contre sur le plan thermique on ne pourra proposer une fibre de lin qui aura les mêmes résistances que la fibre de Kevlar. Ceci limitera l'utilisation éventuelle de la fibre de lin à des matériaux composites non thermiques, indiquant un segment d'utilisation bien précis. Ensuite le lin est livré par la nature sous forme de fibres qui font 6 cm au maximum de longueur, alors que d'autres fibres de renforcement comme le Kevlar sont livrées aux utilisateurs sous forme de filaments continus.

A la question de savoir si la fibre de lin peut entrer dans les marchés futurs des matériaux composites, la réponse est, qu'il serait souhaitable dès maintenant et nécessaire dans un avenir proche que la recherche en collaboration avec l'industrie, entreprenne des études bien dirigées dans cette direction.

CONCLUSIONS GENERALES DU SYMPOSIUM

D. FINLAY MAXWELL
Président de la Commission Recherche Scientifique de Comitextil

Au cours de ce symposium, vous avez eu sur les réalisations scientifiques un exposé d'introduction, quatre sessions détaillées, quatre exposés principaux et une récapitulation de ces derniers. Le message ne peut être plus clair. Nous avons réalisé des progrès importants dans chacun de ces domaines. Indiscutablement, nous avons prouvé qu'avec l'aide de la Commission, nous pouvons entreprendre ensemble des recherches fructueuses en collaborant avec des services de recherche complémentaires, et même au-delà des frontières nationales.

Pourquoi des recherches communes ? Pourquoi ne pas essayer de réaliser ces travaux à l'échelle nationale ? L'évolution constante des technologies impose de plus en plus des recherches multi-disciplinaires sur les textiles.

Prenons l'exemple, hypothétique, de la teinture ou du traitement de surface du lin, ou dans ce cas précis, de tout tissu. Il faut étudier l'effet de toute une série de produits chimiques, probablement de traitements à chaud, y compris l'utilisation de la radio-fréquence pour provoquer un gradient thermique à l'intérieur du tissu ou l'utilisation d'un rayonnement hyperfréquence ou infra-rouge pour stimuler la résurgence des additifs appliqués sous forme de mousse. Il est toujours possible d'utiliser des rayons alpha ou bêta ou même des lasers. Nous pouvons aussi recourir aux techniques d'analyse Kawabata pour optimiser ces traitements.

Notre organisation dispose-t-elle d'un membre compétent dans ces domaines ? Si ce n'est pas le cas, comment faut-il mener ce genre de recherches ? Un des atouts de la recherche communautaire est son habilité à rassembler toujours plus de disciplines - chaque groupe spécialisé étant hautement qualifié en la matière.

De tels programmes sont d'autant plus appréciables qu'ils font fréquemment l'objet d'une évaluation et d'une auto-critique constructives. J'ai appris par un ou deux délégués qu'une partie des travaux avait déjà été exécutée ailleurs. Ce fait se reproduira encore, et plus particulièrement pour les programmes nationaux exécutés indépendamment.

Encore une fois, les programmes communautaires jouent un rôle important puisqu'ils permettent d'éliminer ou du moins d'éviter dans une large mesure les répétitions coûteuses.

C'est pourquoi j'envisage des programmes communautaires pour des projets à plus grande échelle. Les programmes nationaux sont peut-être mieux orientés vers le développement d'une compétence particulière au sein de cet ensemble.

Les deux voies de développement peuvent se compléter.

Nous avons discuté des succès mis en évidence au cours de ce symposium. Qu'entendons-nous faire des résultats ?

Premièrement, nous devons nous assurer de la meilleure distribution possible de ces résultats. Dans une première étape de 6 mois, la Commission reproduira un dossier complet des documents et le fera parvenir à tous les délégués. Ensuite, ce sera à nous de redistribuer les informations reçues à un public aussi large que possible par le biais de nos organisations et fédérations.

Sans faillir, nous devons nous efforcer d'assurer la réalisation de la majorité du travail accompli. Toutefois, il reste des travaux inachevés que nous pourrions juger utile de terminer. Dans ce but, j'aimerais voir la création d'une sorte de "programme d'exécution" encouragé par Comitextil.

J'ai pleinement conscience de l'étendue du travail que notre équipe technique devra accomplir mais je tiens à ajouter que ce fait était manifeste depuis quelques temps.

En effet, résidant à Huddersfield, j'ai personnellement considéré ma présence aux réunions une ou deux fois par mois en qualité de président amateur, comme totalement inappropriée pour répondre aux exigences futures. C'est précisément pour cette raison que je me suis récemment retiré pour laisser la place à un successeur qui effectuera les changements nécessaires. Ces changements exigeront une restructuration d'une important partie de notre temps et de nos ressources. Je crois que cette tâche est réalisable et qu'une organisation adéquate peut être mise sur pied pour veiller au développement. Il nous faut des programmes qui répondent aux aspirations de la Communauté, des programmes de la plus haute technicité qui pourront être produits et menés à bien grâce aux ressources disponibles.

RIASSUNTI IN ITALIANO

Fisiologia e confezionamento degli indumenti

Nuove tecnologie di filatura nell'industria laniera

Qualità delle stoffe e degli articoli a maglia

Valorizzazione del lino

Conclusioni generali del Convegno

FISIOLOGIA E CONFEZIONAMENTO DEGLI INDUMENTI

Presidente : Dott. R. JEFFRIES, Shirley Institute

Nel corso della presente riunione verrano illustrati i lavori di ricerca realizzati in due settori ben determinati: da un lato le caratteristiche fisiologiche dell'indumento e il grado di comfort che è in grado di offrire nonchè la relazione fra la qualità del tessuto e la confezione, dall'altro le tecniche di produzione applicate, lo stile e l'aspetto estetico del capo di vestiario. Entrambi i settori erano contemplati dal secondo programma di ricerca per il settore tessile della CEE, dato l'interesse primario che presentano sia per l'industria tessile che per le aziende produttrici e i commercianti del ramo. Il miglioramento del comfort offerto da tessuti e da indumenti rappresenta la preoccupazione costante dei produttori e dei rivenditori, per cui risulta particolarmente auspicabile la realizzazione di progetti che perseguano tale obiettivo. D'altra parte gli sforzi maggiori si stanno orientando verso la realizzazione di capi di vestiario per i quali si punta prevalentemente sulla sobrietà, sulla qualità e sul taglio nonchè sull'automazione della confezione; in entrambi i settori svolgono un ruolo capitale la qualità, la classificazione e il confezionamento dei tessuti e in definitiva degli abiti.

Sono quindi stati presi in esame cinque progetti, la cui realizzazione è stata ispirata dall'intento di fornire risultati e conclusioni che presentassero un interesse immediato e diretto per l'industria tessile e per il settore della confezione dei paesi della CEE e conseguentemente di mettere a disposizione dei medesimi gli strumenti indispensabili ed idonei ad elaborare prodotti di migliore qualità e a mettere a punto processi produttivi in grado di offrire rendimenti più elevati.

La prima relazione fa il punto sulle ricerche condotte nel settore estremamente complesso del comfort offerto dai capi di vestiario e dei fattori che intervengono in tale materia, vale a dire il tipo e le caratteristiche degli abiti e dei tessuti impiegati nella loro confezione. Sono state prese in considerazione due classi di comfort dei capi di abbigliamento: il comfort "termofisiologico" che corrisponde al mantenimento di un equilibrio termico e igrometrico indispensabile a garantire la costanza di una temperatura funzionale accettabile e il comfort "sensoriale", relativo alle proprietà tattili dei vestiti a contatto diretto con l'epidermide e agli effetti che ne risultano.

Tre sono i progetti aventi per tema il comfort termofisiologico, uno dei quali si è prefisso l'obiettivo di fornire parametri precisi di tale tipo di comfort e di elaborare idonee direttive e metodi di valutazione. E' stato possibile definire modelli di comportamento dell'epidermide e sulla

scorta delle misurazioni termiche e igrometriche è stata proposta una
formula empirica, in grado di fissare un "livello di comfort".

L'impiego di tale livello di comfort consente di rinunciare alle prove su
modelli, assai costose. L'obiettivo di un altro progetto consisteva nel
miglioramento del comfort termofisiologico degli indumenti impermeabili.
La ricerca prendeva in esame una gamma molto ampia di tessuti impermeabili e permeabili al vapor acqueo. Sono state prese in particolare considerazione le caratteristiche combinate di permeabilità e di ventilazione.
Grazie a tale progetto si è riusciti a mettere a punto un metodo per la
valutazione degli effetti prodotti dalla ventilazione e ad elaborare una
serie di direttive per il confezionamento di indumenti impermeabili.
L'obiettivo di un terzo progetto consisteva nel prendere in esame gli
effetti di ordine fisiologico (aumento della temperatura corporea e
sudorazione) indotti dagli indumenti di lana e di cotone o confezionati
con tessuti misti lana o cotone.

Il quarto progetto analizza il problema del comfort sensoriale degli indumenti a contatto con la pelle. Inizialmente si è dovuto ricorrere a
prove accurate ed esaustive su modelli, per individuare le sensazioni e i
fattori fisici che determinano il comfort sensoriale degli indumenti a
contatto diretto con l'epidermide. A conclusione di tali prove, in base
ai risultati acquisiti si è riusciti ad elaborare una serie di metodi
sperimentali semplici facenti capo a criteri oggettivi e soggettivi. Tali
metodi consentono di misurare vari parametri, quali la sensazione di ir-
ritazione, di prurito, di solletico, di pressione, di freddo o di umido a
contatto con la pelle, le azioni delle fibre non amalgamate e i fenomeni
di elettrostatica.

Nella seconda relazione viene affrontato il problema dell'individuazione
e della classificazione delle proprietà fisiche dei tessuti nell'ambito
della confezione e delle possibilità di migliorare i processi di fabbricazione e le proprietà estetiche degli indumenti. E' stata esaminata la
resistenza dei tessuti agli sforzi di trazione multidirezionali in fase
di fabbricazione. Tale ricerca coinvolge l'impiego di tecniche di punta
che non sono necessariamente rispondenti alle esigenze produttive delle
aziende del settore.

I produttori hanno preso parte alla fase di verifica dei risultati delle
prove. Nella seconda parte del progetto è stata esaminata l'interazione
tessuto/macchina in fase di produzione, come pure gli effetti dell'adattamento automatico delle macchine alla qualità dei tessuti. Prove sperimentali sono state condotte in due settori specifici: l'alimentazione del
tessuto nelle operazioni di cucitura (per individuare le proprietà dei
tessuti e le caratteristiche delle macchine preposte all'avanzamento del
tessuto) e la pressatura in linea (per determinare le condizioni di equilibrio fra velocità di produzione, parametri di pressatura e dimensioni
degli indumenti).

414

VALUTAZIONE DEL COMFORT OFFERTO DALLE STOFFE E DAGLI INDUMENTI

Relatore : K.H. UMBACH, Hohenstein

1. - PARAMETRI TERMOFISIOLOGICI CHE DETERMINANO IL COMFORT ALL'UTENZA

Allo stadio attuale, il grado di competitività dei capi di vestiario dipende, oltre che dalle caratteristiche di natura meccanica e tecnologica
e dalla facilità di lavaggio e di smacchiatura, anche dal grado di benessere termofisiologico che offrono all'utente. D'altra parte, l'unica possibilità di tener conto delle buone caratteristiche di indossabilità nel
quadro di una programmazione del processo evolutivo dei capi di
vestiario, consiste nella capacità di valutare quantitativamente il
comfort fisiologico. Il progetto in questione mirava precisamente ad
individuare un metodo analitico in tal senso, in grado di essere
impiegato dall'industria in fase di perfezionamento dei prodotti finiti.

Tale metodo analitico, come risulta dalla figura 1 (vedasi pagina 15),
si articola in varie fasi successive. Per valutare le proprietà termo-
fisiologiche dei tessuti che compongono i capi di vestiario, si utilizza
un modello che riproduce il fenomeno di termoregolazione dell'epidermide
umana (modello epidermide), illustrato in figura 2 (vedasi pagina 15).
In base a tale modello è stato possibile mettere a punto vari metodi
sperimentali, che consentono di determinare quantitativamente le
proprietà dei tessuti, ad esempio l'isolamento termico o la trasmissione
e l'assorbimento dell'umidità, sia in condizioni "normali" di utenza sia
in presenza di sudorazione abbondante ed estesa da parte dell'utente, e
nello stesso tempo di esprimere tali proprietà con parametri specifici
dei tessuti utilizzati. Si ottiene per tale via una serie di tabelle,
ricavate da formule previsionali definite nel corso del progetto, che
consentono di valutare il comfort termofisiologico che un determinato
prodotto tessile può offrire. Esistono formule diverse per i vari gruppi
di prodotti, quali la maglieria intima, le camicie, pantaloni ecc. Le
prove effettuate ricorrendo ad indossatori hanno confermato una perfetta
corrispondenza fra il grado di comfort dei capi di vestiario, calcolato
utilizzando tale modello di previsione, con la sensazione di benessere
soggettivo espressa dall'indossatore (vedasi figura 3, pagina 16).

Per il fatto che la valutazione delle proprietà dei tessuti che determinano il comfort di un dato capo di vestiario fa capo a metodi sperimentali razionali, i risultati del progetto di ricerca offrono all'industria tessile la possibilità di programmare oggettivamente il comfort
dell'indumento, quale elemento strutturale, già nella fase preliminare di
produzione di un nuovo articolo. La conferma riposa sul fatto che gli
esperimenti fondamentali realizzati nel corso del progetto hanno consentito l'elaborazione di direttive strutturali, grazie alle quali taluni

capi d'abbigliamento, quali indumenti sportivi, impermeabili, termo-
isolanti, possono essere progettati garantendo proprietà fisiologiche
ottimali.

Nella seconda fase del sistema analitico, illustrato in figura 1, la
ricerca sugli elementi che compongono i capi di vestiario avviene con
l'ausilio di un manichino mobile, che riproduce fedelmente i fenomeni di
termoregolazione del corpo umano (vedasi figura 4, pagina 16).

Tale metodo fornisce parametri termofisiologici specifici (resistenza
alla trasmissione del calore e del vapore) per la struttura completa del
capo d'abbigliamento, compresi gli strati d'aria componenti il microclima
e aderenti alla superficie esterna dell'indumento.

E' stato possibile mettere a punto un modello di previsione per capi di
vestiario grazie ai risultati di centinaia di prove d'indossabilità controllate nella camera climatica (figura 5, vedasi pagina 18), che costituiscono la fase n°3 del sistema d'analisi di figura 1 e nel corso delle
quali sono stati ricavati valori oggettivi per le funzioni corporee e valutazioni soggettive fornite dai partecipanti alle prove.

Ricorrendo a tale modello, che congloba i parametri specifici degli
indumenti misurati su manichino, è possibile definire, in termini
universalmente comprensivi e in perfetta aderenza con i risultati
sperimentali, il comfort offerto dai capi di vestiario indossati, in
corrispondenza di qualsiasi condizione climatica onerosa e di qualsiasi
attività svolta dall'utente.

Per tale via è consentito attualmente non soltanto definire la gamma di
temperatura o di utilizzazione in cui l'utente, in base ad un determinato livello di attività, è convenientemente protetto dal freddo o dalla
sudorazione, e questo grazie ad un numero limitato di prove di
laboratorio, ma è altrettanto possibile calcolare direttamente le
variazioni nel tempo delle funzioni del corpo e dell'umidità nell'ambito
del microclima, in fase di utilizzazione del capo di vestiario, in
funzione di determinate condizioni climatiche esterne e del tipo di
attività svolto.

Le ricerche svolte nell'ambito di tale progetto hanno confermato che i
valori ricavati per le funzioni corporee e il microclima influenzano
direttamente le sensazioni di comfort termofisiologico della persona.
Anche soltanto limitatamente ai capi di vestiario confezionati con stoffe
è stato possibile elaborare formule biofisiche, in grado di fornire
direttamente il grado di comfort all'utenza, valido per tutta la gamma
dei capi di vestiario. Nell'industria dell'abbigliamento, e per la fase
di progettazione dei futuri indumenti, si è riusciti ad eliminare quasi
del tutto la necessità di ricorrere alle prove di indossabilità, lunghe e
costose (fasi n° 4 e 5 del metodo di analisi).

A conferma dei risultati offerti dal progetto di ricerca, attualmente tali prove sono sostituite, sulla base di un numero limitato di ricerche di
laboratorio, dalla semplice valutazione degli effetti fisiologici prodotti dai capi di vestiario.

2. - COMFORT ALL'UTENZA IN BASE ALLA SENSAZIONE EPITELIALE

L'obiettivo di tale progetto consisteva nel determinare i parametri
strutturali specifici che concorrono a definire il comfort all'utenza in
base al contatto dei capi di vestiario con la pelle. Si è proceduto alla
messa a punto di vari test di laboratorio, di semplice realizzazione, che
consentono di individuare i vari aspetti del comfort legato alle sensazioni provate dal contatto con la pelle. In base ai risultati forniti da
una ricerca bibliografica, da un'indagine condotta su un campione di
1.000 persone interrogate circa le loro abitudini in materia d'abbigliamento e da una serie di prove d'utenza con l'intervento di indossatori
con 22 tipi diversi di capi d'abbigliamento consistenti in T-shirts, è
risultato che i parametri che determinano il comfort all'utenza in funzione delle sensazioni cutanee sono:

2.1. Aderenza localizzata

La pressione esercitata da indumenti eccessivamente a contatto con
la pelle può mascherare le altre sensazioni. E' indispensabile verificare quali siano le pressioni massime tollerabili nelle parti sensibili del corpo. Sono stati messi a punto alcuni apparecchi sperimentali, che consentono di determinare quantitativamente la pressione esercitata dagli indumenti a contatto con il corpo.

2.2. Irritazioni localizzate imputabili alle etichette degli indumenti

Sensazioni di fastidio particolarmente elevate possono essere indotte da ritagli o dagli spigoli delle etichette, specialmente del tipo
autocollante.

2.3. Sensazione di pizzicore

I tessuti di lana sono considerati suscettibili di provocare "pizzi-
core" quando il 3,5% delle fibre hanno un diametro superiore a 30 μm
o lo 0,6% delle fibre hanno un diametro superiore a 40 μm.

2.4. Sensazione di solletico

I tessuti a superficie non rasa provocano sensazioni di sollecito
sgradevoli, in particolare quando la pelle è umida per il sudore.
Fotografando le punte delle fibre in rilievo sulla superficie del
tessuto si riesce a determinare la predisposizione di un tessuto a
provocare solletico.

2.5. Perdita di fibre

Le fibre che si staccano dal tessuto dell'abito e le fibre che aderiscono alla pelle provocano sensazioni di disagio. Con l'ausilio di un apparecchio in cui l'articolo viene sottoposto ad una serie di scosse, è possibile determinare quantitativamente la tendenza di determinate fibre a staccarsi dal tessuto, per mezzo di un "indice di perdita di fibre" comparato a campioni fotografici.

2.6. Tendenza ad aderire alla pelle (per tessuti umidi)

Come risulta dalla figura 8 (vedasi pagina 21), è stato realizzato un apparecchio sperimentale, nel quale la forza necessaria a staccare una striscia di tessuto da un cilindro in plexiglas riproduce la sensazione sgradevole prodotta dai tessuti aderenti alla pelle. Le prove con indossatori hanno confermato una correlazione accettabile fra gli indici di aderenza sperimentali e le valutazioni soggetive degli indossatori.

2.7. Sensazione iniziale di freddo

Anche se la durata della sensazione di freddo, che l'utente prova nella fase iniziale di contatto con gli indumenti che indossa, si limita a pochi secondi, tale sensazione rischia di influenzare sensibilmente e permanentemente la sua valutazione soggettiva circa il grado di comfort di tali capi. Esiste un apparecchio in grado di determinare quantitativamente tale sensazione di freddo iniziale prodotta dai tessuti.

2.8. Sensazione di prurito

Con la misurazione dell'attrito che si produce a contatto fra un tessuto e un soggetto sperimentale, si è in grado di valutare il grado di predisposizione di tale tessuto ad indurre sensazioni di prurito.

2.9. Fenomeni di elettricità statica

Sensazioni cutanee fastidiose sono spesso originate da fenomeni di scariche elettrostatiche, quando una persona si spoglia o compie determinati movimenti del corpo. La misurazione a determinate condizioni dell'intensità di tale scarica che si produce quando ci si spoglia da determinati indumenti, consente di valutare il grado di comfort offerto da tali indumenti.

Riassumendo, il progetto di ricerca ha consentito non soltanto di mettere a punto metodi analitici che consentono di valutare il grado di comfort sensoriale cutaneo per gli indumenti indossati, ma altresì, grazie ai ri-

sultati fondamentali acquisiti, di elaborare direttive tali da fornire
all'industria, con un netto progresso rispetto al passato, elementi utili
alla produzione su basi oggettive di indumenti da indossare a contatto
con la pelle, che offrono un elevato comfort per quanto riguarda le sensazioni a livello epiteliale.

3. - MIGLIORAMENTO DEL GRADO DI COMFORT OFFERTO DAGLI INDUMENTI PROTET-
TIVI

La funzione della maggior parte degli indumenti protettivi consiste nel
difendere l'utente dagli effetti delle intemperie. A livello fisiologico,
il problema da risolvere per queste categorie d'indumenti si identifica
nella necessità di garantire contemporaneamente l'impermeabilizzazione e
una buona trasmissione dell'umidità. Mancando quest'ultima proprietà, il
comfort offerto da tali indumenti protettivi risulta alquanto pregiudica-
to.

Per poter valutare criticamente, in base alle proprietà fisiologiche offerte, i tessuti impermeabili attualmente disponibili, inizialmente si è
fatto ricorso a metodi sperimentali specifici, in grado di determinare,
in corrispondenza di determinate condizioni reali, il grado d'impermeabilizzazione nonchè la permeabilità dei tessuti nei confronti dell'aria e
dell'acqua.

Sulla scorta di una serie di esperimenti realizzati in applicazione di
tali metodi sperimentali è stato confermato che determinati tessuti stra-
tificati con membrana integrante, speciali rivestimenti microporosi e
tessuti a microfibre, sono particolarmente idonei a soddisfare sia le e-
sigenze tecnologiche che i requisiti fisiologici richiesti. Grazie alla
combinazione dei vari processi produttivi, si è rivelata la possibilità
di ottenere proprietà nettamente superiori a quelle offerte dai prodotti
attualmente disponibili. Se inoltre la produzione appropriata di tali
tessuti è integrata da un'idonea confezione che consenta convezione e
ventilazione, nell'ambito del microclima realizzato dal movimento corpo-
reo della persona che indossa l'indumento o dai movimenti dell'aria
ambiente, si è in grado di rendere ottimali le proprietà degli indumenti
protettivi contro l'azione delle intemperie. La possibilità di realizzare
tale idea in pratica è stata confermata nel corso del progetto di
ricerca, grazie ad una serie di prove pratiche con indossatori, nel corso
delle quali è stato messo a punto un metodo che consente la
determinazione quantitativa del tasso di ventilazione negli indumenti. I
risultati delle prove sono inoltre sfociati in un modello matematico, che
esprime i processi di trasmissione del calore e dell'umidità nell'
indumento, per determinare la permeabilità all'umidità, unitamente alla
ventilazione studiata in fase di confezione e i relativi effetti sul
comfort offerto dagli indumenti protettivi in determinate condizioni
climatiche.

**CARATTERISTICHE OTTIMALI DEGLI INDUMENTI
IN FASE DI IDEAZIONE E DI PRODUZIONE**

Relatore : J. DESCHAMPS, CETIH

La parte del programma esaminata in questa sede ha per oggetto lo studio delle caratteristiche fisiche dei tessuti in fase d'ideazione dei modelli di capi di vestiario, per favorire:

- la migliore realizzazione di modelli degli articoli, eliminando al massimo le modifiche necessarie,
- processi di produzione semplificati
- realizzare indumenti di taglio perfetto ed esteticamente validi.

**1. - CLASSIFICAZIONE DEI TESSUTI IN FUNZIONE DELLE RISPETTIVE PROPRIETA'
FISICHE**

I sistemi di classificazione dei tessuti più frequentemente utilizzati ancora attualmente fanno riferimento a determinate categorie d'utilizzazione. Sostanzialmente sono di natura soggettiva e direttamente legati alle informazioni visive e tattili acquisite dagli informatori quando esaminano e palpano il tessuto, anche se talvolta si ricorre a sistemi che utilizzano parametri elementari misurabili, quali il peso del tessuto.

Di fronte allo sviluppo dell'ammodernamento dell'industria della confezione e dell'automazione dei processi produttivi e dell'impiego a breve e a lungo termine dell'informatica in fase di disegno e di produzione con l'ausilio dei computer, risulta indispensabile introdurre una classificazione maggiormente basata su criteri oggettivi, in funzione di parametri misurabili.

Per facilitarne l'applicazione da parte della maggior parte delle industrie del settore, il sistema di classificazione obiettiva adottato deve limitarsi a prendere in considerazione un numero ridotto di parametri misurabili; la scelta ottimale potrebbe consistere nella determinazione di un metodo di classificazione che definisca classi precise di tessuti in funzione di una o di alcune proprietà opportunamente selezionate.

Si è proceduto ad un'indagine sulle proprietà dei tessuti che attualmente costituiscono per le industrie criteri più o meno intuitivi adottati in fase di progetto e di produzione di capi d'abbigliamento, e che sarebbe utile approfondire in modo sistematico.

Sono state studiate le proprietà più significative, che facilitano il confezionamento dei prodotti finiti e favoriscono la realizzazione di indumenti che più si avvicinano alle esigenze degli ideatori, per quanto riguarda la qualità, il comfort e l'estetica.

Per qualsiasi tipo di abiti e di tessuti attualmente in commercio è stata definita la massima gamma di variazioni di tali proprietà.

Per determinare le variazioni realmente significative, sono state esaminate le modalità di variazione di ciascuna di tali proprietà, riferite ad uno stesso tipo di tessuto, ricavato da un'unica pezza o da due pezze diverse.

Si è constatata la necessità di identificare le deformazioni che si verificano nella struttura dei tessuti in fase di trasformazione in capi di vestiario.

2. - IDENTIFICAZIONE DELLE SOLLECITAZIONI E DELLE DEFORMAZIONI A CUI SONO SOGGETTI I TESSUTI IN FASE DI TRASFORMAZIONE IN CAPI DI ABBIGLIAMENTO

Con l'ausilio di documenti fotografici delle modifiche subite dalla struttura dei tessuti in fase di cucitura dei vari elementi che compongono un indumento (cuciture laterali, di spalla, carré, giunture maniche) e delle riprese, si è riusciti ad evidenziare le proprietà delle stoffe che possono maggiormente influenzare la "confezionabilità" dei tessuti.

Le proprietà che determinano la facilità di taglio e di deformabilità in tutte le direzioni sembrano essere le più importanti in tale contesto, per cui si rivela necessario realizzare metodi di misurazione semplici, che consentano ai tecnici del settore la possibilità di sfruttare pienamente tali proprietà.

3. - RICHERCHE ATTE A FORNIRE ALL'INDUSTRIA DEL SETTORE METODI DI MISURAZIONE DELLA RESISTENZA AL TAGLIO E ALLE DEFORMAZIONI MULTIDIREZIONALI DEI TESSUTI

Dal momento che le proprietà di taglio e di deformazione multidirezionale dei tessuti sono sembrate le più importanti fra quelle che entrano in gioco in fase di confezionamento degli indumenti, si è ritenuto utile proporre un metodo di misurazione globale di tali parametri.

Sono stati messi a punto metodi sperimentali in grado di realizzare tali misurazioni.

Si è realizzata una verifica "in situ" delle caratteristiche dei tessuti descritte, onde determinare l'influenza in fase di ideazione e di produzione alle condizioni reali dell'industria della confezione.

4 . – VALUTAZIONE E VERIFICA "IN SITU" DELL'INFLUENZA DELLE PROPRIETA'
   DEI TESSUTI, CLASSIFICATE IN ORDINE D'IMPORTANZA

Onde verificare la validità delle ipotesi formulate, sono stati realizzati da aziende di confezione, con i loro mezzi normali di produzione e nel
rispetto di un apposito capitolato d'oneri e a partire di tipi di tessuto
imposti, alcuni modelli di capi di vestiario (gonne, abiti, soprabiti,
giacche).

I risultati ottenuti sono stati sottoposti all'esame di tecnici della
confezione e di rappresentanti dei consumatori (parametri osservati: rendimento, comfort e valore estetico).

5. – CONCLUSIONE

La presente ricerca costituisce unicamente una fase preliminare nella ricerca di soluzioni al problema dell'influenza delle proprietà fisiche dei
tessuti sulla fase di progettazione dei capi di vestiario.

Anche se attualmente non si è ancora in grado di realizzare interamente
il modello ideale che risponda perfettamente alle esigenze sia di ordine
estetico sia del comfort, si potranno fare decisivi passi avanti in tal
senso qualora si riesca ad assicurare la regolarità delle caratteristiche
dei tessuti forniti all'industria della confezione.

422

SINTESI E CONCLUSIONI DEL PRESIDENTE DELLA SESSIONE

R. JEFFRIES, Shirley Institute

La sessione sulla fisiologia e struttura degli indumenti era suddivisa in
due parti:
- valutazione della comodità dei tessuti e degli indumenti,
- ottimizzazione del disegno e della concezione degli indumenti.

Una prima conclusione che si può trarre dai lavori è che la scelta dei
due argomenti era indovinata: si tratta infatti di due aspetti che hanno
entrambi grande rilevanza per le esigenze presenti e future dell'industria del tessile e dell'abbigliamento, se si vuole che essa conservi anche per l'avvenire le proprie capacità d'innovazione e di competitività.

La comodità dei tessuti e delle confezioni, considerata nella pluralità
dei suoi aspetti, si riduce essenzialmente a un problema di funzionalità,
il quale sia per i produttori che per gli utenti di stoffe e di indumenti
presenta già attualmente un grande interesse, destinato a crescere ancora
con l'andar del tempo, per tutta una serie di motivi: primo, di pari
passo con le esigenze in fatto di proprietà "prottetive" degli abiti si
affineranno anche le richieste in fatto di comodità; secondo, dai consumatori proviene una domanda sempre più spiccata di indumenti sportivi
pratici e comodi, soprattutto per le attività e le situazioni in cui si
produce una sudorazione profusa; terzo, anche per gli abiti normali di
ogni giorno esiste la tendenza ad esigere caratteristiche di maggiore comodità.

Nella sua relazione riguardante l'attività di ricerca sulla comodità degli indumenti, il dott. Umbach ha riferito sulle ricerche intraprese
presso l'Istituto di Hohenstein, l'Istituto Shirley e l'Istituto per gli
studi sulla percezione della TNO rispettivamente nei settori del conforto
termofisiologico, del conforto sensoriale e del conforto degli impermeabili con tessuto semipermeabile ("capace di respirare"): dall'attuazione
del progetto si è ricavato un solido fondamento di conoscenze cui l'industria potrà attingere per l'ideazione di abiti di maggiore comodità per
tutta una serie di situazioni, senza bisogno di ricorrere a lunghe e di-
spendiose prove di resistenza in condizioni reali.

La relazione arriva tuttavia alla conclusione che, nonostante i risultati
eccellenti sinora conseguiti, molto resta ancora da fare in questo
settore. Anzitutto c'è bisogno di più estese conoscenze scientifiche di
base sui vari aspetti della comodità del vestiario: parte del lavoro di
ricerca da effettuarsi in questa direzione dovrà mirare ad approfondire
la comprensione dei lati fisiologici, psicologici e para-medici della
problematica; per citare un solo esempio, sebbene si possiedano conoscen-

423

ze abbastanza estese sulla soglia dolorifica della cute, poco invece si sa sulla sensibilità cutanea al contatto leggero con le stoffe ed i tessuti degli indumenti indossati sul corpo. Un grande bisogno di approfondimento sussiste anche per il concetto di conforto psicologico, a proposito del quale sarebbe lecito porsi ad esempio la domanda seguente: per comodità si deve intendere semplicemente l'assenza di sensazioni di disagio? O non sarebbe ammissibile un'interpretazione "in positivo" del concetto di comodità, che comporti ad es. la creazione di tessuti e indumenti capaci di dare un senso "attivo" di benessere? Qualcuno dirà forse che si tratta di un'idea peregrina, ma anche qui sarà il futuro a dire l'ultima parola. In secondo luogo, occorrerà mettere a punto nuovi materiali dotati di proprietà ancora più spinte sotto l'aspetto della comodità, cioè stoffe e tessuti che, come ho detto prima, contribuiscano a creare una sensazione positiva di conforto, piuttosto che accontentarsi di stoffe che semplicemente non provochino sensazioni di disagio. Terzo: bisogna pensare a nuove forme di indumenti, che consentano di sfruttare appieno le potenzialità e le proprietà dei nuovi materiali "a conforto positivo"; ed inoltre sarà forse anche opportuno cercare nuovi metodi di ventilazione del vestiario, nell'intento di esaltare al massimo le proprietà di conforto globale d'un indumento.

Alla relazione del dott. Umbach è seguita un'interessante discussione, nel corso della quale è stata riconosciuta la grande importanza che il lavoro di ricerca sulle sensazioni di conforto rappresenta per l'insieme dell'industria e dell'abbigliamento, la quale, grazie ai dati raccolti con l'attuazione del progetto, sarà in grado di predire con maggiore sicurezza il comportamento degli indumenti sotto l'aspetto del conforto. Il dott. Manni ha riferito in particolare sui lavori di ricerca appoggiati dalla Lanerossi, nel corso dei quali si è cercato di determinare, mediante la misurazione di parametri fisiologici, il grado di conforto di tessuti di lana in confronto ad altri misti di lana, e di cotone in confronto al lino: grande interesse hanno suscitato le osservazioni sulle caratteristiche tecnico-scientifiche degli indumenti protettivi che consentono il passaggio del vapore acqueo, ma non dell'acqua liquida. In risposta ad una richiesta di chiarimento, il dott. Umbach ha precisato che, sebbene nell'ambito di altre attività di ricerca presso l'Istituto di Hohenstein si siano studiati anche gli effetti che il finissaggio delle stoffe ha sulle sensazioni di conforto, di questo lavoro non ha parlato al Simposio, perché esso non rientra a stretto rigor di termini nel progetto specifico (per quanto riguarda il finissaggio, comunque, conta molto la sua natura, cioè se esso è idrofilizzante o idrofobizzante). Particolare interesse, nel corso della discussione, è stato dedicato alle tecniche utilizzate per il rilevamento obiettivo dei fattori di conforto, le quali costituiscono di fatto il punto chiave dell'intera problematica.

Nella seconda parte della sessione il signor Deschamps ha riferito sul lavoro di ricerca svolto dal CETIH e, in subappalto, dal British Clothing Centre sull'ottimizzazione del disegno e della confezione del vestiario.

Sull'argomento in discussione si è arrivati a due conclusioni d'interesse. Primo: si è convenuto che si tratta di un punto di grande rilevanza per il futuro del settore del tessile e dell'abbigliamento, dato che per garantire l'efficienza e la competitività dell'industria occorrerà approfondire la comprensione del rapporto fra le proprietà di stoffe e tessuti e la tipologia e le modalità di confezione dei vestiti; ma una volta che si saranno acquisite migliori conoscenze sulle connessioni fra tessuti e vestiario, si potranno modificare o almeno integrare le tecniche attualmente in uso per il disegno e la produzione dell'abbigliamento, le quali, basate in gran parte sull'abilità personale e sull'esperienza maturata in anni di lavoro, si sono dimostrate finora soddisfacenti, ma non basteranno certamente più da sole in futuro. Grazie alle nuove informazioni che si acquisiranno sul rapporto tessuti/vestiario si potrà non solo migliorare la qualità e potenziare la produttività conseguibili con gli attuali metodi di produzione, ma anche imboccare con decisione la strada dell'automazione e dell'impiego diffuso della robotica pure nel settore dell'abbigliamento. Il lavoro di ricerca sui cui il sig. Deschamps ha riferito nella sua relazione riguarda appunto lo studio delle proprietà fisiche delle stoffe e dei tessuti in rapporto all'ottimizzazione della tipologia dei vestiti. L'intento è duplice: da una parte, creare capi di vestiario che siano comodi ed efficaci e che nello stesso tempo presentino buone qualità stilistiche ed estetiche, e dall'altra trasformare il processo di produzione in maniera da consentire una più agevole penetrazione dell'automazione e della robotica nel settore dell'abbigliamento (il che implica, come presupposto di base, una comprensione più approfondita delle interazioni fra le stoffe e le macchine di confezione).

La seconda conclusione che si può trarre dalla relazione del sig. Deschamps è che il lavoro di ricerca sull'interfaccia tessuti/vestiario si trova ancora in una fase iniziale: bisogna infatti precisare che, benché le indagini sinora svolte abbiano apportato utilissimi contributi alla conoscenza di quest'aspetto di grande complessità e difficoltà, moltissimo resta tuttavia da fare se si intende costituire la solida base di conoscenze di cui vi sarà bisogno in futuro. Per una buona riuscita delle ricerche in questo difficilissimo settore ci sarà bisogno della piena collaborazione degli scienziati, dei tecnologi e degli industriali: incoraggiante e benaugurante è a questo proposito la collaborazione che il sig. Deschamps ha ricevuto dall'industria per l'esecuzione delle proprie ricerche, e che si spera possa continuare ed anzi approfondirsi in avvenire.

I due aspetti trattati nella sessione dedicata alla fisiologia e alla concezione degli indumenti rivestono un'importanza essenziale per l'industria comunitaria del tessile e dell'abbigliamento, dato che dai progressi che si realizzeranno in questi due settori dipenderà la sua capacità di conservare il proprio potenziale innovativo e la propria competitività.

NUOVE TECNOLOGIE DI FILATURA NELL'INDUSTRIA LANIERA

Presidente : Dr. M. BONA, Città Degli Studi

Ormai da parecchi anni, la filatura del cotone è stata interessata da un' importante innovazione tecnologica, grazie alla messa a punto di tecniche non convenzionali in parziale sostituzione del classico sistema a ring; le principali fra queste tecniche rientrano nelle categoria degli open end, inizialmente del tipo a rotore ma recentemente anche - in prospettiva - ad attrito o a getto d'aria.

E' naturale che tali importanti sviluppi abbiano prima di tutto interessato il settore del cotone (e di alcune fibre sintetiche), sia per obiettive ragioni tecniche che rendono più agevole il trattamento di tali fibre (più corte e più resistenti rispetto alla lana), sia perchè gli investimenti di ricerca e sviluppo dei costruttori meccanotessili si sono innanzi tutto rivolti ad un mercato di potenziale sostituzione che è almeno dieci volte più grande di quello laniero.

Per favorire l'applicazione di nuove tecnologie di filatura anche per la lana, al fine di ridurre o contenere i crescenti costi di fabbricazione, era quindi necessario uno sforzo di ricerca che, partendo dalle esigenze reali dell'industria e con il supporto della medesima, studiasse il comportamento della fibra sottoposta ai nuovi processi non convenzionali, e precisasse le condizioni meccaniche e tessili per una loro eventuale applicazione in condizioni ottimali. Questa ricerca, che ha coinvolto 8 Istituti appartenenti a 5 Paesi della CEE, era già cominciata per iniziativa di alcuni di essi, ma è stata potenziata in maniera decisiva, e soprattutto integrata, grazie al lancio del Secondo Programma Comunitario di Ricerca tessile ed ai relativi finanziamenti.

Inizialmente, il centro del problema consisteva ovviamente nell'esecuzione di sperimentazioni sui diversi modelli di open end a rotore disponibili sul mercato, non trascurando tuttavia altre soluzioni non convenzionali, quali la filatura ad attrito e quella a fuso cavo.

Sotto questo primo aspetto, il programma di ricerca comunitario ha fornito numerose indicazioni di grande interesse (sia per il pettinato che per il cardato), la cui importanza pratica è accentuata dal fatto che, grazie alla collaborazione fra più istituti, è stato possibile sottoporre a prova molti tipi di macchine.

Progressivamente, tuttavia, ci si è resi conto che la natura tecnica del problema si spostava a monte della filatura propriamente detta, investendo la preparazione iniziale delle fibre e la cardatura.

Per quanto riguarda la preparazione, il programma ha permesso di
suggerire interessanti modifiche ai processi convenzionali in uso per le
fibre vergini, come, ad esempio, una depurazione meccanica che può
accompagnare, semplificandolo, il classico carbonizzo; particolare attenzione è stata naturalmente riservata all'impiego di materie prime di
recupero, che è di grande importanza economica per il settore del cardato
europeo e presenta i problemi tecnici più difficili in vista delle
lavorazioni ulteriori.

Nella prospettiva di un futuro impiego delle nuove tecnologie di
filatura, sia partendo da materie prime vergini che soprattutto di
recupero, uno dei risultati di maggiore importanza del programma è senza
dubbio costituito dall'ideazione, dalla messa a punto a livello di prototipo e dalla sperimentazione pratica di carde di nuova concezione, al
fine di produrre un nastro al tempo stesso omogeneo e ben pulito ed
aperto, da alimentare all'open end.

E' auspicabile che i costruttori, che operano nel settore sia dei filatoi
che delle carde convenzionali, prendano gli opportuni contatti con gli I-
stituti di Ricerca, al fine di sviluppare macchine industriali basate sui
nuovi concetti, che potrebbero del resto condurre anche ad
un'interessante evoluzione nel campo della filatura convenzionale.

Un ultimo aspetto del programma di ricerca ha riguardato l'esame del comportamento fisico-chimico della lana durante la lavorazione, condizione
indispensabile per poter utilizzare in maniera ottimale le nuove tecnologie, evitando taluni inconvenienti quali la formazione eccessiva di
polveri e depositi nei rotori, che ne impedirebbero lo sfruttamento
economico:

Fra le caratteristiche della lana che la rendono più difficilmente
filabile con le nuove tecnologie (rispetto ad es. al cotone), deve infatti essere segnalata la sua maggiore fragilità e la necessità di impiegare
additivi che devono essere attentamente scelti e dosati se si vuole
filare con l'open end.

In particolare, quindi, va considerato quale importante risultato del
programma di ricerca la messa a punto di metodi di laboratorio che
potranno servire in pratica per il controllo del processo di fabbricazione, come pure per la selezione delle lane più adatte ad essere
impiegate con le nuove tecnologie.

Le tre relazioni tecniche che seguono forniranno agli ascoltatori i
necessari dettagli, ulteriormente approfondibili in sede di discussione,
sui tre principali temi che sono stati citati: preparazione, filature,
controlli.

Come conclusione generale, riteniamo di dover sottolineare che, grazie all'eccellente spirito di collaborazione dimostato dagli Istituti che hanno preso parte al programma comunitario, si sono realmente poste le basi per una decisiva evoluzione della tecnologia di filature anche nel settore laniero; anche se, naturalmente, alcuni importanti problemi restano da risolvere, come il trasferimento delle nuove idee in macchine e processi industriali (il che richiede la collaborazione fra Istituti ed Aziende meccanotessili).

Un tema importante per ricerche future è infine costituito da uno studio accurato dell'impiego dei nuovi filati, la cui struttura è comunque, in maggiore o minore misura, diversa da quella tradizionale: ciò comporta infatti modifiche alle lavorazioni successive di tessitura e finissaggio e d'adattamento al mercato dei nuovi prodotti.

PREPARAZIONE DEI MATERIALI RECUPERATI PER LA FILATURA
DI TIPO NON CONVENZIONALE

Relatore : Dott. P. ARTZT, Istituto Tecnica della Filatura

Rispetto alla fibra di cotone, la fibra di lana è una materia prima molto costosa. Nel calcolo del costo di fabbricazione dei filati, il costo delle materie prime costituisce l'elemento predominante. Ne consegue che qualsiasi aumento del rendimento della materia prima si ripercuote positivamente, con legge esponenziale, sul calcolo dei costi. E' quindi compito della ricerca individuare quei processi che siano in grado di preparare adeguatamente i cascami di lana, per poterli trasformare in fibre utilizzabili con il massimo rendimento.

Gli istituti di ricerca CELAC (Belgio), WIRA (Regno Unito) e ITF (Germania) hanno cooperato alla realizzazione di ricerche il cui obiettivo consisteva nell'utilizzazione economica delle fibre e nella semplificazione dei processi di trattamento.

I cascami più importanti ricavati dalla trasformazione della lana sono:

1. I cascami dei nastri della carda:
   sono essenzialmente formati da fibre lunghe, come risulta dal diagramma di figura 1 (vedasi pagina 88) per la lana in fiocchi.
   La percentuale di impurità presenti può raggiungere l'80%.

2. Le pettinacce:
   le lunghezze delle fibre sono distribuite come indicato in figura 2 (vedasi pagina 88), distribuzione simile a quella del cotone. Il tenore di sostanze vegetali può raggiungere il 20%. Le particelle vegetali sono strettamente incorporate alle fibre.

Attualmente questi cascami sono sottoposti in generale ad un processo di carbonizzazione, nel quale si verifica l'ossidazione delle impurità a base di cellulosa.

La carbonizzazione offre i seguenti vantaggi:

1) un elevato grado di depurazione,
2) un elevato rendimento;

a cui si accompagnano per contro i seguenti svantaggi:

1) l'impianto richiede elevati investimenti,
2) notevole consumo d'energia,

3) utilizzazione di grandi quantità di prodotti chimici,
4) rischi d'inquinamento dell'ambiente,
5) controlli sofisticati di un processo complesso,
6) effetti negativi sulle proprietà delle fibre di lana (fragilità),
7) percentuale elevate di fibre corte nella lana carbonizzata.

I processi meccanici di depurazione della lana a fiocchi offrono un rendimento di circa il 50%. Come illustrato in figura 3 (vedasi pagina 90), alcuni processi utilizzati per la depurazione del cotone sono stati opportunamente modificati e adattati alla lana.

In linea di massima un rendimento del 50% nella depurazione della lana in fiocchi con un tenore d'impurità normale è più che sufficiente, dal momento che le fibre sono sottoposte ad ulteriori processi di depurazione, in fase di cardatura e di filatura a rotore. Esiste la possibilità di utilizzare le fibre di lana depurate in misti con fibre sintetiche o cotone.

Una varietà molto interessante risulta combinando:

1) un processo di depurazione meccanica preventiva seguito da
2) una fase di carbonizzazione di breve durata

La durata del processo reale di carbonizzazione delle piccole particelle residue d'impurità può essere ridotta a seconda dei casi, anche per rendere meno fragili le fibre. Con il processo di depurazione meccanica preventiva si ottiene un velo di carda privo di nodi.

Le fibre depurate sono state incorporate a fibre di poliestere, acrilico e cotone e trasformate in filati sull'impianto ad attrito (Dref 2) e a rotore.

Dalle prove è risultato che, utilizzando un'anima a filo continuo, la produzione di filati con impianti di filatura ad attrito offre taluni vantaggi.

Sulle macchine a turbina, si sono ottenuti filati rotor da 30 tex a partire dal 50% di pettinacce di lana depurate.

Grazie alla lunghezza delle fibre, i cascami dei nastri di carda sono quelli che offrono il miglior rendimento. E' inoltre possibile ottenere nella fibra depurata lunghezze di fibra superiori, poichè con una depurazione meccanica periodica si riesce ad eliminare una percentuale superiore di fibre corte.

Nel corso di prove comparative fra il sistema di filatura su ring e il sistema wrap a partire da cascami di carda carbonizzati, rendimenti nettamente superiori e filati con migliori caratteristiche sono stati ottenuti a partire da fibre depurate con processi meccanici. Le fibre depurate meccanicamente rappresentano pertanto per la filatura materia prima di qualità superiore, poichè tali fibre sono trasformabili in prodotti diversi da quelli ottenuti a partire dalle fibre sottoposte a carbonizzazione.

Per la produzione di filati ordinari destinati alla fabbricazione di coperte, tessuti d'arredamento e tappeti, la fase di stiratura a monte della filatura rotor può essere soppressa. E' opportuno far ricorso ad una carda per lana a due tamburi con uno sgrossatore.

All'uscita della carda il velo si suddivide e si ottengono nastri del peso unitario adeguato per essere introdotti direttamente nei filatoi rotor. Il tenore di sostanze grasse residue nella lana va limitato allo 0,5% circa. In tal caso non è necessario ricorrere ad una nuova lubrificazione, onde evitare difficoltà dovute alla presenza di residui sugli elementi della macchina.

Molteplici sono le possibilità d'impiego degli scarti di confezione. Le maggiori difficoltà si incontrano a livello di cernita di tali scarti in funzione del colore e della composizione delle fibre. Una ricerca avente per tema vari aspetti del problema ha fornito interessanti informazioni sugli attuali settori di utilizzazione, sui paesi d'origine e i processi di trasformazione di tali fibre.

I progetti di ricerca realizzati in cooperazione dagli istituti CELAC, ITF Denkendorf e WIRA hanno fornito risultati complementari e indicano vie nuove nel settore di preparazione degli scarti, in vista di più redditizie riutilizzazioni.

NUOVI PROCESSI PER LA CARDATURA E LA FILATURA

Relatore : J.P. BRUGGEMAN, I.T.F. - Nord

Com'è noto, la cardatura trasforma le materie prime, opportunamente condizionate attraverso i processi illustrati dai nostri colleghi, in un nastro iniziale che deve presentare caratteristiche ottimali per quanto riguarda la purezza (assenza di impurità e di sostanze vegetali), la regolarità del titolo e l'omogeneità della composizione.

Nel processo di filatura a partire dal cardato tradizionale, la carda fornisce un velo suddiviso in stoppini che alimentano il filatoio continuo. Il processo open-end ha indotto la soppressione del divisore della carda continua, sostituendolo con un'uscita a 2 o 4 nastri, che successivamente vengono filati direttamente sulla macchina a fibre liberate. Tale processo, molto interessante per i fili per tappeti, è praticamente inapplicabile per i fili più fini, a causa della mancanza di un qualsiasi sistema di calibrature del titolo dei vari nastri e per le fibre di uno stesso nastro.

La ricerca e l'industria hanno quindi optato per le carde tipo cotone, che forniscono un nastro il cui titolo consente l'utilizzazione sui filatoi open-end.

Nonostante i risultati incoraggianti ottenuti, si sono evidenziate talune anomalie, fra cui:

- la presenza di un tenore troppo elevato di impurità vegetali,
- un mediocre parallelismo delle fibre,
- la presenza di un numero eccessivo di nodi.

Per ovviare a tali lacune, i laboratori del CELAC in Belgio, di TECNOTESSILE in Italia, della WIRA in Gran Bretagna hanno sollecitato la cooperazione dei costruttori e degli industriali del settore, con l'obiettivo di mettere a punto nuovi tipi di macchinari per la lana, che tengano conto delle esigenze di:

- purezza del nastro,
- diminuzione del numero di nodi,
- lavorabilità delle fibre,
- il parallelismo delle fibre,
- l'omogeneità del nastro.

L'idea di accoppiare il grado di lavorabilità dei cilindri lavoratori-spogliatori della carda da lana con la cardatura fibra per fibra e la pulitura della carda da cotone tramite l'azione dei cappelli si è concretizzata in alcuni tipi di carde miste:

- la carda mista monotamburo con due scardassi e una sezione di cappelli (carda SACM - figura 1 ,vedasi pagina 105),
- la carda tandem a 2 tamburi con relativi cappelli (figura 2, vedasi pagina 105)
- la carda tandem a 2 tamburi di cui uno equipaggiato con 4 gruppi di cilindri lavoratori-spogliatori e l'altro con cappelli (figura 3, vedasi pagina 106),
- la carda a 3 tamburi di cui 2 con cappelli forniti da file di punte variabili (figura 4, vedasi pagina 106).

Anche se, com'è noto, il comportamento di un determinato materiale dipende dal sistema di cardatura adottato, tuttavia i risultati delle ricerche condotte dai laboratori sui misti di lana riciclata e poliestere, come pure sulle lane vergini open-top o sulle pettinacce carbonizzate o meno o infine su materiali tipo Prato, hanno confermato che le carde miste a 2, e specialmente a 3 tamburi, che alternano l'azione dei lavoratori-spogliatori e dei cappelli, sono in grado di fornire un nastro utilizzabile per la filatura open-end.

In funzione dei diagrammi delle lunghezze delle fibre ottenibili all' uscita della carda, è opportuno prevedere un numero minimo di operazioni di stiratura, controllate o meno, onde migliorare a breve e a lungo termine il grado di omogeneità. In presenza di diagrammi non compatibili, è indispensabile prevedere un sistema di autoregolazione sulla carda stessa. L'andamento delle curve CBL (figura 5, vedasi pagina 107) evidenzia gli eventuali miglioramenti da apportare alla qualità del nastro, con conseguente maggiore facilità di lavorazione in fase di filatura.

FILATURA

L'obiettivo principale perseguito dagli istituti partecipanti alla ricerca è consistito nello studio delle applicazioni del processo open-end, e la gamma dei materiali esaminati testimonia pienamente l'impegno profuso.

Vanno citati:

- la Dref 2
- la San Giorgio
- la BD 200
- l'ITG 300
- l'Autocoro
- la RU II

I risultati confermano che tale processo incontra difficoltà estreme per la filatura della lana pura, salvo presumibilmente della Dref. Per contro le previsioni sono ottimistiche per il misto lana-fibre sintetiche in

433

proporzioni variabili, per il quale si possono ottenere titoli fino a 40
tex o 33 tex in funzione dei diagrammi, purchè si prendano alcune
precauzioni, ad es.:

- la scelta del cilindro sgrossatore e relativa velocità
- la scelta del coefficiente di torsione
- la scelta dell'ugello, in funzione delle opzioni precedenti.
- la scelta del diametro della turbina e relativo profilo.

E' particolarmente indispensabile controllare il tasso di grasso presente
sul nastro. Il limite accettabile per evitare avvolgimenti sullo sgrossatore è di circa 0,5-0,8%. D'altra parte il tasso di polvere contenente
elevati tenori di squame nella turbina è imputabile prevalentemente all'
azione meccanica dello sgrossatore.

Va sottolineata la maggior efficacia dello sgrossatore a lamelle o selettore rispetto agli sgrossatori a denti o ad aghi. L'arresto della produzione dell'ITG 300 ha senz'altro contribuito a ridurre l'interesse per il
sistema.

Le difficoltà indotte dalla presenza di polveri hanno sollecitato gli
sforzi dei ricercatori, intesi a mettere a punto un simultatore in grado
di selezionare i lotti che presentano i rischi maggiori d'intasamento
delle turbine.

Infine i filati open-end presentano caratteristiche diverse dai filati
prodotti sul filatoio continuo, in particolare per quanto riguarda la
struttura, la morbidezza e le proprietà meccaniche; ciò nonostante sono
normalmente utilizzati in maglieria, nel settore dell'abbigliamento e
nelle stoffe per arredamento. Soltanto per i velluti a coste e i tessuti
piatti garzati si raccomanda una scelta prudente dei misti e dei diagrammi dei componenti.

Va ricordata la ricerca specifica effettuata dall'UMIST sui filatoi a fu-
si cavi, che ha fornito una serie di filati con caratteristiche meccaniche interessanti e che offre un comfort nei capi d'abbigliamento confezionati con tali filati del tutto identico a quello dei prodotti tradi-
zionali.

I risultati conseguiti nell'ambito di questo ampio programma di ricerca
sono stati resi possibili unicamente grazie alla stretta collaborazione
fra i vari laboratori della Comunità, a sua volta sostenuti dagli industriali del settore e interessati alla ricerca, che desideriamo ringra-
ziare in questa sede per la loro fattiva collaborazione.

ANALISI FISICHE E CHIMICHE

Relatore : J. KNOTT, Centexbel

Com'è noto, i parametri fisici quali la lunghezza e il diametro della lana, la regolarità dello stoppino, non sono i soli fattori che entrano in gioco nella filatura a turbina open-end.

Molto importante è l'azione dei parametri fisico-chimici, che determinano essenzialmente le condizioni superficiali della fibra, a causa della formazione di polveri che si depositano sulle pale della turbina. Gli obiettivi della ricerca in collaborazione fra il Deutsches Wollforschungsinstitut di Aachen, l'Institut Textile de France Section Nord e il Centexbel- Verviers consistevano:

1) nell'elaborazione di nuove prove o nella modifica di prove già operanti, per raccogliere dati previsionali sul comportamento delle lane utilizzate nel processo di filatura open-end,

2) nell'esame dei fenomeni di formazione delle polveri in fase di produzione del filato,

3) nello studio delle azioni svolte:
   - dalla percentuale di grasso
   - dalla presenza di oleanti
   - dai trattamenti subiti preventivamente dalla lana (carbonizzazione, tintura) sulle proprietà di filatura.

1. - MESSA A PUNTO DI PROVE SPECIFICHE

1.1. Determinazione del tenore di polvere presente nella lana a fiocchi, nello stoppino e nel filato

La depolverazione delle materie prime si effettua tramite vibrazioni ad alta frequenza (30 Hz, ampiezza 5 mm) delle fibre in sospensione etanolica.
La quantità di materiale è dosata per via gravimetrica. Oggetto della ricerca è stata parimenti la cinetica del processo di depolverazione.

1.2. Determinazione del "tippy" per la lana

Una prova particolarmente utile risulta quella che prevede la determinazione della percentuale di fibre che presentano fenomeni di ossidazione alle estremità, in considerazione delle notevoli sollecitazioni meccaniche sopportate dalle fibre nel corso della filatura open-end.

Sono stati presi in esame due metodi:

1) il primo, impiegato unicamente a livello di laboratorio, fa ricorso al microscopio fotometrico a scansione rapida. L'apparecchio registra la luminosità degli elementi colorati (parti ossidate delle fibre).

2) il secondo si basa sull'osservazione del fatto che le punte delle fibre ossidate presentano un tenore più elevato di cisteina, con conseguente predisposizione a fissare in modo selettivo determinati coloranti cationici in presenza di soluzione ad acidità pH 2.

Tale metodo si presta ottimamente per la lana a fiocchi, poichè consente di determinare rapidamente il grado di trasformazione delle punte delle fibre.

1.3. Analisi delle polveri

I metodi analitici a disposizione permettono di determinare, tramite filtrazione micrometrica, ad esempio la composizione "morfologica" delle polveri (punte a residui di fibre, pellicole, corteccia).

Vari parametri che influiscono sul processo di filatura open-end sono stati esaminati ricorrendo ai metodi descritti, senza dimenticare altri metodi chimici consueti, quali la determinazione del grasso neutralizzabile con soluzione di biclorometano, il dosaggio della cisteina per via chimica o al microscopio a raggi infrarossi per riflessione interna multipla, nonchè il dosaggio degli amminoacidi con il metodo Moore e Stein, la determinazione della solubilità alcalina, il dosaggio dei gruppi amminici terminali con la ninidrina, ecc.

2. - EVOLUZIONE DELLE VARIE FASI DI TRASFORMAZIONE NEL CORSO DELLE OPERAZIONI DI LAVORAZIONE, A PARTIRE DALLA PREPARAZIONE FINO ALLA FILATURA

Per poter individuare più da vicino i fenomeni che danno origine alla presenza di sostanze estranee, responsabili delle anomalie che si riscontrano nel processo open-end, ci è parso utile evidenziare le caratteristiche del prodotto in corrispondenza delle varie fasi di lavorazione, a partire dalla lana a fiocchi.

Le modifiche a cui è soggetta la lana a fiocchi sono state vizualizzate sottoponendola all'azione del blu di metilene. Dopo le operazioni di cardatura, pettinatura, la frottatura, lo stiro e la filatura, si è proceduto all'analisi del prodotto e delle polveri residue. E' stato del pari effettuato un esame comparativo dei filati ricavati dal filatoio open-end con selettore e sgrossatore, nonchè dei filati prodotti con i processi di tipo tradizionale.

I risultati riportati nella tabella 1 ribadiscono l'importanza del ruolo svolto dalla carda in merito all'eliminazione delle polveri. L'operazione di frottatura del materiale è una causa non trascurabile della formazione di quantità supplementari di polveri.

I filatoi open-end a selettore sottopongono la lana a sollecitazioni inferiori che non i filatoi a sgrossatore; tale osservazione è confermata dal rapporto corteccia/pellicola, inferiore nelle polveri prodotte da tale tipo di macchina.

Si ricorda infine che le lane perfettamente carbonizzate riducono l'insudiciamento delle turbine, data la riduzione delle sostanze vegetali presenti e delle polveri.

Tabella 1

Andamento delle percentuali di polvere nel corso
del processo di filatura

(* = media su 4 campionamenti)

Campione	% di polveri (*)
Lana a fiocchi	1,06 - 0,91
Lana all'uscita della carda	0,55 - 0,46
Lana a monte della frottatura	0,43 - 0,43
Lana a valle della frottatura	0,52 - 0,51
Lana a monte della pettinatura	0,40 - 0,35
Nastro all'entrata open-end	0,50 - 0,51
Filato open-end sfibratore	0,76 - 0,76
Filato open-end selettore	0,59 - 0,61
Filato tipo tradizionale	0,53 - 0,51

SUMMARY AND CONCLUSIONS BY THE CHAIRMAN OF THE SESSION

M. BONA
Città Degli Studi

La sessione si è svolta mercoledì pomeriggio in maniera molto soddisfacente, alla presenza di circa 70 delegati.

Conformemente al programma, sono stati presentati tre rapporti, che riassumevano il lavoro svolto da 8 Istituti appartenenti a 5 Paesi-membri : preparazione delle materie di recupero per la filatura non convenzionale (Dr Artzt), contraenti : Institut für Textiltechnik, WIRA, CELAC; nuove tecnologie di cardatura e filatura (J.P. Bruggeman), contraenti : ITF-NORD, Centexbel, Celac, WIRA, UMIST, Tecnotessile; Analisi fisiche e chimiche (Prof. Knott), contraenti : Centexbel, Deutsches Wollforschungsinstitut, ITF-Nord.

Le relazioni sono state seguite con grande attenzione dai presenti, e hanno dato luogo ad una discussione molto animata e molto concreta. Ciò ha permesso di chairire meglio determinati argomenti tecnici, fra i quali si può citare :
- L'alternativa fra il carbonissaggio classico dei sottoprodotti e i nuovi processi di epurazione meccanica. Se per delle "blousses" poco inquinate da materie vegetali, il trattamento meccanico può condurre a risultati soddisfacenti, per quelle molto cariche un carbonissaggio ben condotto sembra ancora essere la migliore soluzione. Al fine di evitare la feltratura durante la neutralizzazione, è possibile soprimere la battitura (è stato dimostrato che, dopo cardatura, le materie non battute prescutano meno bottoni e praticamente la stessa pulizia di quelle carbonizzate secondo il procedimento classico). Occorre anche approfondire il confronto fra i vari procedimenti, in condizioni industriali, dal punto di vista economico.
- Il problema delle polveri, molto critico per il funzionamento delle turbine. In effetti, le polveri si formano, in seguito ad alterazione meccanica, durante tutto il processo di fabbricazione, in maniera più o meno accentuata a seconda dello stato della materia prima e delle condizioni di lavoro. Di conseguenza, è altrettanto importante studiare come esse possano essere eliminate, che approfondire il meccanismo della loro formazione. Un'osservazione interessante ha attirato l'attenzione sulle polveri di origine minerale che, pur non essendo in quantità rilevanti, possono esercitare un ruolo importante, ad es. per il loro effetto abrasivo.
- Il confronto, per quanto riguarda gli organi di entrata degli open end a turbina, fra cardino e selettore. Il primo sistema, che è di gran lunga il più impiegato, esercita un'azione più brutale sulle fibre, e molti dati presentati nella relazione e durante la discussione hanno provato che ne risulta un maggiore tasso rottura delle fibre stesse (le conferme sono per ora indirette, perchè devono ancora essere eseguite delle misure dirette di lunghezza sul filato). E' particolarmente spiacevole il constatare che la sola Ditta che finora produceva un open end con selettore l'ha ora ritirato dal mercato, il che sembra confermare l'insufficiente interesse che i costruttori dimostrano per il settore laniero : si spera che i risultati del presente Simposio,

che si raccomanda di portare a loro conoscenza in tutti i modi possibili, possano contribuire a mutare tale situazione.

- E' stata attirata l'attenzione sulle proprietà dei prodotti fabbricati con i nuovi filati, in confronto con i tessuti e i capi di maglieria classici, in particolare per quanto riguarda i fili ricoperti (fuso cavo). Una serie di campioni, fabbricati con le varie tecniche sperimentate dai diversi Istituti, hanno del resto consentito di rendersi conto di tali proprietà in modo molto concreto.

- Sono state anche fornite indicazioni per ricerche future, fra le quali la più interessante riguarda lo studio di additivi appositamente messi a punto per eliminare gli attuali inconvenienti che la lana presenta in rapporto all'open end : si tratta in sostanza di individuare i punti deboli della fibra (il che è ora possibile con precisione, grazie a questi lavori di ricerca) e di affrontarli in maniera sistematica. La filosofia è dunque quella di adattare la fibra alla macchina, e non solamente la macchina alla fibra : si tratta di un cambiamento, rispetto all'ottica tradizionale, che deriva dall'accresciuta importanza dei costi di fabbricazione rispetto a quelli della materia prima.

Su un piano più generale, infine, è importante segnalare due interventi che, a titoli diversi, possiedono un significato che va al di là dell'aspetto puramente tecnico :

- il contributo portato alla discussione da un rappresentante spagnolo, presente in qualità di osservatore, che ha confermato alcuni punti enunciati dagli autori dei rapporti. Ciò permette di sperare che l'eccellente spirito di collaborazione sperimentato durante l'esecuzione del presente programma, sarà rafforzato ed esteso in una Comunità allargata.

- il commento del Presidente di Interlaine, quindi del più alto rappresentante in carica dell'industria laniera comunitaria, il quale si è congratulato con gli Instituti per il lavoro svolto, e ha tenuto a sottolinearne l'interesse pratico, sul quale si riprometti di ritornare in occasione della prossima assemblea della sua organizzazione.

QUALITA' DELLE STOFFE E DEGLI ARTICOLI A MAGLIA

Presidente : J. STRYCKMAN, Centexbel

L'industria della magliera copre un ottavo della produzione europea nel settore tessile, da cui la conferma della sua importanza in tale settore a livello europeo. La sua competitività, necessaria per consolidare il mercato interno ed espandere la sua penetrazione nel mercato esterno, dipende dalla capacità di produzione della sua organizzazione industriale e dalla qualità dei prodotti offerti.

All'articolo di maglieria si richiede funzionalità, affidabilità e buone qualità estetiche. Si tratta di un settore indotto a vendere qualità, immaginazione e buon gusto a prezzi minimi e il cui successo è legato alla capacità d'adattamento immediato alle esigenze del mercato e a quelle ancora più impegnative della moda. E' sollecitato a migliorare costantemente e a rinnovare i suoi processi produttivi e i suoi prodotti, nel momento stesso in cui il consumatore si fà sempre più esigente per quanto riguarda la polivalenza di impiego e la garanzia di qualità.

Gli articoli a maglia sono prodotti in prevalenza nell'ambito di piccole e medie aziende che, per poter rispondere agli imperativi di cui sopra, devono far capo ad un supporto logistico esterno, rappresentato essenzialmente dagli istituti e centri di ricerca specializzati.

Proprio per consolidare tale supporto logistico, attraverso l'approfondimento delle conoscenze tecnologiche, otto istituti di ricerca si sono impeganti, riunendo e coordinando le loro capacità e sotto l'egida della Commissione delle Comunità Europee, ad avviare un programma di ricerca che intende fornire ai produttori gli strumenti necessari per garantire la qualità dei prodotti e stimolare l'interesse dei consumatori a favore degli articoli europei.

Garantire la qualità degli articoli di maglieria rappresenta tuttavia un problema complesso, che esige:

- la conoscenza delle proprietà richieste all'articolo,

- la disponibilità di metodi di misurazione di tali proprietà,

- la conoscenza dei valori limite prescritti dalle norme, per poter valutare il livello di qualità,

- la determinazione dei parametri che influenzano la qualità e l'incidenza della variazioni dei medesimi.

L'obiettivo primario del programma è consistito nella determinazione del-
l'influenza dei parametri, dei materiali e delle condizioni di produzione
sulla qualità degli articoli, in funzione dell'impiego a cui è destinata
la maglia e del tipo di articolo.

L'inalterabilità della forma dell'articolo durante il periodo d'uso e dopo lavaggi periodici costituisce un criterio importante e generale per
tutte le categorie di articoli a maglia, che tutti i produttori sono te-
nuti ad osservare. A tali caratteristiche si sommano altri criteri qualitativi, che formano un insieme di proprietà da prendere in considerazione
per definire la qualità di un determinato articolo e per valutarne il livello, conformemente alla funzione che deve assolvere.

Il programma di ricerca è sfociato nell'elaborazione di una serie di direttive che disciplinano i tipi e le caratteristiche dei prodotti di par-
tenza, le condizioni di produzione e i trattamenti successivi a cui sottoporre gli articoli di maglieria, a cui i produttori devono conformarsi
a garanzia della qualità dei loro prodotti.

La presentazione dei risultati segue tre orientamenti complementari, vale
a dire:

- in base ai criteri di qualità degli articoli a maglia, dal punto di
  vista del consumatore,

- ai vari aspetti della qualità in fase di produzione delle stoffe e degli articoli lavorati a maglia,

- alla valutazione oggettiva della qualità.

Complessivamente i risultati ottenuti costituiscono un riferimento molto
utile per l'industria del settore, che è chiamata a controllare, migliorare e mantenere costante la qualità dei suoi prodotti.

CRITERI DI QUALITA' DEGLI ARTICOLI A MAGLIA

Relatore : H.J. SUURMEIJER, T.N.O.

Si è inteso osservare quali siano i criteri adottati dai consumatori, dai commercianti e dagli industriali per valutare la qualità degli indumenti a maglia. Le rispettive opinioni, metodi di misurazione e valutazione del grado d'indossabilità delle maglie sono state registrate nel corso di un sondaggio effettuato prima nel Regno Unito e successivamente su un campione di 1400 capi indossati dai rispettivi acquirenti in Belgio, Danimarca, Francia e Paesi Bassi.

L'esame delle specifiche tecniche in base alle quali si fabbricano gli indumenti confermano che i grandi distributori e gli industriali titolari di un marchio di fabbrica sono, in linea di massima, d'accordo sugli standard minimi prescritti dalle norme. Tuttavia, per il fatto che i loro metodi di collaudo sono differenti, salvo per quanto riguarda la solidità dei colori, tale accordo risulta più apparente che reale.

Le specifiche d'acquisto dei piccoli distributori esaminate risultano piuttosto sommarie e spesso parti importanti delle specifiche fornite dai produttori sono trascurate e comunicate verbalmente. Ad eccezione dei produttori di indumenti su misura, pochi sono quelli che nelle loro specifiche tenevano conto della lunghezza del punto, mentre tale parametro è essenziale in fase di controllo della qualità.

Precise esigenze sono denunciate in merito al miglioramento delle prove di elasticità e di controllo delle cuciture degli indumenti. Differenze notevoli si riscontrano del pari nella definizione delle taglie, per quanto riguarda i rapporti fra misure del corpo e dimensioni del capo di vestiario nuovo e dopo il lavaggio.

Nella maggior parte dei paesi della CEE non sono disponibili classi di dati antropometrici di tipo universale, che forniscano tutti i parametri indispensabili e tengano conto delle differenze a livello regionale o delle variazioni dovute all'età o alla classe sociale.

In fase di ideazione e di realizzazione di indumenti di maglia che ben si adattano al numero limitato di taglie attualmente disponibili alla maggioranza dei consumatori, le tabelle e le denominazioni delle taglie nella CEE non consentono di sfruttare appieno le proprietà specifiche dei tessuti a maglia.

Date le differenze esistenti fra le proporzioni fisiche dei gruppi di popolazioni nell'ambito della CEE, occorrerebbe mettere a disposizione dei produttori e dei distributori idonee tabelle armonizzate delle taglie, e-

laborate in base a ricerche antropometriche sui gruppi di popolazioni nei
mercati locali e nazionali, per potere offrire alla maggior parte dei
consumatori indumenti a maglia confortevoli e perfettamente corrispondenti alle taglie richieste.

In fase di ideazione e di realizzazione di tali indumenti si può far capo
alle conoscenze disponibili in merito alle proprietà strutturali dei tessuti a maglia e ai processi produttivi realizzati in base a tale bagaglio
tecnologico.

In base ad una serie di prove di nuovo tipo con l'ausilio di strumenti di
misura proposti dalla TEFO di Göteborg nonchè alla realizzazione di mani-
chini speciali da parte della ITF Maille e infine all'esame accurato di
oltre 1.000 capi di maglieria intima reperibili in commercio, è stata
confermata la possibilità di valutare l'indossabilità degli indumenti a
maglia nell'ambito del numero limitato di taglie disponibili.

Tale valutazione può disporre dei risultati delle prove dimensionali cri-
tiche sugli indumenti soggetti a carichi minimi o massimi ammissibili e
della comparazione di tali risultati con i dati antropometrici corrispondenti minimi e massimi registrati nell'ambito delle taglie.

Per i T-shirts si è constatato che tali tipi d'indumenti richiedono esami
più approfonditi, poichè le fibre di questi tessuti di jersey, a coste o
uniti, possono deformarsi con l'uso. Le misurazioni effettuate su indumenti stesi su una superificie piana non sono sempre indicative delle loro caratteristiche di impiego e di indossabilità.

Partendo dalle esperienze acquisite con l'ausilio di sagome piatte per
calzini e calze, ecc., sono state realizzate 26 sagome che riproducono
l'altezza del busto, le forme della testa, del torace, delle spalle e
della parte superiore del braccio. Tali forme riproducono il 95% dei dati
antropometrici della popolazione maschile adulta in Francia.

I criteri proposti per valutare le caratteristiche e l'indossabilità dei
T-shirts sono:

- in senso trasversale: si osserva se gli indumenti sono troppo stretti,
   ovvero comodi, aderenti o troppo ampi;

- in senso longitudinale: la lunghezza minima, definita sperimentalmen-
   te, è indicata da una linea sulla sagoma.

Quando si infilano gli indumenti sulle sagome, si verifica se si hanno
difficoltà di introduzione a livello della testa.

Tali criteri sono stati collaudati nel corso di esperimenti di laboratorio e presso gli stabilimenti industriali. Le capacità dimensionali dei

tessuti a maglia, con gradi diversi di aderenza al corpo, sono state studiate simulando le deformazioni degli indumenti durante l'uso in pratica.

I nuovi metodi sperimentali e gli strumenti di misurazione attuali contribuiscono

- a migliorare la garanzia e il controllo della qualità,

- a rendere più efficaci le specifiche per la produzione e il commercio
  di indumenti,

- a migliorare la casistica delle taglie,

- a realizzare una migliore concordanza fra le maglie e le taglie degli
  utenti.

ASPETTI DELLA QUALITA' NELA FABBRICAZIONE DEI TESSUTI E DEGLI ARTICOLI A MAGLIA

Relatore : M. BALLAND, I.T.F. - Maille

I risultati presentati in questa sede costituiscono una sintesi delle attività di ricerca svolte presso la CENTEXBEL, il DANISH TEXTILE INSTITUTE, l'INSTITUT FUR TEXTILTECHNIK e l'INSTITUT TEXTILE DE FRANCE-MAILLE.

Ci proponiamo di discutere i vari argomenti seguendo il ciclo di produzione, a partire dalla fibra fino al prodotto finito.

Iniziamo pertanto con un esame del meccanismo dei costi di produzione delle stoffe e dei costi imputabili alle anomalie. I costi per i materiali sono pari al 56% per la maglia lineare e al 64% per la maglia circolare. L'incidenza sui costi delle anomalie è mediamente del 6%, di cui il 60% è dovuta al filato. I processi di concatenamento del filato offrono un netto miglioramento.

I filati utilizzati in maglieria vanno sottoposti ad un trattamentto alla paraffina. Le paraffine attualmente impiegate sono formate da catene di idrocarburi a 21-38 atomi, con proprietà fisiche e chimiche dipendenti da tali catene. Inoltre, poichè lo spessore degli strati dipende dal tipo di filato, risulta praticamente impossibile realizzare industrialmente un coefficiente d'attrito costante e predeterminato. I risultati più soddisfacenti si ricavano con un tasso variabile dallo 0,4% all'1%.

E' quindi consigliabile controllare ed adattare la lunghezza del filato su qualsiasi tipo di telaio.

Per tal motivo sono state effettuate ricerche su vari prototipi di telai rettilinei, fra i quali il più recente è formato da un misuratore dell'alimentazione del filato, un posizionatore del carrello, un'unità centrale comprendente un "single board computer" e un servocomando regolato dalla tensione del filato.

Le prove a livello industriale hanno fornito risultati positivi, e offrono la possibilità di ridurre le variazioni di lunghezza del filato assorbito ad un valore dell'ordine dell'1%. Si stanno esaminando i rendimenti offerti dal montaggio del dispositivo sui vari tipi di macchine. E' ormai assodato che tali dispositivi sono destinati ad affermarsi rapidamente e consentiranno di realizzare sensibili economie sul materiale.

Si è proceduto a studiare il problema della qualità delle maglie, di lana o di misti lana/acrilico in base a criteri basati sul tatto, l'aspetto, il pilling e la distorsione delle maglie.

Per quanto riguarda il tatto e l'aspetto, i risultati migliori sono offerti dai filati di lana al 100% con le fibre più sottili. Inoltre le lane migliori sono risultate quelle non sottoposte a trattamento anti-feltratura.

Risultati analoghi sono stati riscontrati per i misti lana/acrilico, nonostante le differenze di spessore delle fibre e la presenza o meno di proprietà anti-pilling dei misti.

Dopo il lavaggio non si sono riscontrate differenze degne di nota fra le lane migliori e i misti nel loro complesso.

In merito al pilling, i risultati più mediocri si registrano per le lane a fibre sottili e in particolare per le maglie lineari. Solo per quest'ultime, e in misura non rilevante, si riscontra una differenza fra lana sottoposta a trattamento anti-feltratura e lana non trattata, con caratteristiche migliori a favore delle lane trattate. Tra i misti, il risultato migliore si è avuto con l'acrilico 5,6 Dtex. Presentano qualità superiori soltanto alle lane fini non trattate.

Per quanto riguarda la distorsione delle maglie di jersey (Maschenver-werfungen), è stato chiaramente dimostrato che tale anomalia è dovuta alla presenza di fibre di lana molto spesse in tale zona. Tale fenomeno aumenta la rigidità a flessione del filo e può provocare sia una distorsione che un'ondulazione.

Infine l'attorcigliamento è un'anomalia tipica delle maglie di jersey, che può essere resa meno evidente da taluni trattamenti di appretto, anche se attualmente l'unico procedimento pratico consiste nell'alternare fili a torsione S con fili Z.

Le variazioni dimensionali delle maglie sono uno dei fenomeni maggiormente oggetto di ricerche allo stadio attuale.

Il concetto di "stato rilassato" è accettato da tutti i laboratori e i metodi per determinare tale stato sono molto simili. Si basano sull'essiccamento con centrifuga, preceduto da un numero di lavaggi variabile da 3 a 5 o da un trattamento statico immersione in acqua, con successiva azione meccanica su nastro vibrante.

E' stata confermata l'importanza del tasso d'umidità nella stoffa all'uscita dalla centrifuga, sulle dimensioni delle maglie di cotone; a rigor di logica dovrebbe essere conservato, dopo essicamento, un tasso d'umidità prestabilito, preferibilmente tra il 5% e il 6%.

I fenomeni osservati allo stato rilassato sono stati registrati, con le stesse leggi, anche per il cotone e si è dimostrato che anche le maglie jersey in tessuto acrilico, a costa 1:1 e 2:2 rispondono alle leggi di Munden, di cui sono stati calcolati i coefficienti.

Dopo aver sottoposto a lavaggi successivi campioni di maglie di cotone in fase di produzione, si constata chiaramente che la maggior parte delle variazioni dimensionali si verificano dopo il primo lavaggio, e successivamente le variazioni sono alquanto limitate e si stabilizzano dopo il decimo lavaggio.

Nel senso della lunghezza il restringimento verificatosi dopo il primo lavaggio continua nello stesso senso, mentre nel senso della larghezza, dopo il restingimento prodotto dal primo lavaggio, si constata la formazione di afflosciamenti.

Le dimensioni dei tessuti a maglia variano nel corso di tutto il ciclo di finissaggio. Tuttavia i fenomeni di restringimento, relativamente elevato, in fase di lavaggio, registrati nelle maglie gregge, non vengono neutralizzati in fase di tintura, candeggio o di qualsiasi trattamento ad umido, a secco o di finissaggio. E' ovvio che qualsiasi dilatazione della maglia che si produca ad una determinata fase di lavorazione presuppone un restringimento maggiore in fase di lavaggio.

Nel gruppo di macchine di finissaggio, la più importante dal punto di vista della stabilità è l'essicatoio. Le macchine che offrono i risultati migliori sono le centrifughe continue.

Per le maglie di cotone le leggi di Munden sono di difficile applicazione a causa della differenza di dimensioni esistenti allo stato rilassato fra le maglie gregge e le maglie trattate. In linea di massima, allo stato rilassato le maglie trattate sono più lunghe del 5% e più strette del 4% rispetto alle maglie gregge. Tale problema potrà probabilmente essere risolto quando saranno disponibili banche di dati in materia.

E' stato esaminato il trattamento di maglie in acrilico per pull-overs, ricorrendo ad un metodo di laboratorio atto a misurare il "fissaggio" dei filati. Innanzittutto è stato dimostrato che un trattamento alla pressa di tipo tradizionale non è in grado di fissare le dimensioni di una maglia che risulti già deformata rispetto allo stato rilassato. Soltanto il 30-50% di tali deformazioni possono essere fissate. Ne consegue che, prima di essere trattati alla pressa, i pannelli vanno rilassati e quindi "fissati" alla pressa in presenza di una miscela aria-vapore a temperature variabili a seconda dei tipi di acrilico. Il Leacril richiede 75°C, il Courtelle 75-80°C, il Dralon 80°C e il Crylor 85°C. Si noti che a temperature inferiori a 70°C le maglie non vanno più stirate.

Le prove con indossatori confermano i risultati di laboratorio e confermano l'importanza dello stato iniziale della maglia per le successive evoluzioni dimensionali in fase di uso e lavaggio nonchè l'importanza dei trattamenti.

La stabilità maggiore si osserverà nelle maglie "fissate" alle condizioni che più si avvicinano allo stato rilassato.

In ogni caso, tenuto conto della tendenza al restringimento nel senso della lunghezza e al rilassamento nel senso della larghezza, è opportuno prevedere correzioni inverse dell'ordine del 5% nei due sensi.

Un'attenzione particolare va riservata ai bordi a costa, le cui maglie non devono essere troppo allentate e nello stesso tempo vanno trattate in condizioni per quanto possibile prossime allo stato rilassato.

Rispettando tali precauzioni si possono realizzare pull-overs in acrilico non eccessivamente deformabili all'uso.

Per concludere va osservato che esistono metodi di controllo del grado di lavorabilità per gli articoli di maglieria. E' stata in particolare sottolineata l'importanza delle dimensioni reali dell'ago e dello stato di ammorbidimento, quest'ultimo misurabile dalla resistenza alla penetrazione dell'ago nella maglia.

Va notato che la qualità degli indumenti a maglia dipende da ogni singola fase del ciclo di produzione e che i risultati delle ricerche illustrati in questa sede possono contribuire a migliorarla.

VALUTAZIONE OGGETTIVA DELLA QUALITA'

Relatore : E. FINNIMORE, Deutsches Wollforschungsinstitut

La qualità dei tessuti e degli indumenti a maglia costituisce il tema del programma di ricerca presentato in questa sede.

Per assicurare la competitività dell'industria tessile europea è indispensabile conservare e migliorare costantemente il livello di qualità. La realizzazione di tale obiettivo è legata alla disponibilità di metodi di valutazione oggettiva della qualità, in grado di offrire, ad esempio, un controllo più efficace della qualità in fase di produzione, l'elaborazione di norme internazionali e l'ideazione di tessuti ed indumenti perfettamente rispondenti ad esigenze specifiche. La presente relazione illustrerà vari aspetti della valutazione oggettiva, come risulta dalle ricerche condotte da tre istituti europei.

All'atto dell'acquisto, il consumatore valuta la qualità dei calzini in base alla loro calzabilità, con conseguente maggior comfort all'uso. Presso l'ITF Maille, Francia, è stata condotta una ricerca al fine di realizzare un sistema di controllo oggettivo della taglia dei calzini (demi-bas) e delle corrispondenti caratteristiche di aderenza.

In fase d'uso, i calzini si trovano in uno stato di dilatazione, di cui deve tener conto qualsiasi metodo di misurazione del relativo comfort. Le forme da esposizione utilizzate non si rivelano adatte allo scopo. Per motivi pratici l'industria deve preferibilmente ricorrere a sagome piatte. E' stato quindi necessario definire un metodo in grado di trasformare una superificie piana in una superficie equivalente all'area esterna di un volume (piede e gamba).

Per realizzare tali sagome si è proceduto per fasi successive, prima trasformando un calzino infilato in una sagoma rigida estraendolo dalla gamba con opportuni tagli, quindi procedendo ad incisioni in determinati punti per realizzare una sagoma piatta e quindi trasformando la sagoma in un'altra a profilo continuo e di area equivalente. E' stata così realizzata una serie di 12 sagome che comprendono le taglie 24-46 della maglieria francese.

90 tipi diversi di calzini sono stati esaminati con l'ausilio di tali sagome e parallelamente di indossatori. In figura 1 (vedasi pagina 200) sono illustrati i criteri adottati per valutare il comfort offerto dai calzini.

Fra i 90 campioni esaminati soltanto in tre casi si sono registrati discordanze tra la prova oggettiva e la prova reale. E' stato dimostrato

che in questi tre casi le dimensioni del piede e della gamba si scostava-
no dal valore standard della taglia del piede considerata. Le prove sulle
sagome hanno fornito risultati positivi in cinque maglifici. Tali sagome
sono state immesse sul mercato e il metodo fa ormai parte della normativa
francese. Si spera che su tale base si possa pervenire ad una normativa
europea.

Una delle cause più importanti di difettuosità riscontrabili negli indumenti a maglia industriali è dovuta alle mediocri prestazioni alla cucitura di determinati tessuti a maglia, con conseguente presenza di inconvenienti dovuti alla formazione di smagliature e di cavità in fase di cucitura.

Presso l'Institut für Textiltechnik Denkendorf nella Germania Federale è
stata messa a punto una prova sulla confezionabilità dei tessuti a maglia, basata su criteri oggettivi. In base ai risultati di prove preli-
minari si è constatata una correlazione quasi perfetta tra la forza di
penetrazione dell'ago e la presenza di anomalie a causa della formazione
di anelli.

Si è quindi proceduto a realizzare un sistema computerizzato per la regi-
strazione rapida della forza di penetrazione dell'ago, che presenta il
vantaggio di poter essere montato su macchine per la produzione in serie.

La prova viene effettuata su campioni rappresentativi a monte dell'operazione di taglio, il che presenta il vantaggio di poter lavorare la pezza
intera con dispositivi ausiliari di cucitura. Con questo metodo è stato
chiaramente dimostrato che un buon appretto del tessuto a maglia migliora
sensibilmente la loro confezionabilità. Con processi di appretto adeguati
è possibile ridurre fino al 93% la forza di penetrazione dell'ago per i
tessuti composti da fibre sintetiche al 100%. Tale metodo consente quindi
di effettuare ricerche obiettive al fine di realizzare trattamenti
ottimali di appretto, che contribuiscono direttamente a migliorare la
qualità del prodotto.

Uno dei criteri più importanti che influiscono sulla decisione all'acquisto da parte del consumatore è la sensazione al tatto, unitamente alle
qualità estetiche apparenti del tessuto. La sensazione al tatto resta
tuttavia un elemento molto difficile da definire, valutare e misurare.

In questi ultimi anni molte ricerche sono state condotte al fine di realizzare metodi di misurazione oggettiva della sensazione al tatto. I vantaggi offerti da tali metodi sono: la possibilità da parte degli acquirenti e dei fornitori di presentare specifiche in codice e di quantifica-
re gli effetti prodotti dai vari trattamenti d'appretto. La soluzione che
ha risvegliato il maggior interesse a livello mondiale è quello proposta
dai giapponesi, vale a dire il metodo KES-F di Kawabata.

Presso il Deutsches Wollforschungsinstitut di Aachen in Germania si è cercato di estendere tale sistema di misurazione della sensazione al tatto anche ai tessuti a maglia. In tale contesto occorre tener presente che la gamma dei tipi di maglie è più vasta rispetto alle stoffe.

Sono misurati i parametri rappresentativi della piegatura, della rasatura, del limite elastico a trazione, alla compressione, le proprietà superficiali e strutturali (peso e spessore). Inizialmente le ricerche si proponevano di valutare l'effetto prodotto sui parametri obiettivi della sensibilità al tatto per una serie di maglie di lana e di misto lana/acrilico da vari caratteristiche, quali lo spessore, il tipo di maglia, la fibra tessile, il materiale, il grado di finizione e di appretto. E' stata sottolineata l'importanza del finissaggio a vapore per ottenere una morbidezza ampia e vellutata.

Il secondo tema della ricerca consisteva nella valutazione qualitativa degli effetti indotti da elementi ammorbidenti sulle maglie di lana e misti lana. Attraverso la misurazione oggettiva dei parametri di rigidità, si può individuare un livello ottimale di utilizzazione che favorisce la presenza di effetti ammorbidenti nella lana trattata per resistere ai fenomeni di ritiro (fig. 1).

Tali effetti di concentrazione sono stati rilevati anche in sede di valutazione soggettiva. Un certo parallelismo è stato notato fra i sistemi di valutazione oggettiva e soggettiva in merito ai trattamenti di appretto, anche se il giudizio degli esperti è stato influenzato dal tipo di maglia utilizzato.

I dati quantitativi acquisiti nel corso di tale ricerca possono servire d'aiuto al produttore nella scelta dei materiali e delle migliori condizioni di appretto al fine di realizzare il grado di qualità richiesto.

RIASSUNTO E CONCLUSIONI DEL PRESIDENTE DELLA SESSIONE

J. STRYCKMAN, Centexbel

Due questioni importanti riguardanti la qualità dei tessuti lavorati a
maglia sono state messe in evidenza dalle questioni sollevate nelle presentazioni: da un lato le dimensioni e la stabilità dimensionale e
dall'altro le misure oggettive per determinare la comodità degli
articoli. La discussione è qui di seguito brevemente delineata.

Per quanto riguarda la questione se le dimensioni stabilizzate di un articolo lavorato a maglia finito possano essere determinate in anticipo
sulla base dei parametri del filo e della struttura del lavoro a maglia,
esiste una correlazione tra questi parametri dipendente dal tipo di maglia e di finitura. Con la banca di dati Starfish, creata dall'Istituto
Internazionale per il Cotone, le dimensioni di un articolo finito possono
essere calcolate partendo dalle caratteristiche del tessuto non candeg-
giato. Le equazioni sono disponibili per il cotone in tre tipi differenti
di maglia: Jersey, costa 1:1 e punto incatenato.

Le dimensioni di un articolo dipendono in gran parte dalla lunghezza di
filo utilizzata in ogni maglia. E' dunque molto importante controllare e
verificare questo parametro. Sulla macchina circolare questo è fatto dagli alimentatori positivi. Non vi erano precedentemente dispositivi per i
telai rettilinei, ma due metodi sono stati ora messi a punto e sperimen-
tati.

Nell'uno, lo scopo è di mantenere una tensione costante ad un dato punto
non appena il filo è inserito nella macchina. Nell'altro, la lunghezza di
filo utilizzata in una certa sezione per ogni giro di maglie è posta in
raffronto tramite microelaboratore con una lunghezza desiderata predeter-
minata per il giro. Se le due lunghezze sono diverse, un segnale corregge
l'alimentazione del filo nella macchina. Entrambi i sistemi permettono di
produrre tessuti di lunghezze identiche con una tolleranza dell'1% sulla
lunghezza di filo utilizzata.

La velocità dei telai rettilinei moderni non impedisce l'utilizzazione
del sistema. L'accelerazione del dispositivo di aggancio potrebbe essere
un problema in uno dei sistemi ma la misurazione ha luogo nel letto degli
aghi.

Nell'altro sistema, il sensore reagisce immediatamente alle differenze di
tensione. L'effetto dell'accelerazione del filo all'inizio del giro è an-
nullato tramite la disattivazione simultanea dei dispositivi di tensione.

Incerare il filo prima del lavoro a maglia è un mezzo meno sofisticato
per regolare la lunghezza del filo utilizzato: ma per certi articoli prodotti su commissione costituisce la sola possibilità. Tuttavia, non tutte
le cere reagiscono nello stesso modo. La loro composizione, in altre pa-
role le diverse lunghezze delle catene di idrocarburi, esercitano un'influenza sul punto di fusione, che dipende inoltre dalla quantità di cera
utilizzata: esso dovrebbe essere tra lo 0,4 e l'1%, a secondo della natura del filo. E' necessario dunque sapere come la cera utilizzata si comporta ed applicarla in base alla natura del filo.

La questione se la cera influisca sulla sensazione di chi indossa tessuti
lavorati a maglia ci porta ad affrontare un altro aspetto, quello della
comodità, della gradevolezza e della loro misurabilità oggettiva. In primo luogo, l'inceratura cambia la sensazione provocata dal tessuto, ma il
progetto non ha esaminato a fondo questo aspetto. La misurazione oggettiva di questa proprietà è ancora in uno stadio alquanto primitivo e molto
ancora deve essere fatto in questo campo.

I soli parametri coperti dal programma erano il tipo di maglia e la fibra; un grande numero di campioni è stato già analizzato. Due tipi di filo sono stati esaminati: lana pura e un misto acrilico-lana. Le misurazioni oggettive sono state effettuate utilizzando gli strumenti giapponesi Kawabata, specialmente progettati per misurare la qualità tattile. I
risultati delle misure oggettive sono stati paragonati alle valutazioni
soggettive effettuate da esperti scelti per la loro lunga esperienza con
materiali in tessuti o lavorati a maglia. La correlazione era nel complesso abbastanza buona, nonostante gli errori occasionalmente fatti dagli esperti quando la differenza tra i campioni non era grande.

Il concetto di "morbidezza" del Kawabata è una combinazione di due elementi: morbidezza della superficie e solidità. Manca per queste due misure in Europa una terminologia adatta.

Gli strumenti Kawabata sono stati utilizzati molto di più per i materiali
intessuti che per quelli lavorati a maglia. I parametri dunque devono essere adattati a questi ultimi. Tuttavia, la valutazione soggettiva dovrebbe gradualmente cedere il passo alla misurazione oggettiva degli articoli, poichè gli esperti non sono sempre coerenti, contraddicendosi re-
golarmente anche in un breve spazio di tempo. Per i tessuti lavorati a
maglia, parametri quali l'isteresi di taglio, di spessore o di peso e la
regolarità della superficie sarebbero più adatti. Ciò dovrebbe diventare
chiaro con l'esperienza.

L'importanza di Kawabata e della misurazione oggettiva risiede inoltre
nel fatto che essa rende facile valutare l'effetto dei parametri del materiale e del processo di fabbricazione sulle qualità tattili di un tessuto, cosa che per ragioni pratiche è più difficile, se non impossibile,
con i metodi soggettivi. Il sistema Kawabata potrebbe essere utilizzato

in Europa, ma l'attrezzatura è relativamente costosa e le imprese piccole o di medie dimensioni non potrebbero sostenerne l'acquisto.

In Giappone, sono utilizzate 150 macchine, molte delle quali negli istituti di ricerca o nelle agenzie di acquisto che lavorano in collaborazione con l'industria. E' dunque attualmente utile, e sarà essenziale nell' immediato futuro, per le industrie tessili e dell' abbigliamento, fare uso di questa tecnica. Le imprese potrebbero utilmente servirsi di laboratori centralizzati in grado di acquistare e utilizzare l'attrezzatura e, dall'esperienza guadagnata nelle misurazioni oggettive, offrire un servizio prezioso all'industria.

Il tema "qualità dei tessuti e degli articoli lavorati a maglia" è stato esaminato da tre angolature differenti. La prima ha avuto per oggetto i criteri qualitativi per gli articoli lavorati a maglia, ed il lavoro, effettuato da quattro istituti, ha mostrato che la qualità, l'adattamento, la comodità e la soddisfazione del consumatore dovevano essere migliorati. Per fare questo, si è auspicata la creazione di un metodo standard europeo per la misurazione delle dimensioni del corpo umano, tenendo conto del fatto che anche all'interno di ciascun paese, erano suscettibili di verificarsi differenze regionali nelle forme del corpo. Banche di dati di queste misure devono essere create ed eventualmente aggiornate utilizzando il metodo standard approvato.

Il secondo aspetto, anch'esso trattato da quattro istituti, è stato l'esame della qualità nella fabbricazione di tessuti ed articoli lavorati a maglia.

La questione principale era determinante per la qualità delle merci lavorate a maglia: la stabilità dimensionale sia nella fabbricazione che nell'uso da parte del consumatore. I risultati del lavoro permetteranno un migliore controllo di quest'aspetto per gli articoli di lana, di misto lana-acrilico e di cotone. Un risultato particolarmente interessante è stata la creazione di dispositivi per il controllo automatico della regolarità delle dimensioni degli spazi nei telai rettilinei. Tuttavia, molto resta ancora da fare nel campo della stabilità dimensionale.

L'aspetto da ultimo esaminato sarà in avvenire di importanza crescente nel giudizio di qualità dei tessuti lavorati a maglia: la valutazione oggettiva di qualità. Vi sono tre considerazioni importanti: le qualità tattili, l'adattamento e la cucibilità. Quest'aspetto è stato esaminato da tre istituti. Una metodologia recente ha fornito alcuni risultati iniziali per i tessuti lavorati a maglia in un settore in cui l'esperienza con i prodotti intessuti è già estesa. Queste proprietà sono influenzate da un numero particolarmente grande di parametri. Il programma dunque ha dovuto essere limitato a quelli ritenuti determinanti e ad una gamma limitata di materie prime, in particolare metodi di filatura, strutture di lavoro a maglia e processi di finitura.

Il lavoro ancora da fare in questo campo può, per molti anni a venire, continuare a fornire gli argomenti per molti diversi programmi di ricerca.

In qualità di Presidente di questa sessione, vorrei concludere esprimendo due speranze.

La prima è indirizzata agli istituti impegnati in questo progetto di ricerca. Essi hanno accumulato una vasta quantità di risultati che dovrebbero essere utilizzati. Ciò può essere fatto soltanto se tutti questi risultati sono applicati effettivamente all'industria. E' dunque essenziale che ogni istituto nella sua sfera di attività si accerti che i suoi risultati siano fatti propri dall'industria e fornisca il sostegno tecnico per assicurarsi che le informazioni siano correttamente utilizzate nella pratica.

Nell'interesse di quest'applicazione della ricerca, la Commissione delle Comunità Europee farà ogni sforzo per stimolare il trasferimento di informazioni e per assicurarsi che sia fatto un uso completo dei suoi risultati.

E' stato raccolto un volume enorme di informazioni, ma il numero di parametri in questione è sterminato e la diversità dei prodotti lascia un notevole campo d'azione per le indagini future. Abbiamo visto attraverso la cooperazione che i risultati dei nostri sforzi possono essere moltiplicati.

In secondo luogo, spero che questa collaborazione europea tra gli istituti, che hanno costituito un eccellente gruppo di lavoro durante gli ultimi tre anni, continui. Ciò sarebbe incoraggiato se la Commissione europea dovesse appoggiare richieste di sovvenzioni provenienti dalla grande industria europea della maglieria per permettere l'effettuazione di progetti di ricerca.

L'adempimento di queste due speranze sarebbe indubbiamente un passo avanti cruciale nella salvaguardia del futuro dell'industria tessile europea.

455

VALORIZZAZIONE DEL LINO

Presidente : M. VAN LANCKER, Centexbel

Il lino è una delle rare materie prime tessili vegetali prodotte nell'ambito della Comunità Europea.

Riportiamo alcune cifre che definiscono la posizione del lino in generale e nell'industria tessile in particolare:

- il lino costituisce il 2% delle fibre tessili utilizzate,

- nella CEE la produzione di filacce (lino stigliato e stoppe) era di 79.881 ton nel 1984 (137.000 ton nel 1962 e 82.100 ton nel 1982); la produzione di filati di lino di tipo tradizionale di oltre 33.500 ton nel 1984 (80.000 ton nel 1962 e 30.400 ton nel 1982).

- la bilancia commerciale risulta ampiamente attiva:
  . circa 145 milioni di ECU nel 1984 per la Francia
  . circa 70 milioni di ECU nel 1984 per il Belgio.

Lo squilibrio esistente da vari anni tra la produzione di filacce e il fabbisogno della filatura tradizionale ha spinto gli operatori economici a rivolgersi ai mercati extracomunitari e l'industria a cercare nuovi sbocchi.

Da vari decenni le industrie di trasformazione del lino avevano registrato un'evoluzione negativa e tutti i settori interessati, facenti parte della C.I.L.C. (Confederazione Internazionale del Lino e della Canapa) hanno elaborato un programma di ristrutturazione e d'espansione, che prevedeva un impegno considerevole nel campo della ricerca, della promozione e della commercializzazione.

Il programma di ricerca, denominato "EUROLIN", si proponeva l'obiettivo di migliorare i processi di trasformazione della paglia in prodotti finiti, per:

- economizzare la materia prima,
- instaurare condizioni di lavoro più vantaggiose,
- valorizzare le proprietà intrinseche del lino,
- ridurre i costi di produzione,
- realizzare prodotti finiti più rispondenti alle esigenze dei consumatori.

Con riferimento ad un'analisi della situazione dell'industria del lino nella CEE alla fine degli anni '70, dei problemi e delle lacune risultanti in base alla valutazione delle attività intraprese e dei risultati ottenuti, in vista di un secondo programma la CEE è stata invitata a presentare un progetto di aiuto finanziario per la ricerca nel settore tessile, al fine di accelerare l'esecuzione del progetto EUROLIN già avviato.

Vari fattori confermavano l'opportunità di tale passo, al fine di fornire un nuovo impulso a tale programma:

- un rinnovato interesse da parte dei consumatori per le fibre naturali,
- il progresso delle fibre miste,
- l'avvento di nuovi processi in filatura,
- la necessità di reinvestire a breve termine da parte dell'industria del lino,
- la disponibilità di un ampio mercato per l'esportazione.

Il programma di ricerca presentato alla CEE è stato elaborato dalla Commissione Scientifica e Tecnica della C.I.L.C., per opera del Comité Billaux, comitato di ricerca e sviluppo. Di tale comitato fanno parte i centri di ricerca sul lino con sede nella CEE: ATPUL, ITL, ITF-Nord e ITF-Boulogne in Francia, Centexbel e OVLT in Belgio, TNO e IBVL in Olanda, LIRA in Gran Bretagna, Textil-Forschung Bielefeld in Germania e la Stazione Sperimentale per la Cellulosa in Italia.

Ciascuno di tali centri è specializzato in un settore particolare per il quale è stato invitato a partecipare; tuttavia nell'ambito del programma generale si è reso indispensabile istituire un coordinamento dei vari contributi individuali. L'intervento della Comunità ha accentuato l'esigenza di mantenere la cooperazione ad alto livello, la cui direzione è stata assunta dalla Commissione Scientifica e Tecnica della C.I.L.C. e dal Comitato Billaux. Tale orientamento costituiva una garanzia per incanalare i lavori verso le reali esigenze dell'industria e per far conoscere e applicare rapidamente a livello di stabilimenti i risultati scientifici e tecnici ottenuti.

L'aggiornamento degli obietivi ha consentito una rigorosa definizione degli obiettivi tecnologici sottoelencati, obiettivi assurti a temi di programmi di ricerca nel secondo programma di ricerca e sviluppo tessile e dell'abbigliamento con la denominazione comune "Valorizzazione del lino".

OBIETTIVI TECNOLOGICI:

1. - Approvvigionamento delle materie prime ed estrazione delle fibre

Per garantire la continuità dell'approvvigionamento di materie prime idonee, la ricerca deve orientarsi verso il miglioramento delle colture, del rendimento delle fibre, della resa dei sottoprodotti, la presentazione dei materiali e la riduzione della manodopera attraverso l'automazione.

2. - Trasformazione della fibra in filato

In questa fase occorre adattare la fibra per poter utilizzare materiale ad elevata resa o riccorere ai moderni processi di filatura: filatura OE, ad attrito, Novacore, ecc.

3. - Nobilitazione

Occorre collaudare e mettere a punto nuovi e migliorati processi e dispositivi e parallelamente risolvere i problemi d'indole generale indotti dalla nobilitazione del lino.

4. - Norme di qualifica dei prodotti

Tale procedura si rivela indispensabile, onde individuare le caratteristiche del lino che consentono di valutarne la predisposizione alle diverse applicazioni. Il controllo oggettivo della qualità va integrato da norme precise e da capitolati d'oneri ben definiti.

5. - Morfologia delle fibre

Si impone la necessità di approfondire la conoscenza della morfologia delle fibre di lino nonchè dell'influenza esercitata dalle condizioni di produzione e di lavorazione, per poter sfruttare tutte le risorse potenziali offerte dalle fibre di lino.

Il programma "EUROLIN", di cui il programma "Valorizzazione del lino" costituiva una parte integrante, viene portato avanti presso gli istituti di ricerca unitamente a nuove attività di promozione a livello internazionale.

Grazie a tali attività di ricerca e di promozione, l'industria del lino della CEE potrà disporre di un tessuto tecnologico e socio-industriale moderno e rispondere pertanto alle esigenze della strategia industriale delle CEE.

Le nuove qualifiche già si stanno imponendo a tutti i livelli, in particolare nella ricerca di nuovi sbocchi e nella produzione di manufatti di ottima qualità. L'industria del lino si sta uniformando a tali esigenze e si prepara ad offrire prodotti in grado di soddisfare le richieste della clientela sui mercati mondiali.

PRODUZIONE DELLA FIBRA DI LINO

Relatore : W.W. FOSTER, LIRA

Abbiamo riportato in tabella l'elenco delle attività svolte in cinque
centri di ricerca, nel quadro del programma EUROLIN. L'obiettivo di tali
ricerche consisteva nel migliorare il rendimento globale e la qualità
delle fibre di lino e nello stesso tempo di ridurre i costi di trasformazione.

ATTIVITA'	Centri di ricerca				
	ITL	LIRA	Wagen-ingen	ATPUL	Centex-bel
1. Prove sul terreno					
– irrorazione prodotti chimici anti-allettamento	x				
– irrorazione prodotti chimici essic./maceraz. sul terreno	x				
– irrorazione prodotti chimici essic./maceraz. in silo		x			
– taglio/maceraz. sul terreno			x		
– sistemi di cernita paglie	x				
– materiali inquinanti				x	x
2. Prove in stabilimento					
– miglior. della stigliatura				x	
– sist. di prod. del nastro				x	
– trattamento nastri:					
. enzimi		x			
. prodotti chimici		x		x	
– processi per fibre corte				x	
– lino totale				x	

1. – PROVE SUL TERRENO

Nel corso degli ultimi cinque anni si è esteso il ricorso a sistemi di
irrorazione di prodotti chimici per evitare l'allettamento, ormai pienamente affermati sul mercato.(Nella foto 1 - vedasi pagina   - a sinistra
lino trattato e in centro lino non trattato). Nel 1984 è stato sottoposto
a trattamento il 10% della produzione di lino francese. Tali prodotti,
normalmente impiegati per rallentare la crescita, comportano l'inconveniente di ridurre l'altezza della pianta e di abbassare il rendimento in
semi. Gli steli ridotti sono più spessi di quelli normali. Tali sistemi

d'irrorazione vanno quindi utilizzati unicamente quale ultima risorsa. Tre sono i prodotti praticamente inoffensivi per tale impiego: Cerone, Atheverse, Terpal. Tutte le varietà di lino e in qualsiasi fase della loro crescita sono sensibili a tali prodotti chimici la cui azione è accentuata dall' aggiunta di emolienti.

Sono proseguite le ricerche sulle modalità d'impiego delle irrorazioni di essicazione e macerazione del lino direttamente sui campi (lino trattato nel centro della foto 2 - vedasi pagina238). L'operazione è stata anche effettuata per accelerare l'essicazione e la macerazione normale sul terreno. Data la siccità persistente verificatasi negli ultimi due anni, i risultati sono stati piuttosto mediocri e le informazioni raccolte limitate.

Sono state realizzate macchine speciali per la raccolta del lino macerato sul terreno prima dell'imballaggio, previa eliminazione delle erbacce e dei sassi e migliorando l'essicazione (foto 3 - vedasi pagina240). Attualmente sono operanti circa 200 macchine. Inoltre si può montare sulle macchine imballatrici un'attrezzatura supplementare funzionante secondo lo stesso principio, in grado di eliminare fino all'85% di sassi prima dell' imballaggio del lino (foto 4 - vedasi pagina241).

Si è d'altra parte cercato di mettere a punto processi speciali, tali da rendere superfluo l'impiego di macchine specifiche per la raccolta del lino, quali ad esempio le estirpatrici, e utilizzare unicamente le macchine agricole consuete. Riassumendo, si pensa di procedere al taglio del lino, senza estirpazione, e quindi sottoporre le piante tagliate alla macerazione e all'essicazione sul terreno stesso, però già imballato, e quindi successivamente sottoposto a cernita, ottenendo fibre corte e pulite (vedasi schema allegato). Già fin d'ora si può presumere che le fibre, rispetto a quelle prodotte con macchinari ordinari, potranno essere offerte a prezzi concorrenziali e perfettamente idonee alla preparazione di misti.

Sono stati studiati metodi atti ad individuare la presenza di tracce di filamenti di polipropilene nel lino, anche se non si dispone ancora di un processo in grado di eliminare tale inquinante. Ampie ricerche sono state condotte per identificare la grande varietà di frammenti vegetali presenti nella stoppa e che producono effetti negativi nel tessuto finale. E' ormai confermato che conviene di gran lunga eliminare le erbacce all'atto della raccolta del lino, piuttosto che eliminare successivamente o neutralizzare i residui.

2. - PROVE IN STABILIMENTO

Sono state apportate modifiche essenziali in fase di progettazione di impianti di stigliatura (foto 5, vedasi pagina 242) come pure degli avantreni delle carde e delle sezioni dei battitoi, che offrono un

rendimento più elevato. Altre modifiche interessano processi di trasformazione più efficienti ed economici, che offrono qualità e rendimenti
più elevati: un dispositivo per stoppe ristagliate che produce materiali
sotto forma di nastri e un apprettante per la formazione di nastri di
lino totale. In parte i risultati ottenuti sono molto incoraggianti.

Progressi sensibili si sono registrati nei processi di nobilitazione per
via chimica e biochimica delle fibre di lino di mediocre qualità, in particolare con l'impiego di enzimi, che migliorano la qualità del nastro.

NUOVI PROCESSI DI PREPARAZIONE E DI FILATURA

Relatore : M. VAN LANCKER, Centexbel

Nella tabella seguente sono elencate le attività svolte nei quattro centri di ricerca, nel quadro del programma EUROLIN. Obiettivo dei lavori era la ricerca volta a migliorare i processi di trasformazione della fibra di lino in prodotti finiti, dal punto di vista tecnico ed economico, al fine di:

- economizzare materia prima,

- valorizzare le proprietà intrinseche del lino,

- ridurre i costi di produzione,

- realizzare prodotti più rispondenti alle esigenze dei consumatori.

Attività	Centri di ricerca			
	Biele-feld	Centex-bel	I.T.F.	LIRA
1. Rivalutazione dei materiali di mediocre qualità	x	x		
2. Problemi indotti dalle nuove fasi di produzione:				
- resa ottimale della stigliatura		x		
- preparazione e cernita delle fibre			x	x
- trattamento dello stoppino separazione delle fibre			x	
- rendimento economico di una nuova linea di produzione			x	x
3. Nuovi processi di filatura:				
- filatura a turbina	x			
- filatura ad attrito		x		
- Novacore			x	
- filatura senza torsione				x
4. Modifiche dei prodotti e dei processi			x	x

Annualmente parte del raccolto di lino resta inutilizzato per la filatura di tipo tradizionale, mentre molti scarti prodotti in fase di preparazione alla filatura potrebbero essere recuperati. I lavori di ricerca hanno consentito di individuare e mettere a punto cicli di trasformazione specifici. Sono stati collaudati nuovi prodotti ricavati da tali materiali e avviati nuovi processi industriali.

Ricerche sono state effettuate sulla stigliatura delle fibre lunghe per trasformarle nell'ambito di un ciclo alternativo, denominato "sistema ad alta produttività", e sulla stigliatura quale fase di rettifica del diagramma delle lunghezze. La ditta Piersen & Digneffe ha realizzato un prototipo di stigliatrice, mentre le officine Houguet-Duesberg-Bosson ha messo a punto una macchina che ricalca le caratteristiche di quella studiata nel corso della ricerca, modificandola alle peculiarità della fibra di lino.

La ricerca sistematica nel settore della preparazione, della separabilità e del trattamento dello stoppino prima della filatura ha fornito risultati che consentono alle industrie del ramo di rendere ottimale la lavorazione meccanica e chimica della fibra e di orientare adeguatamente i futuri investimenti. L'apparecchio "Tenometro", messo a punto in fase di ricerca, permette di selezionare in modo perfetto il grado di dissoluzione della fibra dopo il trattamento chimico. L'industria del lino ha collaudato positivamente tale apparecchio di controllo e lo ha ormai preso in dotazione (foto 1 - figura 1 - vedasi pagina 247).

Vista l'esigenza di rinnovare gli impianti di filatura, sono stati esaminati i problemi economici legati ad una nuova linea di produzione, in cooperazione con un costruttore di macchine tessili. Pur essendo convinta dei miglioramenti a livello di produzione, l'industria del ramo resta tuttavia ancora esitante di fronte a tali innovazioni, a causa degli elevati termini d'ammortamento.

Sono stati collaudati nuovi processi di filatura, ad esempio i processi OE, il core-yarn e i filati senza torsione. In alcuni casi è prevista una trasformazione specifica del materiale, ad esempio per il Novacore (foto 2 e 3 - vedasi pagina   ), in altri le fibre di lino e i relativi sottoprodotti sono adattati ai nuovi processi di trasformazione. La ricerca ha studiato un nuovo tipo di filatoio per il lino a fibre corte. Tale processo si è attualmente affermato a livello industriale in vari Stati membri.

La ricerca e la messa a punto di nuovi filati ha indotto ad una serie di modifiche dei prodotti e dei processi e alla realizzazione di nuovi prodotti, sia nell'industria della maglia che in tessitura, e al servizio di settori disparati, quali l'abbigliamento sportivo, la confezione, l'arredamento, ecc. Taluni di tali prodotti si trovano attualmente in commercio nella maggior parte dei Paesi membri.

SETTORE DELLA NOBILITAZIONE DEL LINO

Relatore : M. SOTTON , ITF

Nella tabella che segue sono elencate le attività più importanti svolte
da quattro centri di ricerca nel quadro del programma EUROLIN. I campi di
ricerca sono in parte complementari e si situano a vari livelli, nell'intento di fornire, attraverso l'approfondimento delle conoscenze in materia, risposte valide ai problemi d'indole generale connessi con la nobilitazione del lino e nello stesso tempo miglioramenti tecnici o vere
innovazioni per quanto riguarda i prodotti e i processi che interessano i
diversi settori del lino nei Paesi membri.

Attività	Centri di ricerca			
	Centexbel	ITF	LIRA	Laboratori
1. Studio della struttura interna delle fibre, per la messa a punto di nuove tecnologie		x		
2. Trattamenti alcalini essenziali per la nobilitazione delle stoffe	x		x	x
3. Innovazioni in materia di nobilitazione: - dei tessuti per la confezione - dei tessuti per arredamento: . poltrone, parati, biancheria . per la riduzione dell'infiam- mabilità dei sedili imbottiti	x		x	x
4. Miglior rendimento: - dei trattamenti in bagni rapidi (schiume) - degli appretti meccanici		x	x	

Il programma di ricerca EUROLIN ha apportato tutta una serie di conoscenze fondamentali, atte a colmare le lacune esistenti sulla fibra "lino".
Tale patrimonio costituisce fin d'ora una fonte di progresso e d'innovazione. La fibra di lino originale costituisce un vero e proprio "Kevlar
cellulosico", con struttura particolarmente anisotropa e cristallina,
scarsamente reagente, particolarmente fragile alle sollecitazioni di compressione e al taglio, che provocano profondi sfaldamenti strutturali. La
presenza di tali dislocazioni sono all'origine della scarsa resistenza

del lino alle abrasioni. Tali fenomeni possono essere favorevolmente
sfruttati (appretti meccanici) e bastano da soli per giustificare la necessità di ricorrere ai trattamenti alcalini preventivi, necessari per la
nobilitazione delle stoffe e che, omogeneizzando le fibre elementari, le
rendono più facilmente aggredibili da parte dei reattivi.

Ampie ricerche sono state condotte nei Paesi membri sui trattamenti alcalini preventivi, essendosi registrata una lacuna nelle conoscenze in materia da parte delle industrie di trasformazione e, in tale fase, il lino
va inoltre trattato in modo diverso dal cotone. I risultati delle ricerche hanno contribuito all'elaborazione di raccomandazioni rigorose ad uso
degli industriali e che peraltro differiscono sensibilmente da un Paese
all'altro, a seconda dei prodotti a cui si riferiscono (confezione, cami-
ceria, arredamento, ecc.).

Il problema della nobilitazione dei tessuti destinati alla confezione è
stato esaminato in stretta collaborazione con le industrie di finissaggio. I risultati confermano l'importanza essenziale del trattamento alca-
lino preventivo, realizzato per ottenere una buona sensibilità al tatto e
la stabilità dimensionale, a favore della qualità finale del prodotto.
Per tale via si annullano tutte le differenze iniziali esistenti nei materiali, nei filati e nei tessuti. Sul piano tecnico, e con il pieno con-
senso dell'industria del settore, una delle conseguenze più importanti
consiste nella fissazione di un capitolato d'oneri per i tessuti destinati alla confezione che prevede l'applicazione di etichette che ne garan-
tiscono la qualità.

Una procedura analoga è stata seguita per i marchi di garanzia dei tessuti di lino destinati all'arredamento, in particolare per le poltrone e i
parati.

Vanno ricordati i progressi realizzati nel campo degli appretti anti-mac-
chia: raccomandazioni relative ai prodotti più efficaci (fluorocarbonati)
e alle tecniche d'applicazione (cilindri spalmatori piuttosto che vapo-
rizzazione o appretti schiumogeni). Per i tessuti impiegati nel rivesti-
mento di poltrone si è pervenuti ad un compromesso accettabile fra le
esigenze di aumento della resistenza all'abrasione (appretti vinilici o
al silicone) e della qualità estetica e la facilità di lavorazione.

E' stata avviata un'importante ricerca sulla produzione di stoffe di lino, in grado di ottemperare alle nuove norme di legge sulla resistenza al
fuoco delle poltrone imbottite rivestite di tessuto, che interessano sia
l'arredamento privato che pubblico. Si è constatato che le esigenze
dell'arredamento domestico possono essere facilmente soddisfatte ricorrendo a tessuti di lino o a fibre miste lino-sintetico.

In merito alle forniture di tessuti per locali pubblici, si prestano ottimamente i tessuti lino/nylon (80/20), a condizione che sul rovescio

siano rivestiti con uno strato di materiale ignifugo: tale misto lino/
nylon è attualmente oggetto di collaudi a livello industriale.

Le proprietà dei tessuti sottoposti a trattamento antipiega in fase
schiumosa sono analoghe a quelle dei tessuti trattati con processi tradi-
zionali, salvo una notevole riduzione dei costi di essicazione.

466

SINTESI E CONCLUSIONI DEL PRESIDENTE DELLA SESSIONE

M. VAN LANCKER, Centexbel

Le relazioni presentate e le discussioni che ne sono seguite hanno toccato una serie di argomenti di grande interesse per l'industria sia del lino che del tessile in generale; in particolare sono stati esaminati a fondo i punti seguenti:

- processi alternativi di macerazione;
- uso di nuove tecniche di produzione,
- classificazione dei prodotti
- struttura morfologica del lino.

A proposito del tempo che ancora ci vorrà prima che siano stati messi definitivamente a punto i nuovi processi di macerazione (chimica e biochimica), va osservato che, al loro attuale stadio di sviluppo, questi processi non sono ancora abbastanza evoluti da consentire di affrontare il problema delle fibre coltivate in condizioni climatiche sfavorevoli. Mediante la macerazione chimica, una volta che siano stati perfezionati i processi relativi, per il momento ancora non pienamente soddisfacenti, sarebbe possibile coltivare il lino anche in zone caratterizzate da condizioni generali poco propizie; i risultati conseguiti finora permettono comunque di prevedere che una macerazione del lino con mezzi chimici sarà fattibile in un futuro non lontano, il che consentirà di ricavare fibre tessili di ottima qualità anche da piante coltivate in situazioni meteorologiche avverse.

Oggetto di intenso interesse nella discussione sono stati i nuovi procedimenti di lavorazione, con particolare riguardo alla rottura per stiramento delle fibre lunghe e la filatura delle fibre corte. Per quanto riguarda il primo dei due procedimenti, è stato affermato che il numero dei punti di rottura nella stiratura delle fibre lunghe dipende in larga misura dalla lunghezza massima accettabile del materiale in questione, e dai risultati delle prove eseguite si direbbe che sia più vantaggioso lavorare con una terza zona.

I procedimenti utilizzati per le fibre corte sono in genere analoghi a quelli impiegati per il cotone, con l'avvertenza che, col lino, si deve prestare particolare attenzione alla cernita del materiale da lavorare ed alla produzione della fibra. In determinate condizioni, per lavorare le fibre corte è possibile utilizzare le macchine per la cardatura della lana. Va comunque detto che la serie delle prove industriali finora compiute è assai limitata.

Per i settori che non lavorano in prima linea col lino, e cioè che usano
il lino per lo più in combinazione con altre fibre, c'è ovviamente bisogno di metodi di analisi che forniscano dati ed elementi di giudizio
obiettivi.

Sul problema della classificazione sistematica delle materie prime e dei
prodotti sono ormai in corso da anni ricerche miranti all'elaborazione di
metodiche di analisi e di procedure adeguate. Attualmente si dispone di
tre prove, due delle quali basate sulla simulazione, cui si può ricorrere, per quanto si debba dire che non sono considerate del tutto affidabi-
li: una consente di valutare la resa in pettinato; con la seconda si può
stabilire la divisibilità meccanica del pettinato, mentre con la terza,
basata su di una tecnica a flusso d'aria, si misura la finezza della fibra. Nessuna delle tre comunque può dare risultati pari a quelli di un'
analisi organolettica effettuata da esperti. Per la classificazione del
prodotto (cioè filamenti, tessuto, ecc.) si fa normalmente uso di prove
standard; tuttavia, in certi casi (come ad es. la misura dell'elasticità
delle fibre o il grado di polimerizzazione) si deve ricorrere a prove
specifiche o specialmente adattate per il lino.

Particolarmente interessante è stata l'esposizione delle nuove acquisi-
zioni sulla struttura morfologica della fibra del lino, a proposito della
quale sono state fatte due importanti constatazioni: primo, le differenze
di reazione fra la fibra di lino e quella di cotone ai trattamenti alcalini e, secondo, le proprietà meccaniche della fibra vergine di base del
lino, le quali sono analoghe a quella del Kevlar.

Trattando la fibra di lino con soda caustica (processo di mercerizzazione
o caustificazione), si incontrano resistenze derivanti dalla struttura
altamente cristallina della fibra e da una serie di altri fattori. A differenza di quanto succede con le fibre del cotone, dove la soda caustica
penetra con grande omogeneità, la penetrazione della soda dentro ed attraverso le fibre di lino è caratterizzata da discontinuità marcate: comunque, gradualmente l'intera fibra del lino si converte in cellulosa
mercerizzata, ed a questo punto gli effetti sono gli stessi che per le
fibre di cotone, cioè ne risultano aumentate sia la penetrabilità ai prodotti chimici che la stabilità. Un'importante caratteristica del trattamento delle fibre di lino con soda caustica è che vengono conservati i
nodi e le dislocazioni della fibra. Diverso è il risultato che si ottiene
se, in funzione di agente mercerizzante, si usa l'ammoniaca liquida: in
questo caso il processo di trasformazione della fibra è più lento e meno
drastico che con la soda.

Data la particolarità delle proprietà meccaniche possedute dalle fibre
vergini, si può pensare ad una loro eventuale utilizzazione per la fabbricazione di materiali compositi con matrice organica o minerale. Le
difficoltà legate ad un eventuale impiego del lino in sostituzione delle
fibre di rinforzo del tipo Kevlar in materiali compositi derivano dal

fatto che la fibra di lino, a meno che non si modifichino le tecniche di estrazione attualmente in uso, non possiede proprietà meccaniche intrinseche comparabili; se inoltre si considera che non esistono fibre a base di lino con una resistenza al calore pari a quella del Kevlar, se ne deve concludere che nello speciale settore in discussione l'utilizzazione del lino dovrà semmai limitarsi alla fabbricazione di materiali compositi per applicazioni specifiche in condizioni non termiche. In più bisogna tener presente che, mentre le fibre di rinforzo come il Kevlar sono disponibili in commercio sotto forma di filamenti continui, il lino si presenta invece, nella forma naturale, con fibre di 6 cm di lunghezza al massimo.

In conclusione, per rispondere al quesito se nel futuro il lino potrà conquistarsi un posto sul mercato dei materiali compositi, occorrerebbe che già da ora la ricerca, in collaborazione con l'industria del settore, si occupasse del problema, il quale è comunque destinato ad assumere una rilevanza cruciale forse a breve scadenza.

CONCLUSIONI GENERALI DEL CONVEGNO

D. FINLAY MAXWELL
Presidente del Comitato per la Ricerca Scientifica del Comitextil

Sull'argomento dei risultati della ricerca scientifica si sono svolte, nell'ambito di questo Convegno, una rassegna "in anteprima" delle novità, quattro sessioni con esposizione di relazioni approfondite, quattro presentazioni di scoperte di interesse saliente, ed una rassegna riassuntiva di queste quattro presentazioni. A questo punto il messaggio che esce dal Convegno dovrebbe essere chiaro a tutti: abbiamo conseguito successi rilevanti in ciascuno dei settori esaminati ed abbiamo dimostrato al di là di ogni dubbio che, con l'appoggio della Commissione, siamo in grado di condurre a termine con esito positivo progetti comuni di ricerca, realizzati da più Istituti o centri operanti in collaborazione complementare fra di loro e al di sopra dei confini nazionali.

Qualcuno può chiedersi perché si dia preferenza alla ricerca collettiva al posto dell'attività realizzata da ciascun paese per proprio conto in ambito nazionale. La risposta è che il ritmo sempre più rapido dell'evoluzione tecnologica impone al settore del tessile la necessità d'una ricerca pluridisciplinare.

Tenterò di chiarire meglio la problematica citando un caso ipotetico, ad esempio la tintura o il trattamento superficiale d'un tessuto di lino o di una stoffa qualsiasi: si tratta, come ognuno sa, di processi per i quali c'è bisogno di studiare gli effetti di tutta una serie di sostanze chimiche, probabilmente anche di trattamenti termici compresi entro una gamma assai ampia di temperature, ed eventualmente l'impiego di radiofrequenze per creare un gradiente termico diretto verso l'interno del tessuto o, in via alternativa, di micro-onde o di radiazioni infrarosse per stimolare lo spostamento verso l'esterno di ausiliari tessili applicati sotto forma di schiume; esiste inoltre la possibilità di ricorrere alle radiazioni alfa o beta o anche al laser; e per ottimizzare tutti questi trattamenti è possibile utilizzare le tecniche d'analisi di Kawabata.

Ora viene la domanda: esiste nel nostro settore un'organizzazione che da sola disponga di capacità specializzate in tutti questi comparti? Se no, come si vuole allora che possa realizzarsi questo genere di ricerca? Uno dei maggiori punti di forza della ricerca comunitaria risiede appunto nella possibilità di riunire una gamma sempre più vasta di differenti discipline, dato che ogni centro o unità specializzata possiede un elevato grado di competenza nel proprio ramo specifico.

Un altro vantaggio dei programmi comunitari è costituito dal fatto che essi sono più frequentemente sottoposti a valutazioni e riesami critici e autocritici. Da taluno dei delegati è stata mossa l'obiezione che parte dei lavori presentati in questa sede riguardano in qualche punto ricerche già svolte altrove: ma questo è un inconveniente che si verificherà sempre, soprattutto là dove vi siano programmi nazionali che procedano parallelamente e in certo grado indipendentemente l'uno dall'altro. Qui appunto sta un altro vantaggio dei programmi comunitari, i quali possono, fra l'altro, servire ad eliminare od almeno a ridurre in misura significativa i doppioni, che sono fonte di inutili sprechi.

In linea di massima, personalmente direi che i programmi comunitari di ricerca si prestano in particolare per l'attuazione dei progetti più impegnativi ed interessanti una più vasta gamma di conoscenze, mentre i programmi nazionali sono forse più adatti all'approfondimento specialistico di aree più ristrette: con ciò voglio dire che le due direttrici di ricerca possono essere rese complementari fra di loro.

Dopo che nel corso del Convegno sono stati illustrati e discussi i successi riportati nella nostra attività di ricerca, resta ora da rispondere alla domanda: che cosa si intende fare dei risultati ottenuti?

La prima cosa da fare è curare la massima diffusione dei risultati conseguiti: a tal fine, come primo passo, la Commissione provvederà entro sei mesi alla stampa e alla distribuzione degli atti del Convegno, copia dei quali sarà inviata a tutti i delegati; e da quel punto subentrerà l'iniziativa personale di ognuno, che deve farsi parte diligente per diffondere entro la più vasta cerchia possibile, nell'ambito della propria organizzazione o Federazione, le conoscenze che ci siamo scambiati in questa sede.

La seconda cosa da fare sarà di dare attuazione pratica alla maggior parte dei risultati teorici emersi dal lavoro di ricerca che è stato portato a termine sinora. In più, ci sono le ricerche ancora in corso, le quali - o almeno quelle che si riterrà meritino l'impegno richiesto - andranno portate a compimento: a questo proposito, cioè per orientare e seguire i lavori di ricerca che restano da completare, mi permetterei di suggerire al Comitextil l'impostazione di una specie di "programma di attuazione".

Si tratta di una serie di compiti che, mi rendo conto benissimo, avranno come inevitabile conseguenza un ulteriore aumento del carico di lavoro gravante sul nostro personale tecnico, per quanto debba dire che questa sia una tendenza in atto già da qualche tempo, di cui ho fatto personale esperienza, e da cui ho anzi tratto, per quanto mi riguarda, le necessarie conseguenze: appunto perché ho compreso che, restando nella mia sede di lavoro di Huddersfield, avrei potuto fare solo il presidente, per così dire, a tempo perso, cioè avrei potuto partecipare al massimo ad una o due riunioni al mese e dedicare all'Associazione una quota del mio tempo che sarebbe stata affatto insufficiente in rapporto alle future esigenze, ho deciso di ritirarmi, lasciando la carica a disposizione di un successore il quale potrà introdurre le riforme che saranno ritenute necessarie.

Nell'ambito di queste riforme bisognerà, fra l'altro, pensare ad una nuova ridistribuzione del tempo e delle risorse a nostra disposizione. Personalmente penso che sia senz'altro possibile attuare una riorganizzazione lungo queste linee, la quale consenta la realizzazione di nuovi programmi su di una base di continuità. L'esigenza essenziale del momento è l'impostazione e la realizzazione di programmi che rispondano ai bisogni della Comunità e che possiedano un altissimo contenuto tecnico: da parte mia sono sicuro che, mettendo assieme le risorse a nostra disposizione, riusciremo a far fronte anche a tale esigenza.

L I S T   O F   P A R T I C I P A N T S

ALIBERT, G.
ITF Section Nord
B.P. 637
F - 59656 VILLENEUVE D'ASCQ CEDEX

ALLO, G.
Conseiller scientifique
Institut pour l'Encouragement de la
Recherche scientifique dans l'In-
dustrie et l'Agriculture (IRSIA)
6, rue de Craver,
B - 1050 BRUXELLES

AMOY, Ms C.
Attachée à la Direction des
Affaires économiques
Union des Industries textiles
10, rue d'Anjou
F - 75008 PARIS

ANDERSON, A.
Production Director
James Johnston & Co. of Elgin Ltd
Newmill
UK - ELGIN, Morayshire IV30 2AF

ARNAUD, P.
Filature de Garrot - Ets. Arnaud
B.P. 10
F - 81210 ROQUECOURBE

ARTZT, P.
Forschungsinstitut für Textil-
technik
Körschtalstrasse 26
Reutlingen-Denkendorf
D - 7036 DENKENDORF

BACART, J.
Professeur - Institut supérieur
industriel de de l'Etat
8, rue de Séroule
B - 4800 VERVIERS

BAETENS, E.
Chef de Projet
CENTEXBEL
St. Pietersnieuwstraat 41,
B - 9000 GENT

BAKER, B.
Quality Control Manager
ASDA Stores
Asda House, Britannia Road
UK - MORLEY, LEEDS, West Yorkshire

BALLAND, M.
ITF Maille
270, rue du Faubourg Croncels
F - 10042 TROYES Cedex

BARELLA, A.
Investigador
Consejo Intertextil Español
Gran Via de las Corts Catalanes 670
E - 08010 BARCELONA

BARENBRUG, M.
Technician Garment Sizing
C & A Nederland -
Dutch Delegate AEIH Technical
Committee
Postbus 249
NL - 1012 LW AMSTERDAM

BARRE, P.
Secrétaire du Bureau Marketing
Confédération internationale du Lin
et du Chanvre
27, boulevard Malesherbes
F - 75008 PARIS

BASCOUR, G.
ITCB-Belgium
Sterrekundelaan 14, Bus 27
B - 1030 BRUSSELS

BAUER, M.
Forschungsinstitut für
Textiltechnik
Körschtalstrasse 26
Reutlingen-Denkendorf
D - 7036 DENKENDORF

BELLY, M.
Chercheur
CENTEXBEL
8, rue de Séroule
B - 4800 VERVIERS

BENEYTO CASTELLO, J.V.
Ingegnero
Instituto del Textil
Plaza Emilio Sala, 1
E - ALCOY-ALICANTE

BERNOCCHI, M.
Vicepresidente
Cassa di Risparmi e Depositi di
Prato
Via degli Alberti 2
I - 50047 PRATO

BEUKER, F.
Dr.med.
Wissenschaftlicher Berater
Universität Düsseldorf
Institut für Sportwissenschaften
Universitätsstr. 1
D - 4000 DÜSSELDORF 1

BLANCHARD, Ms V.
LA REDOUTE
57, rue Blanchemaille
F - 59200 ROUBAIX

BLANKENBURG, G.
Professor
Deutsches Wollforschungsinstitut
Veltmanplatz 8
D - 5100 AACHEN

BLUM, C.
Directeur général
COMITEXTIL
24, rue Montoyer
B - 1040 BRUXELLES

BOLATTI
European Investment Bank
L - 2950 LUXEMBOURG

BONA, M.
S.P.A. Città degli Studi di Biella
Centro di Formazione Professionale
Strade Orema, Biella, Via per Ivrea
I - 13051 BIELLA VERCELLI

BONGERS, J.
FENECON
Kon. Wilhelminaplein 13
NL - AMSTERDAM

BOURDEAU, P.
Director for Environment, Raw
Materials and Materials Technology
Commission of the European Communi-
ties, DG Science, Research and
Development (DG XII/G)
200, rue de la Loi
B - 1049 BRUSSELS

BRACH, J.
Directeur adjoint - CELAC
Zoning du Petit Rechain
69H, ave du Parc
B - 4655 CHAINEUX

BRADY, H.
Head of Department of Technology
Scottish College of Textiles
Netherdale
UK - GALASHIELS, Selkirksh. TD1 3HF

BRADY, P.R.
Commonwealth Scientific and Indus-
trial Research Organisation
P.O.Box 21
Australia - BELMONT, Victoria 3216

BRUGGEMAN, J.P.
ITF Nord
B.P. 637
F - 59656 VILLENEUVE D'ASCQ Cedex

BRÜNN, G.W.
Advisor/Consultant - Forschungsge-
meinschaft Bekleidungsindustrie eV
Mevissenstrasse 15,
D - 5000 KÖLN 1

BRUNNSCHWEILER, D.
Technical Director
Cosmopolitan Textile Co Ltd,
Road Five, Industrial Estate
UK - WINSFORD, Cheshire, CW7 3QU

BUCHER, Ms
ITF Maille
270, rue du Faubourg Croncels
F - 10042 TROYES Cedex

BÜHLER, G.
Forschungsinstitut für Textil-
technik
Körschtalstrasse 26
Reutlingen-Denkendorf
D - 7036 DENKENDORF

BUTI, A.
Sindaco Revisore
Cassa di Risparmi e Depositi di
Prato
Via degli Alberti 2
I - 50047 PRATO

CANDRIES, Ms J.
Ingénieur textile - Adj. Direct.
COMITEXTIL
24, rue Montoyer
B - 1040 BRUXELLES

CEDNÄS, Ms M.
Swedish Institute for Textile
Research
P.O. Box 5402
S - 402 29 GÖTEBORG

CISELET, J.C.
Directeur
ITCB-Belgium
Sterrekundelaan 14, Bus 27
B - 1030 BRUSSELS

COCCHI, R.
Tecnotessile
Centro di Ricerca S.A.
Via Valentini 14
I - 50047 PRATO

COENEN, R.
Vorstand, Dipl.-Ing.
Gesamttextil / Frkt.
Postfach 20 51
D - 4432 GRONAU-EPE

COGLIANDRO, C.
Ets. Cogliandro
110, rue de l'Epine
F - 59200 TOURCOING

COIMBRA, S.
Président directeur général
Lainière de Picardie
Buire-Courcelles
F - 80200 PERONNE

COOKE, W.D.
Department of Textile Technology
UMIST
P.O.Box 88
UK - MANCHESTER M60 1QD

CONROUX, J.J.
Marketing Manager
Institut Battelle
7, route de Drize
CH - 1227 CAROUGE/GENEVE

COSTERMANS, J.-M.
Secrétaire général
MAILLEUROP
24, rue Montoyer
B - 1040 BRUXELLES

COUSIN, J.
Secrétaire général
Confédération internationale du Lin
et du Chanvre (CILC)
27. boulevard Malesherbes
F - 75008 PARIS

CRISP, S.
Head of Consumer Hazards and Tex-
tiles Subdivision - Laboratory of
the Government Chemist
Cornwall House, Waterloo Road,
UK - LONDON SE1 8XY

DANON, G.
Styliste - CETIH
14, rue des reculettes
F - 75013 PARIS

DAVID, C.
Commission of the European
Communities, DG Internal Market and
Industrial Affairs (III/C/2)
200, rue de la Loi
B - 1049 BRUSSELS

DE MAUTORT, L.
European Investment Bank
L - 2950 LUXEMBOURG

DE RUYTTERE, F.
Industriel
FIBELIN - NV/SA BENTEX
40, Hooiwege
B - 9940 SLEIDINGE

DE SCHRYNMAKERS DE DORMAEL, P.
Filateur de Lin - FIBELIN
Ets Stanislas Cock
Kleine Dam 17
B - 9100 LOKEREN

DE WILDE, Ms P.
Secrétaire
COMITEXTIL
24, rue Montoyer
B - 1040 BRUXELLES

DEGROOTE, K.
Textielbedrijf De Leie NV
Gentsesteenweg 131
B - 8500 KORTRIJK

DELFOSSE, P.
Chef de Projet
CENTEXBEL
Zoning de Petit Rechain
69H, avenue du Parc,
B - 4655 CHAINEUX

DESCHAMPS, J.
CETIH
14, rue des Reculettes
F - 75013 PARIS

DHONT, J.
Conseiller technologique
CENTEXBEL
St. Pietersnieuwstraat 41
B - 9000 GENT

DIAS, M.
Secrétaire/Membre du Comité Econo-
mique
Région Nord - Syndicats TEXTILES
CGT
153, rue de Maufait
F - 59100 ROUBAIX

DIRICQ, Ms. M.C.
LA REDOUTE
57, rue Blanchemaille
F - 59200 ROUBAIX

DOMICENT, Ms C.
Secrétaire général
Association européenne des Indus-
tries de l'Habillement (AEIH)
24, rue Montoyer
B - 1040 BRUXELLES

DU PRÉ, W.
Wetenschappelijk Medewerker
Vezelinstituut TNO
Postbus 110
NL - 2600 AC DELFT

DYANT, J.P.
Président directeur général
Ets. E&H Dyant, Filature de Laine
B.P. 288
F - 38203 VIENNE CEDEX

EDME, J.
ITF Nord
B.P. 637
F - 59656 VILLENEUVE D'ASCQ Cedex

FEDDERSEN, A.
Chefkonsulent - Textilindustrien
Bredgade 412
DK - 7400 HERNING

FESTA BIANCHET, O.
Président - Interlaine
19, rue du Luxembourg, Bte 14
B - 1040 BRUXELLES

FINLAY MAXWELL, D.
Former President Rech. Scient.Com.
COMITEXTIL
John Gladstone & Co Ltd
Wellington Mills - Lindley
UK - HUDDERSFIELD HD3 3HJ

FINNIMORE, Ms. E.
Project Leader
Deutsches Wollforschungsinstitut
Veltmanplatz 8,
D - 5100 AACHEN

FOCHER, B.
Stazione Sperimentale per la
Cellulosa, Carta e Fibre Tessili
Vegetali ed Artificiali
Piazza Leonardo da Vinci 26
I - 20133 MILANO

FONTAINE, J.C.
Secrétaire général
EUROCOTON
24, rue Montoyer
B - 1040 BRUXELLES

FOSTER, W.W.
Lamber Industrial Research Associa-
tion
Lisburn
IRL - Co ANTRIM BT27 4RJ

FRANCK, R.R.
Director
International Linen Promotion Ltd
31, Great Queen Street
UK - LONDON WC2B 5AA

FRANKE, E.D.
Comité international de la Rayonne
et des Fibres synthétiques (CIRFS)
29, rue de Courcelles
F - 75008 PARIS

FUGIGLANDO, P.
R/D Tessili
Centro Sperimentale SNIA FIBRE
Via Friuli 55
I - 20031 CESANO MADERNO

FUNDER, A.
Institutsleiter
Textil-Forschung Bielefeld eV
Postfach 3008
D - 4800 BIELEFELD 1

GALLICO, L.
Direttore
CNR Istituto Ricerche e Sperimenta-
zione Laniera "O.Rivetti"
Piazza Lamarmora 5
I - 13051 BIELLA

GARCIA-SORROCHE, J.A.
Chief of Laboratory
Indyuco SA
Tomas Breton 62,
E - Madrid 28045

GOEBEL, L.
TNO Faser Institute
Postbus 671
NL - 7500 ER ENSCHEDE

GREENWOOD, P.F.
Development Manager
International Institute for Cotton
Technical Research Division
Kingston Road, Didsbury
UK - MANCHESTER M20 8RD

GRIGNET, J.
Directeur des Recherches - Verviers
CENTEXBEL
Zoning du Petit Rechain
69H, Avenue du Parc
B - 4655 CHAINEUX

GROS, L.
Président de la Commission
Recherche scientifique de
COMITEXTIL
17, avenue de Saxe
F - 75007 PARIS

GROSHENS, P.
Directeur général adjoint
Lainière de Picardie
Buire-Courcelles
F - 80200 PERONNE

GSCHWANDTNER, H.
Dipl.-Ing.
Geschäftsführender Gesellschafter
Blaas Textilwerke Ges.mbH
Postfach 68
A - 9560 FELDKIRCHEN/KÄRNTEN

GUILLOTIN, Ms M.
Ingénieur
Département qualité-confectionabi-
lité des matières premières
CETIH
14, rue des reculettes
F - 75013 PARIS

HAGEGE, R.
Directeur de Laboratoire
ITF - Boulogne
B.P. 79
F - 92100 BOULOGNE BILLANCOURT

HANSEN, J.
Civilingeniør - Dansk Beklædnings
og Textil Institut
Postbox 80
DK - 2630 TÅSTRUP

HANSS, W.
Referent für Bekleidungsphysio-
logie, Vorsitzender des Verbandes
Mieder- u. Badebekleidung e.V. Köln
Franz-Albert-Str, 4 A
D - 8000 MÜNCHEN 50

HARKER, R.P.
Industrial Liaison Manager
WIRA Technology Group
Wira House - West Park Ring Road
UK - LEEDS LS16 6QL

HAY, Ms. A.
Research Assistant
Scottish College of Textiles
Netherdale
UK - GALASHIELS, Selkirksh. TD1 3HF

HAZEL, B.G.
Secretary General CRIET
Comités réunis de l'Industrie de
l'Ennoblissement textile dans les
CEE
272, Royal Exchange
UK - MANCHESTER M2 7ED

HICKMAN, W.S.
Project Leader
Interox Research & Development
P.O. Box 2
UK - WIDNES, England

HONEYMAN, J.
Expert CEE
55 Kingston Road
UK - NANCHESTER M20 8SB

HOSPIED, B.
Directeur technique
EUROMOTTE S.A.
49, Place Motte
B - 7700 MOUSCRON

INNES, Ms. J.
Loan Officer
European Investment Bank
L - 2950 LUXEMBOURG

JACQUEMART, J.
ITF - Boulogne
B.P. 79
F - 92105 BOULOGNE BILLANCOURT

JEFFRIES, R.
Technical Group Manager
Shirley Institute
856, Wilmslow Road, Didsbury
UK - MANCHESTER M20 8RX

JULIA, C.
Technicien
ITF SUD/LECARIM
Boulevard du Thoré
F - 81200 MAZAMET

JOMBART, P.
LA REDOUTE
57, rue Blanchemaille
F - 59200 ROUBAIX

KEMPTON, R.T.S.
Chairman
T.W. Kempton Ltd
Burleys Way
UK - LEICESTER LE1 3TR

KINAST, U.
Dipl.-Ing.
Füssener Textil AG
Mühlbachgasse 2-4
D - 8958 FÜSSEN/ALLGÄU

KIRCHDÖRFER, Ms. E.
Bekleidungsphysiologisches Institut
Hohenstein e.V.
Schloss Hohenstein
D - 7124 BÖNNIGHEIM

KNOTT, J.
Chargé de direction de recherche
CENTEXBEL
8, rue de Séroule
B - 4800 VERVIERS

KROENER, H.H.
Director General
Industrievereinigung Chemiefaser
e.V.
Karlstr. 21
D - 6000 FRANKFURT 1

LAKIN, W.
Secretary General
INTERLAINE
19, rue du Luxembourg, Bte 14
B - 1040 BRUXELLES

479

LAMBRECHTSEN, A.P.
Chairman
FENETEXTIEL
GAMMA HOLDING NV
P.O. Box 80
NL - 5700 AB HELMOND

LARCY, D.
Chercheur
CENTEXBEL
Sint Pietersnieuwstraat 41
B - 9000 GENT

LARSEN, A.
Consultant
COMITEXTIL
Terrestlaan 3
B - 1900 OVERIJSE

LAWRENCE, C.A.
Lecturer
University of Manchester, Institute
of Science and Technology
P.O. Box 88
UK - MANCHESTER M60 1QD

LEACH, H.
President
COMITEXTIL
25 Sefton Drive - Worsley
UK - MANCHESTER M28 4NG

LECLUYSE, L.
LINTEXCO
Nijverheidslaan
B - 8630 GULLEGEM

LENOIR, J.
Chercheur
CENTEXBEL
Zoning du Petit Rechain
69H, Avenue du Parc
B - 4655 CHAINEUX

LEROY, P.
Secrétaire général
Fédération de la Maille belge
Fibelin (Fédération des Filateurs
de Lin et de Chanvre de Belgique)
Rode Beukendreef 14,
B - 9831 ST. MARTENS-LATEM (DEURLE)

LESSING, L.
Zignani Tessile SpA
Via Ita Marzotto 8,
I - 30025 FOSSALTA DI PORTOGRUARO V

LEUTSCHER, H.J.
Scientist
Intituut voor Bewaring en Verwer-
king van Landbouwprodukten (IBVL)
P.O. Box 18
NL - 6700 AA WAGENINGEN

LINSTER, R.
Commission of the European
Communities, DG Personnel and
Administration (DG IX/D/2)
JMO - B1/39
L - 2920 LUXEMBOURG

LOMAX, G.R.
Senior Research Officer
Shirley Institute
Didsbury
UK - MANCHESTER M20 8RX

LOMBARDI, Ms M.
Confederazione Nazionale
Artigianato
Via Sante Prassede 24
I - 00184 ROMA

LOTENS, W.A.
Project leader TEX 23 NL(N)
Institute for perception TNO
NL - 3769 ZG SOESTERBERG

LÜTTEKE, W.
Geschäftsführer
Unternehmen der Maschen-Industrie
Wolfgang März Strickwarenfabrik
GmbH
Warngauerstr. 54-58
D - 8000 MÜNCHEN

LYSY, R.
Project Leader
Colgate Palmolive R & D
Avenue du parc industriel
B - 4411 MILMORT (Herstal)

MACH, D.
Prokurist - Leiter
Qualitätskontrolle und
Anwendungstechnik
LENZING AG - Sparte Fasern
A - 4860 LENZING

MALVAUX, C.
Ing. technicien - CELAC
69 H, avenue du Parc
B - 4655 CHAINEUX

MANNI, E.
Professore
LANEROSSI S.P.A.
Via Pasubio 149
I - 36015 SCHIO

MARIOTTE, P.
Ingénieur de Laboratoire
CETIH
14, rue des reculettes
F - 75013 PARIS

McCANN, G.
Senior Lecturer
Department of Technology
Scottish College of Textiles
Netherdale
UK - GALASHIELS, Selkirksh. TD1 3HF

MECCHIA, L.
Responsabile Controlli e Processi
Chimici Lino
Zignango Tessile SpA
Via Ita Marzotto 8,
I -30025 FOSSALTA DI PORTOGRUARO VE

MECHEELS, J.
Direktor
Bekleidungsphysiologisches Institut
Hohenstein e.V.
Schloss Hohenstein
D - 7124 BÖNNIGHEIM

MENAULT, J.
Direction technique - RHOVYL
F - TROUVILLE EN BARROIS

MERLIER, P.
Adjoint à la Direction
Fédération de la Maille belge
Rode Beukendreef 14,
B - 9831 ST. MARTENS-LATEM (DEURLE)

MEURIS, I.
Professeur, Expert CEE
2, rue des Mouettes
B - 7500 TOURNAI

MIEGE, R.
Commission of the European Commun-
ities, DG Information Market and
Innovation (XIII/A/2) - JMO B4/102
L - 2920 LUXEMBOURG

MULLIER, J.M.
Directeur de Filature
Ets D. LEUREN - Filature de Lin
222, rue de Lille
F - 59223 RONCQ

MUNOZ GONZALES, Ms R.
Ingegnero tecnico
Instituto del Textil
Plaza Emilio Sala, 1
E - ALCOY-ALICANTE

NAHON, G.
Délégué Technico-Commercial
Interox SA
c/o Solvay & Cie SA
44, rue du Prince Albert
B - 1050 BRUXELLES

NICOLAY, D.
Commission of the European Commun-
ities, DG Information Market and
Innovation (XIII/A/2)
L - 2920 LUXEMBOURG

NYS, J.
Ingénieur
Rhone Poulenc Spécialités chimiques
Département Silicones
47, rue de Villieurs
F - 92527 NEUILLY-SUR-SEINE Cedex

ORSI, B.
Directeur
Linificio Canapificio Nazionale
Villa d'Alme
I - BERGAMO

PAPASTATHOPOULOS, S.
Commission of the European Commun-
ities, DG Internal Market and
Industrial Affairs (III/C/2)
200, rue de la Loi
B - 1049 BRUXELLES

PARRINI, V.
Presidente Tecnotessile
Università di Firenze - Tecno-
tessile, Centro di Ricerche SpA
Prato, Dipartimento di Chimica
Organica
Via Gino Capponi 9,
I - 50121 FIRENZE

481

PATERSON, N.M.I.
Lecturer
Department of Technology, Scottish
College of Textiles
Netherdale
UK - GALASHIELS, Selkirksh. TD1 3HF

PERALES HERNANDEZ, M.
Director
Instituto del Textil
Plaza Emilio Sala, 1
E - ALCOY-ALICANTE

PITZ, J.
Professeur
Institut supérieur industriel
de l'Etat
8, rue de Séroule
B - 4800 VERVIERS

PONCHEL, J.
ITF Nord
B.P. 637
F - 59656 VILLENEUVE D'ASCQ Cedex

PRATI, G.
Direttore
Stazione Sperimentale per la Cellu-
losa, Carta e Fibre Tessili Vege-
tali ed Artificiali
Piazza Leonardo da Vinci 26
I - 20133 MILANO

PRICE, C.D.
Head of Technology
British Clothing Centre
Clayton Wood Rise
UK - LEEDS LS16 6RF

PRUNIER, J.
ATPUL
Station expérimentale
Lagny le sec
F - 60330 LE PLESSIS BELLEVILLE

RADAELLI, G.M.
Direttore Commerciale Linificio e
Canapificio Nazionale SpA
Associazione Nazionale Lino Canapa
e Fibre dure
Corso Venezia 36
I - 20121 MILANO

RAE, A.
Consultant
British Clothing Centre
Clayton Wood Rise
UK - LEEDS LS16 6RF

RAMELLA, E.
Amministratore
Pettinatura e Filatura di Lino
Italiana Lini SpA
Via Per Pralungo 4
I - 13060 COSSILA S. GIOVANNI
              BIELLA

RANKIN, A.G.
Research Manager
Lambeg Industrial Research Associa-
tion
The Research Institute Lambeg
UK - LISBURN, Co Antrim, N Ireland
              BT27 4RJ

ROUSSEAU, L.
Directeur
CELAC
69 H, avenue du Parc
B - 4655 CHAINEUX

RUTSAERT, P.A.
Head of Division for Textiles and
Clothing - Commission of the European Communities, DG Internal Market and Industrial Affairs (III/C2)
200, rue de la Loi
B - 1049 BRUSSELS

SALMON MINOTTE, J.
Secrétaire section tissage
Confédération internationale du Lin
et du Chanvre
27, bvd Malesherbes
F - 75008 PARIS

SATTA, A.
Ingénieur Recherche - ITF-Boulogne
B.P. 79
F - 92105 BOULOGNE BILLANCOURT

SCHICK, E.
Dipl.-Ing. - Industrieverband Garne
Schaumainkai 87,
D - 6000 FRANKFURT AM MAIN

SCOTT, C.C.
Principal Scientific Officer
Department of Trade and Industry
Room 607, Ashdown House
123 Victoria Street
UK - LONDON SW1

SEUFZGER, Ms B.
Entwicklungsingenieur
W.L. Go& Co GmbH
Wernher-von-Braun-Strasse 18
D - 8011 PUTZBRUNN BEI MÜNCHEN

SMIRFITT, J.A.
Information Officer
HATRA
7 Gregory Boulevard
UK - NOTTINGHAM NG7 6LD

SMISSAERT, L.
Chef de Projets
CENTEXBEL
St. Pietersnieuwstraat 41,
B - 9000 GENT

SØRENSEN, T.
Dansk Textil Institut
Gregersensvej 5
DK - 2630 TÅSTRUP

SOTTON, M.
ITF - Lyon
rue de Collongue
F - ECULLY 69

SPERO, G.
Président général
Confédération internationale du lin
et du chanvre
27, boulevard Malesherbes
F - 75008 PARIS

STANDISH, J.M.
Fabric Development Manager
Levi Strauss
189, Avenue Louise
B - BRUXELLES

STEVENS, Ms J.C.
Technical Manager
International Institute for Cotton,
Technical Research Division
Kingston Road, Didsbury
UK - MANCHESTER M20 8 RD

STRYCKMAN, J.
Directeur
CENTEXBEL
24, rue Montoyer
B - 1040 BRUXELLES

SULTANA, C.
Directeur
Institut Technique Agricole du Lin
5, rue Cardinal Mercier
F - 75009 PARIS

SUURMEIJER, H.J.
Vezelinstituut TNO
Schoemakerstraat 97
NL - 2600 AC DELFT

SWAAP, H.,
Ministry of Economic Affairs
Postbus 20101
NL - 2500 EC DEN HAAG

TALL, P.
Senior Technical Executive
Marks & Spencer Plc
Baker Street
UK - LONDON W1A 1DN

TENT, H.
Director
Commission of the European
Communities, Joint Research Centre
200, rue de la Loi
B - 1049 BRUSSELS

TEMPESTINI, F.
Membro del Comitato di gestione
Cassa di Risparmi e Depositi di
Prato
Via degli Alberti 1
I - 50047 PRATO

THEBAULT, P.
Association générale des Produc-
teurs du Lin
8, rue Cardinal Mercier
F - 75009 PARIS

TIQUET, F.
Professeur
Insitut supérieur industriel
de l'Etat
8, rue de Séroule
B - 4800 VERVIERS

483

UMBACH, K.U.
Bekleidungsphysiologisches Institut
Schloss Hohenstein
D - 7241 BÖNNIGHEIM

VAN BODENGRAVEN, J.
Secretaris
FENECON
CC 2.08.02
Koningin Wilhelminaplein 13
NL - 1062 HH AMSTERDAM

VAN CAUSENBROECK, F.
Directeur
Comité central belge de l'Ennoblis-
sement textile
Martelaarslaan 43,
B - 9000 GENT

VAN DORT, J.M.
Product Development Manager
ten Cate Over-All Fabrics bv
Sluiskade NZ 14
NL - 7602 HR ALMELO

VAN LANCKER, M.
Directeur des Recherches - Gent
CENTEXBEL
St. Pietersnieuwstraat 41,
B - 9000 GENT

VANDER STOCKT, G.
ITCB-Belgium
Sterrekundelaan 14, Bus 27
B - 1030 BRUSSELS

VANNESTE, G.
LINTEXCO
Nijverheidslaan
B - 8630 GULLEGEM

VERCHER PEREZ, Ms R.
Ingegneros tecnicos
Instituto del Textil
Plaza Emilio Sala, 1
E - ALCOY-ALICANTE

VERDAN, M.
Senior Research Chemist
Ste Du Pont de Nemours
146, Rte du Nant d'Avril
CH - 1217 MEYRIN/GENEVE

VERRET, R.
President
Werner International Management
Consultants
523, ave Louise, bte 29
B - 1050 BRUXELLES

VERHASSELT, J.
Directeur
Ets AGACHE, CILC
B.P. 13
F - 59840 PERENCHIES

VERROKEN, A.
Research Manager CRIET
c/o SA UCO
Bellevue 1
B - 9218 LEDEBERG

WARD, C.D.
Senior Project Officer
HATRA
7 Gregory Boulevard
UK - NOTTINGHAM NG7 6LD

WINDLE, L.
Senior Research Officer
Wira Technology Group
Wira House - West Park Ring Road
UK - LEEDS LS16 6QL

WOHLFART, H.
Vorsitzender
Forschungskuratorium Gesamttextil,
Frankfurt
Schoeller Eitorf AG
Postfach 140
D - 5208 EITORF

WURM, J.G.
Head of Division - Materials
Technology
Commission of the European
Communities, DG Science, Research
and Devlopment (DG XII/G/4)
200. rue de la Loi
B - 1049 BRUSSELS

485